NEWS^{99}

The Proceedings of the International Symposium on

Nuclear Electro-Weak Spectroscopy for Symmetries in Electro-Weak Nuclear-processes

NEWS^{99}

The Proceedings of the International Symposium on
Nuclear Electro-Weak Spectroscopy for Symmetries in Electro-Weak Nuclear-processes

in honor of Professor Hiro Ejiri

Osaka, Japan 9 – 12 March, 1999

Editors

Hiroshi Toki
Research Center for Nuclear Physics, Osaka University, Japan

Tadafumi Kishimoto
Department of Physics, Osaka University, Japan

Kichiji Hatanaka
Research Center for Nuclear Physics, Osaka University, Japan

Keiji Takahisa
Research Center for Nuclear Physics, Osaka University, Japan

World Scientific
New Jersey • London • Singapore • Hong Kong

Published by

World Scientific Publishing Co. Pte. Ltd.

P O Box 128, Farrer Road, Singapore 912805

USA office: Suite 1B, 1060 Main Street, River Edge, NJ 07661

UK office: 57 Shelton Street, Covent Garden, London WC2H 9HE

British Library Cataloguing-in-Publication Data
A catalogue record for this book is available from the British Library.

NEWS 99
Proceedings of the International Symposium on Nuclear Electroweak Spectroscopy

ISBN 981-238-125-2

This book is printed on acid-free paper.

Printed in Singapore by Uto-Print

Organization of the Symposium:

International Advisory Committee

J. Arvieux (Orsay)
P. Barnes (Carnegie-Mellon)
J. Cameron (Indiana)
R. Chrien (BNL)
J. Deutch (Luvin)
A. Fasseler (Tubingen)
E. Fiorini (Milan)
A. Gal (Jerusalem)
M. Harakeh (KVI)
E. Henley (Washington)
Ph. Hubert (Bordeaux)
H. Kamitsubo (SPring-8)
T. Kirsten (Max-Plank)
K. Kubo (Tokyo Metropolitan)
S. Kullander (TSL)
V. Lobashev (Moscow)
A. Masaike (Kyoto)
A. Morales (Zaragoza)
H. Ohtsubo (Osaka)
A. Sanda (Nagoya)
O. Schult (Julich, IKP)
Z. Sun (Beijing)
A. Suzuki (Tohoku)
A. Thomas (Adelaid)
W. Weise (Munichr)
K. Yazaki (Tokyo)

Organization Committee

H. Toki (RCNP) Chair.
T. Kishimoto (Osaka) Co-Chair.
K. Hatanaka (RCNP) Scientific Secretary
M. Fujiwara (RCNP)
T. Fukuda (KEK)
A. Hosaka (Numazu)
Y. Kuno (KEK)
K. Maeda (Tohoku)
T. Motobayashi (Rikkyo)
K. Muto (TIT)
Y. Nagai (TIT)
T. Nakano (RCNP)
M. Nomachi (RCNP)
T. Noro (RCNP)
H. Ohsumi (Saga)
K. Okada (Kyoto Sangyo)
K. Sato (RCNP)
T. Sato (Osaka)
T. Shibata (KEK)
K. Takahisa (RCNP)

OPENING ADDRESS

T. Kishimoto
President of Osaka University

Good morning, ladies and gentlemen. It is my pleasure to welcome you to Osaka University on the occasion of the International Symposium on Nuclear Electro-Weak Spectroscopy. I am extremely pleased to see so many active scientists get together from all over the world to discuss the research frontiers on Nuclear Physics. I understand that the Research Center for Nuclear Physics of Osaka University is playing a central role in pushing forthfully the research frontiers in Nuclear Physics.

In this occasion, I would like to mention about Osaka University. Osaka University was established in 1931 as one of the seven Imperial Universities. The first president of the university was a famous physicist, Dr. Hantaro Nagaoka. Thus, Osaka University has a strong tradition in Nuclear Physics. Osaka University has developed under the motto "Live locally and grow globally", and at present we have 10 schools and more than 20 institutes and research centers and welcome more than 5,000 undergraduate and graduate students every year. I am proud of the research activities of this university not only in the nuclear physics but also in all the fields of sciences including life sciences, engineering, and human sciences.

I have heard that a part of this Symposium is devoted to Prof. H. Ejiri, who is the director of RCNP for 6 years and will finish his term as the Director and also as a Professor in the end of this month. I want to express my sincere thanks to him for his contributions to the University. He made a lot of suggestions for the education and the research policies of the University. In addition, I am aware of the activities of RCNP in expanding its field to Quark Nuclear Physics by establishing LEPS (laser electron photon source) in SPring8, to Lepton Nuclear Physics by creating the low back ground laboratory in Oto village and to Computational Nuclear Physics by sharing the supercomputor in Osaka University. Professor Ejiri played the major role for the establishments of these facilities and in the research fields.

I have heard also that the subjects of this Symposium are those Prof. Ejiri has been working throughout in his research life. I am sure that this Symposium will provide vital information to those who attend here and also the discussions among the participants will clear up many of the questions individuals have in their daily research. I hope all the participants of this Symposium find it exciting and also enjoyable by making mutual discussions. Please enjoy your stay in Osaka University.

Thank you.

CONTENTS

QUARKS AND HADRONS BY ELECTRO WEAK PROBES

SYMMETRIES IN FLAVOR NUCLEAR PHYSICS

SYMMETRIES IN NUCLEAR STRUCTURES BY ELECTROMAGNETIC SPECTROSCOPY

Double Beta Decay in Gauge Theories

J.D. Vergados
Institute of Theoretical Physics, University of Tuebingen, D–72076 Tuebingen, Germany

and

Theoretical Physics Section, University of Ioannina, GR–451 10, Ioannina, Greece; Electronic address: vergados@cc.uoi.gr

Neutrinoless double beta decay is a very important process both from the particle and nuclear physics point of view. ¿From the elementary particle point of view it pops up in almost every model, giving rise among others to the following mechanisms: a) The traditional contributions like the light neutrino mass mechanism as well as the $j_L - j_R$ leptonic interference (λ and η terms). b) The exotic R-parity violating supersymmetric (SUSY) contributions. ¿From the nuclear physics point of view it is challenging, because: 1) The nuclei, which can undergo double beta decay, have complicated nuclear structure. 2) The energetically allowed transitions are suppressed (exhaust a small part of all the strength). 3) Since in some mechanisms the intermediate particles are very heavy one must cope with the short distance behavior of the transition operators. 4) The intermediate momenta involved are quite high and one has to consider momentum dependent terms of the nucleon current. Taking the above effects into account from the experimental limits on the interesting nuclei A = 76, 82, 96, 100, 116, 128, 130, 136 and 150, we have extracted new limits on the various lepton violating parameters. In particular we get a stringent limit on the R-parity violating parameter $\lambda'_{111} < 4.0 \times 10^{-4}$

1 Introduction

The old exotic neutrinoless double beta decay ($0\nu\beta\beta-decay$) is very interesting since it violates lepton number by two units. It has been resurrected with the advent of gauge theories which favor Majorana neutrinos and through the pioneering work of Kotani and his group[1] it was brought again to the attention of the nuclear physics community.

From a theoretical point of view it is the most likely, if not the only, process capable of deciding whether or not the neutrino is a Majorana particle, i.e. it coincides with its own antiparticle[2,3,4,5,6,7]. It is expected to occur whenever one has lepton number violating interactions. Lepton number, being a global quantity is not sacred, but it is expected to be broken at some level.

¿From a nuclear physics point of view calculating the relevant nuclear matrix elements it is indeed a challenge.

¿From the experimental point of view it also very challenging, since the recent superkamiokande results have given the first evidence of physics beyond the Standard Model (SM). Work in this direction[8] was one ofa the many great accomplishments of the person honored in this symposium, Professor H. Ejiri.

The best known possibility is via the exchange of a Majorana neutrino between the two decaying neutrons[2,3,4,5,9,7,6]. Nuclear physics dictates that we study the light and heavy neutrino components separately. In the presence of only left-handed currents for light intermediate neutrinos the obtained amplitude is proportional $\eta_\nu =< m_\nu > /m_e$xi, where $< m_\nu >$ is a well known average neutrino mass (see, e.g., Refs[4,3]). In the case of heavy Majorana neutrino components the amplitude is proportional to the parameters[4,10] η_N^L and η_N^R, which in turn are proportional to the average inverse neutrino mass, i.e. it is again suppressed.

It is also possible to have, in addition, interference between the leptonic left and right currents, $j_L - j_R$ interference. In this case the amplitude in momentum space becomes proportional to the 4-momentum of the neutrino and, as a result, only the light neutrino components become important. One now has two possibilities. First the two hadronic currents have a chirality structure of the same kind $J_L - J_R$. Then ch leads to the lepton violating parameter $\lambda = \kappa\eta_{RL}$ with $\kappa = (m_L/m_R)^2$, i.e. given by the ratio of the squares of the masses of the L and R gauge bosons. Second the two hadronic currents are left-handed, which can happen via the mixing of the two bosons. The relevant lepton violating parameter is $\eta = \epsilon\eta_{RL}$ which is now proportional to this mixing ϵ. Both of these parameters, however, are proportional to the mixing between the light and heavy neutrinos[4].

In recent years supersymmetric models are taken seriously and semirealistic calculations are taking place. Hence in principle calculable one can use $0\nu\beta\beta$-decay to constrain some of the R-parity violating couplings, which cannot be specified by the theory[11,12,13,16,17]. Recent review articles[10,7,6] give a detailed account of the latest developments in this field.

From the above discussion it clear that one has to consider the case of heavy intermediate particles. One thus has to consider very short ranged operators in the presence of the nuclear repulsive core. If the interacting nucleons are point-like one gets negligible contributions. We know, however that the nucleons are not point like and they have structure described by a form factor first considered by Vergados[14] adopted later by almost everybody.

The other approach is to consider the double beta decay of pions in flight between nucleons, like

$$\pi^- \longrightarrow \pi^+, e^-, e^- \quad , \quad n \longrightarrow p, \pi^+, e^-, e^- \qquad (1)$$

first considered by Vergados[15] and revived by the Tuebingen group[16,17].

One may explicitly consider six quark clusters in the nucleus[18].

The other recent development is the better description of nucleon current by including momentum dependent terms, such as the modification of the axial current due to PCAC and the inclusion of the weak magnetism terms[19].

Granted that one takes into account all the above ingredients in order to obtain quantitative answers for the lepton number violating parameters from the the results of $0\nu\beta\beta$-decay experiments, it is necessary to evaluate the relevant nuclear matrix elements with high reliability. The most extensively use methods are the SMC (for a recent review see[6]) and QRPA(for a recent review see[7,6]). The SMC is forced to use few single particle orbitals, while this restriction does not apply in the case of QRPA. The latter suffers , of course, from the approximations inherent in the RPA method. So a direct comparison between them is not possible.

The above methods of calculation offer some optimism in our efforts for obtaining nuclear matrix elements accurate enough to allow us to extract reliable values of the lepton violating parameters from the data. We will iprovide a review in the case of the targets of experimental interest (^{76}Ge, ^{82}Se, ^{96}Zr, ^{100}Mo, ^{116}Cd, ^{128}Te, ^{130}Te, ^{136}Xe, ^{150}Nd).

2 The R-parity violating contribution to $0\nu\beta\beta$ decay.

Even though the intermediate neutrino mecahanism is the most conventional we will not elaborate further in this paper since it is by now well understood. We will elaborate on a relatively topic which is supersymmetric R-parity violating contributions to $0\nu\beta\beta$ decay. R-parity is defined as +1 for ordinary particles and -1 for their superpartners. R-parity violation has recently been seriously considered in SUSY models. It allows additional terms in the superpotential given by:

$$W = \lambda_{ijk} L_i^a L_j^b E_k^c \epsilon_{ab} + \lambda'_{ijk} L_i^a U_j^b D_k^c \epsilon_{ab} + \lambda''_{ijk} U_i^c U_j^c D_k^c \tag{2}$$

where a summation over the flavor indices i,j,k and the isospin indices a,b is understood (λ_{ijk} is antisymmetric in the indices i and j) The last term has no bearing in our discussion, but we will assume that it vanishes due to some discreet symmetry to avoid too fast proton decay. The λ's are dimensionless couplings not predicted by the theory.

In the above notation L,Q are isodoublet and E^c, D^c isosinglet chiral superfields, i.e they represent both the fermion and the scalar components. The second term in the superpotential could lead to neutrinoless double beta decay [11,12 16]. Typical diagrams at the quark level are shown in Fig.1. Note that as intermediate states, in addition to the s-leptons and s-quarks, one must consider the neutralinos, 4 states which are linear combinations of the gauginos and higgsinos, and the colored gluinos (supersymmetric partners of the gluons).

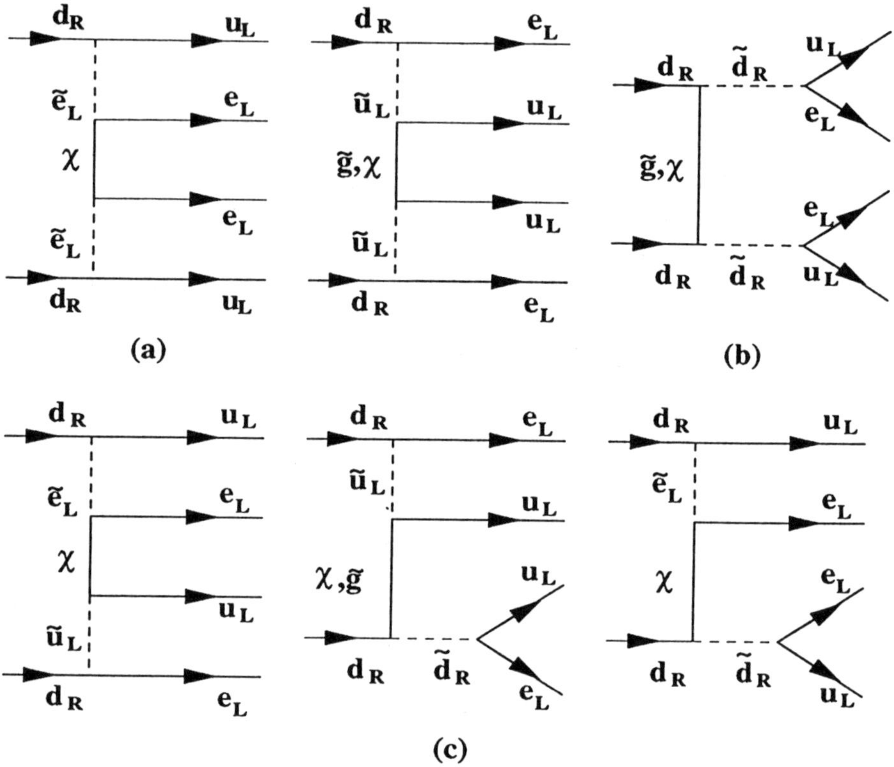

Figure 1: The R-parity violating contribution to $0\nu\beta\beta$ decay mediated by s-fermions and neutralinos (gluinos).

The effective lepton violating parameter η_{SUSY}, extracted from experiment, assuming that pion exchange mode dominates, as the authors of Ref[16,7] claim, is given by

$$\eta_{SUSY} = (\lambda'_{111})^2 \frac{3}{8} (\chi_{PS}\ \eta_{PS} + \eta_T) \tag{3}$$

with $\eta_{PS}(\eta_T)$ associated with the scalar and pseudoscalar (tensor) quark couplings given by

$$\eta_{PS} = \eta_{\tilde{\chi},\tilde{e}} + \eta_{\tilde{\chi},\tilde{q}} + \eta_{\tilde{\chi},\tilde{f}} + \tilde{\eta}_{\tilde{g}} + 7\eta'_{\tilde{g}} \tag{4}$$

$$\eta_T = \eta_{\tilde{\chi},\tilde{q}} - \eta_{\tilde{\chi},\tilde{f}} + \tilde{\eta}_{\tilde{g}} - \eta'_{\tilde{g}} \tag{5}$$

They find $\chi_{PS} = (2/3)$, but stricktly speaking[10] it depends on ratios of nuclear matrix elements. the parameters $\tilde{\eta}$ depend on the SUSY model and they are not going to be discussed further (see Ref. [10]). It has been found [17,10] that the pionic contribution (1) dominates. One may evaluate the relevant amplitudes in two approaches. First using harmonic oscillator wave functions, but adjusting the parameters to fit the π decay experiments[10]. Second using an elementary particle approach[17]

Thus we can write the effective operator for the 1π and 2π mode in the form

$$ME_k = (\frac{m_A}{m_p})^2 \; \alpha_{k\pi} \; \frac{m_p}{m_e} \; [M_{GT}^{k\pi} + M_T^{k\pi}] \tag{6}$$

Where the two above matrix elements are the usual GT and T matrix elements with the additional radial dependence given by

$$F_{GT}^{1\pi} = e^{-x} \quad , \quad F_T^{1\pi} = (3 + 3x + x^2) \; \frac{e^{-x}}{x} \tag{7}$$

$$F_{GT}^{2\pi} = (x-2)e^{-x} \quad , \quad F_T^{2\pi} = (1+x) \; e^{-x} \tag{8}$$

$$\alpha_{1\pi} \;=\; - \; c_{1\pi} \; \rho \; , \alpha_{2\pi} \;=\; c_{2\pi} \; \rho \; , \; \rho = \frac{1}{48 f_A^2} \; (\frac{m_\pi}{m_p})^4 (\frac{m_p}{m_A})^2 \tag{9}$$

and

$$c_{1\pi} = 109 \; (P) \; , \; c_{2\pi} = 198 \; (P) \; , \; c_{1\pi} = 132 \; (FKS) \; , \; c_{2\pi} = 170 \; (FKS) \tag{10}$$

where P=pesent, FKS is Ref. [17]. In our work the matrix elements employed take the form

$$ME_{eff} = ME_2 \quad , \quad \chi_{PS} = \frac{2}{3} \; (4\frac{ME_1}{ME_2} + 1) \tag{11}$$

It has been found[17,10] that χ_{PS} is close to the value of 2/3.

3 The effective nucleon current

As we have already mentioned the operators sre defined at the quark level, while the actual computations are done at the nucleon level. One has to be careful in applying this procedure in the case of $0\nu\beta\beta$ decay. one has to tackle two problems. The first is the question of the correct treatment of short range two nucleon correlations, in particular if the exchanged particles are heavy. The

Table 1: The lifetimes predicted for $0^+ \to 0^+$ $0\nu\beta\beta$-decay in various mechanisms (light neutrino, heavy neutrino, λ and η terms and SUSY contribution) for suitable input of lepton violating parameters and available nuclear calculations.For the definitions of the references see text.

$(\beta\beta)_{0\nu} - decay : 0^+ \to 0^+$ transition

$T_{1/2}^{0\nu-theor}(\langle m_\nu\rangle, \langle\lambda\rangle, \langle\eta\rangle, \langle\eta_N\rangle, \langle\eta_{SUSY}\rangle)$

	^{48}Ca	^{76}Ge	^{82}Se	^{96}Zr	^{100}Mo	^{116}Cd	^{128}Te	^{130}Te	^{136}Xe	^{150}Nd
Ref.	10^{24}	10^{24}	10^{24}	10^{24}	10^{24}	10^{24}	10^{25}	10^{24}	10^{24}	10^{22}
$\langle m_\nu\rangle = 1eV, \langle\lambda\rangle = 0, \langle\eta\rangle = 0, \langle\eta_N\rangle = 0, \langle\eta_{SUSY}\rangle = 0$										
R	12.8	34.8	4.80						24.2	
H	6.34	3.36	1.16				0.80	0.32		
E1		4.60	1.84				0.90	0.48		
E2		28.0	11.2				3.00	1.32	6.60	
S		8.12	2.86				3.60	1.66		
M		4.66	1.20		2.54		1.54	0.98	4.42	6.74
T		4.32	1.22		0.52		1.96	1.08	2.80	8.90
P1	5.00	7.20	3.00	1.22	7.80	9.40	3.80	1.72	6.60	
P2	56.0	36.0	5.60	54.0		9.80	30.0	4.20	5.60	
S1		17.9			0.50	1.44	2.18		17.5	
P		4.22	1.08	1.61	0.46	0.99	2.53	1.46	10.1	8.78
$\langle m_\nu\rangle = 0, \langle\lambda\rangle = 10^{-6}, \langle\eta\rangle = 0, \langle\eta_N\rangle = 0, \langle\eta_{SUSY}\rangle = 0$										
R	7.45	50.2	3.25						22.2	
S		7.75	1.14				14.8	0.89		
M		7.35	0.99		0.95		13.5	0.95	4.90	3.73
T		8.02	1.07		0.55		21.1	1.18	3.47	6.71
P1	2.71	8.90	2.08	0.94	30.6	39.1	22.7	1.34	2.73	
P2	27.9	41.2	4.39	27.7	10.3	10.8	165	2.22	4.42	
$\langle m_\nu\rangle = 0, \langle\lambda\rangle = 0, \langle\eta\rangle = 10^{-8}, \langle\eta_N\rangle = 0, \langle\eta_{SUSY}\rangle = 0$										
R	6.42	27.2	6.24						22.2	
S		36.7	11.1				10.7	5.92		
M		7.35	0.99		0.95		13.5	0.95	4.90	3.73
T		2.25	0.65		0.28		0.67	0.44	1.21	3.39
P1	15.11	3.10	6.51	1.48	3.44	19.2	1.20	0.62	1.23	
P2	43.2	22.8	5.16	7.95	102	83.2	1.90	1.05	0.96	
$\langle m_\nu\rangle = 0, \langle\lambda\rangle = 0, \langle\eta\rangle = 0, \langle\eta_N\rangle = 10^{-7}, \langle\eta_{SUSY}\rangle = 0$										
P1	4.95	0.25	3.35	67.1	4.70	23.5	0.78	3.03	1.42	
P2	124	0.59	7.23	671	1.47	33.6	1.27	1.31	1.01	
P		15.4	4.10	8.10	0.97	8.40	8.51	4.54	3.94	40.6
$\langle m_\nu\rangle = 0, \langle\lambda\rangle = 0, \langle\eta\rangle = 0, \langle\ eta_N\rangle = 0, \langle\eta_{SUSY}\rangle = 10^{-8}$										
F		3.3	0.86	0.71	0.30	0.85	0.93	0.45	1.2	3.2
P		4.5	1.0	1.4	0.59	1.4	1.5	0.71	2.0	5.2

second involves the correct estimate of the momentum dependent corrections in the nucleon current.

The first effect leads to the introduction of nucleon formfactors of a dipole shape

$g_V(\vec{q}^{\,2}) = 1.00/(1+\vec{q}^{\,2}/\Lambda_V^2)^2$, $g_A(\vec{q}^{\,2}) = 1.25/(1+\vec{q}^{\,2}/\Lambda_A^2)^2$

with $\Lambda_V = \Lambda_A = m_A \approx 0.85$ GeV. We should mention that the the transition operator at short distances can be avoided of the *ββdecay* is induced by pions in flight between the two nucleons , as mentioned above [10]. It can also be avoided, if one employs the presence of six quark clusters inside the nucleus [18].

As we have mentioned the effective nucleon current in addition to the usual V and A terms (P,S,T in SUSY contributions) contains momentum dependent terms [19]. With these modifications of the nuclear current in addition to the well known V-A terms one gets a tensor contribution. The corresponding form factors become $h_F = g_V^2(\vec{q}^{\,2})$ and :

$$h_{GT}(\vec{q}^{\,2}) = g_A^2(\vec{q}^{\,2})\ [\,1\ -\ \frac{2}{3}\ \frac{\vec{q}^{\,2}}{\vec{q}^{\,2}+m_\pi^2}\ +\ \frac{1}{3}\ (\frac{\vec{q}^{\,2}}{\vec{q}^{\,2}+m_\pi^2})^2\,]\ +\ \frac{2}{3}\ \frac{g_M^2(\vec{q}^{\,2})\vec{q}^{\,2}}{4m_p^2},$$

$$h_T(\vec{q}^{\,2}) = g_A^2(\vec{q}^{\,2})\ [\,\frac{2}{3}\ \frac{\vec{q}^{\,2}}{\vec{q}^{\,2}+m_\pi^2} - \frac{1}{3}\ (\frac{\vec{q}^{\,2}}{\vec{q}^{\,2}+m_\pi^2})^2\,]\ +\ \frac{1}{3}\ \frac{g_M^2(\vec{q}^{\,2})\vec{q}^{\,2}}{4m_p^2}, \quad (12)$$

The exact results will depend on the details of the nuclear model, since the new operators have different momentum (radial) dependence than the traditional ones and the tensor component is entirely new. For light neutrinos we can get a crude idea of what is happening by taking the above average momentum $\langle q \rangle$=100 MeV/c. Then we find that the GT ME is reduced by 22%. Then assuming that T matrix element is about half the GT one, we find that the total reduction is 28%. This is in perfect agreement with the exact results.

4 Extraction of the lepton violating parameters

The limits deduced for the lepton-number violating parameters depend on the values of nuclear matrix element, of the kinematical factor and of the current experimental limit for a given isotope.

4.1 Traditional lepton violating parameters

The predicted lifetimes for a given set of lepton violating parameters are given in table 1. In all cases, except those marked as (P), the modification of the nucleon current has not been taken into account, even, though, we expect the nuclear matrix elements entering the light neutrino mass mechanism to be decreased by about 30% , independently of the nuclear model.

The references in table 1 are defined as follows: R=Retamosa *et al*[21], H=Haxton *et al*[22], E1=Engel *et al*[23], E2=Engel *et al*[24], S=Suhonen *et al*[25], M=Muto *et al*[26], T=Tomoda *et al*[5], P1=Pantis *et al*[9], P2=Pantis *et al*[9] (p-n pairing), S1=Simkovic *et al*[27] (and private communication), F=Faessler *et al*[17], P=Present calculation (see Simkovic *et al*[19] for the nuclear Matrix elements). Notice in particular that the present calculation, marked P in the table, involves not only renormalized QRPA , but takes into account the corrections in the hadronic current[19] discussed above (see table 1).

4.2 R-parity induced lepton violating parameters

With the above ingredients and using the nuclear matrix elements of[17,10] we can extract from the data values of η_{SUSY}. Then one can use these values of η_{SUSY} in order to extract values for the R-parity violating parameters λ'_{111}.

One finds[10].

$$\lambda'_{111} = C_{\tilde{\chi}^0}(\eta_{SUSY})^{1/2} \quad (neutralinos\ only) \tag{13}$$

$$\lambda'_{111} = C_{\tilde{g}}(\eta_{SUSY})^{1/2} \quad (gluino\ only) \tag{14}$$

When both neutralinos and gluinos are included we write

$$\lambda'_{111} = C_{\tilde{\chi}^0,\tilde{g}}(\eta_{SUSY})^{1/2} \tag{15}$$

The values of the coefficients C, which have a strong dependence on the SUSY parameters, can be found elsewhere[10]. The extracted limits on λ'_{111} are shown in table 2 in the case of the pion mwchanism (for the nucleon mode see our report[10]).

The quantities λ'_{111} were assumed positive. If not, the absolute value is understood.

5 Conclusions

We have seen that $0\nu\beta\beta$ decay pops up in almost any fashionable particle model. From the experimental limits on the lifetimes (e.g. see[28,7,10] one can set useful limits not only on the light neutrino mass $< m_\nu >$, but in addition on other lepton violating parameters like $< \eta_N^{L,R} >$ or the parameters λ and η . Finally we mention again the limit extracted on the R-parity violating parameter λ'_{111}.

A set of limits, for our choice of nuclear matrix elements, derived from the various nuclear targets is given in Table 2. For ^{76}Ge we have used here

Table 2: Summary of the results presented in this work.

	$\langle m_\nu \rangle$	$\langle \lambda \rangle$	$\langle \eta \rangle$	$\langle \eta_N \rangle$	$\langle \eta_{SUSY} \rangle$	λ'_{111}
(A,Z)	eV	$10\times^{-6}$	$10\times^{-8}$	$10\times^{-8}$	$10\times^{-8}$	$10\times^{-4}$
	Pr	P1	P1	Pr	Pr	Pr
^{76}Ge	0.27	0.56	0.32	0.44	0.31	4.0
^{100}Mo	2.9	26	8.8	6.2	3.2	18
^{116}Cd	5.9	37	26	11	7.6	18
^{128}Te	1.8	5.6	1.3	2.9	1.6	8.1
^{130}Te	13	7.6	5.2	20	10	18
^{136}Xe	49	2.1	1.4	4 5	2.4	9.4
^{150}Nd	8.5	5.6	5.3	16	7.3	23

the unpublished new limit of the Heidelberg- Moscow experiment [29] $T_{1/2} \geq 5.7 \times 10^{25}$ yr.

We see that limits are quite stringent, but they, of course, have uncertainties in them. They come from nuclear physics, especially for the short ranged operators or from particle physics, as, e.g., in the case of supersymmetry.

There is a new experimental proposal for measurement of the $0\nu\beta\beta$-decay of ^{76}Ge, which intents to use 1 ton (in an extended version 10 tons) of enriched ^{76}Ge and to reach the the half-life limit $T_{1/2}^{0\nu-exp} \geq 5.8 \times 10^{27}$ and $T_{1/2}^{0\nu-exp} \geq 6.4 \times 10^{28}$ after one and 10 years of measurements, respectively. From these half-life values one can deduce the possible future limits on the effective light neutrino mass 2.7×10^{-2} eV and 8.1×10^{-3} eV, respectively.

Thus one can say with certainty that the experimental detection of the $0\nu\beta\beta$-decay process would be a major achievement with important implications on the field of particle and nuclear physics as well as on cosmology.

Acknowledgments

I would like to express my appreciation to the Humboldt Foundation for their award and my thanks to the Institute of Theoretical Physics at the University of Tübingen for its hospitality.

References

1. M. Doi, T. Kotani, N. Nishiura, K. Okuda, and E. Takasugi, Phys. Lett. **103B**, 219 (1981); bf 113B, 513(E) (1982).
2. W. C. Haxton and G. S. Stephenson, Prog. Part. Nucl. Phys. **12**, 409 (1984).
3. M. Doi, T. Kotani, and E. Takasugi, Prog. Theor. Phys. (Supp.) **83**, 1 (1985).

4. J.D. Vergados, Phys. Rep. **133**, 1 (1986).
5. T. Tomoda, Rep. Prog. Phys. **54**, 53 (1991).
6. J. Suhonen and O. Civitarese, Phys. Rep. **300**, 123 (1998).
7. A. Faessler and F. Šimkovic, J. Phys. G **24**, 2139 (1998).
8. H. Ejiri et al., Nucl. Phys. **A 611**, 85 (1996).
9. G. Pantis, F. Šimkovic, J.D. Vergados, and A. Faessler, Phys. Rev. C **53**, 695 (1996).
10. J.D. Vergados, hep-ph/990731.
11. R. Mohapatra, Phys. Rev. D **34**, 3457 (1986).
12. J. D. Vergados, Phys. Lett. B **184**, 55 (1987).
13. M. Hircsh, H.V. Klapdor-Kleingrothaus and S.G Kovalenko, Phys. Rev. Lett. **75**, 17 (1995); Phys. Rev. Rev. D **53**, 13297 (1996).
14. J.D. Vergados, Phys. Rev. **C24**, 640 (1981).
15. J.D. Vergados, Phys. Rev. **D25**, 914 (1982).
16. A. Faessler, S. Kovalenko, F. Šimkovic, and J. Schwieger, Phys. Rev. Lett. **78**, 183 (1997); Phys. Atom. Nucl. **61**, 1229 (1998).
17. A. Faessler, S. Kovalenko, and F. Šimkovic, Phys. Rev. D **58**, 115004 (1998); Phys. Rev. D **58**, 055004 (1998).
18. J.D. Vergados, Nuc. Phys. **B250**, 618 (1985).
19. F. Šimkovic, G. Pantis, J.D. Vergados, and A. Faessler, Phys. Rev. C **60**, (1999)3597.
20. F. Šimkovic, G.V. Efimov, M.A. Ivanov, V.E. Lyubovitskij, Z. Phys. A **341**, 193 (1992).
21. J. Retamosa, E. Caurier, and F. Novacki, Phys. Rev C **51**, 371 (1995).
22. W. C. Haxton, G. S. Stephenson and D. Strottman, Phys. Rev. D **25**, 2360 (1982).
23. J. Engel, P. Vogel, X.D. Ji and S. Pittel, Phys. Lett. B **225**, 5 (1991).
24. J. Engel, P. Vogel, and M. R. Zirnbauer, Phys. Rev. C **37**, 731 (1988).
25. J. Suhonen, S.B. Khadkikar ad A. Faessler, Phys. Lett. B **237**, 8 (1990); *ibid.* Nucl. Phys. A bf 529, 727 (1991).
26. K. Muto, E. Bender and H.V. Klapdor, Z. Phys. A **334**, 177 (1988). 1120 (1990).
27. F. Šimkovic, G. Pantis, and A. Faessler, Phys. Atom. Nucl. **61**, 1218 (1998), Prog. Part. Nucl. Phys. **40**, 285 (1998).
28. M. Moe and P. Vogel, Ann. Rev. Nucl. Part. Sci **44**, 247 (1994).
29. Heidelberg-Moscow collaboration, L. Baudis et al., Phys. Lett. **B 407**, 219 (1997); hep-ex/9901021.

FLAVOR CHANGING LEPTON PROCESSES

Yoshitaka KUNO
Institute of Particle and Nuclear Studies (IPNS),
High Energy Accelerator Research Organization (KEK),
Oho 1-1, Tsukuba, Ibaraki, Japan 305

The flavor changing lepton processes, or in another words the lepton flavor changing processes, are described with emphasis on the updated theoretical motivations and the on-going experimental progress on a new high-intense muon source.

1 Introduction

Although the Standard Model has had amazing success in explaining all measurements so far, there are still many fundamental questions in particle physics which the Standard Model cannot address. Many experiments therefore focus on discovery of new phenomena to elucidate the underlying fundamental truth. In the minimal Standard Model, lepton flavor conservation is built in by hand with the assumption of vanishing neutrino masses. In fact, virtually any new physics or interaction beyond the Standard Model would predict lepton flavor violation (LFV) at some level. LFV processes of major current interest are those with muons such as $\mu^+ \to e^+\gamma$, $\mu^- N \to e^- N$ conversion in nuclei, $\mu^+ \to e^+e^+e^-$ and muonium-antimuonium conversion. In general, a search for muon LFV processes with a sensitivity of about 10^{-12} in a branching ratio could explore a mass scale of several hundred TeV, which is not accessible directly by present accelerators.

2 LFV and SUSY

Recently considerable interest in LFV processes has arisen based on supersymmetric (SUSY) extensions to the Standard Model, in particular supersymmetric grand unified theories (SUSY-GUT). In supergravity-mediated SUSY models, radiative corrections in the renormalization group evolution from the GUT scale to the electroweak scale lead to finite mixing in the slepton mass matrix even when it is assumed to be diagonal at the Plank scale. Recently, Barbieri and Hall found that the slepton mixing thus generated is very large owing to the surprisingly large top-quark Yukawa coupling [1]. Then, $\mu^+ \to e^+\gamma$ occurs due to this slepton mixing through loop diagrams. The predicted branching ratio is at a level which near-future experiments can reach. The branching ratio of $\mu^+ \to e^+\gamma$ predicted in SUSY SU(5) models [2] is shown in Figure 1.

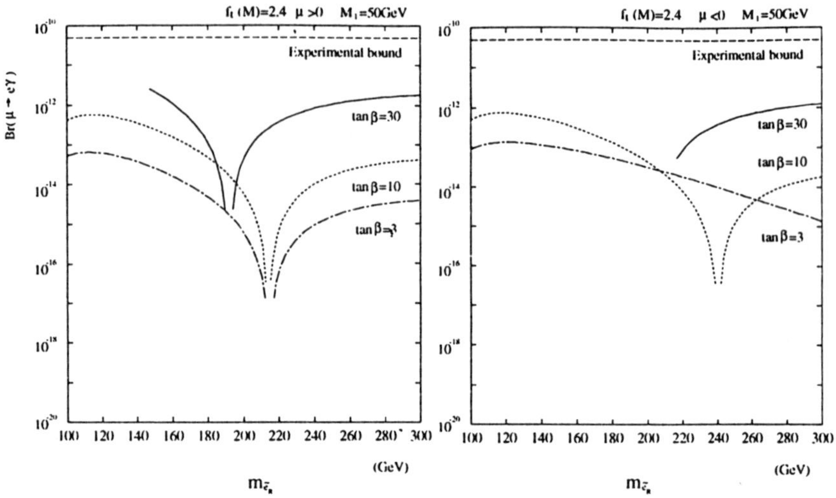

(a) (b)

Figure 1: Predictions of $\mu^+ \to e^+\gamma$ branching ratio in SU(5) SUSY models.

It ranges from 10^{-15} to 10^{-13} for a right-handed slepton mass of 100 to 300 GeV/c^2. The effect of higher dimensional operators in SU(5) SUSY-GUT was considered, and the branching ratio is enhanced for large $\tan\beta$ [3]. The SO(10) SUSY GUT models give an even larger value of 10^{-13} to 10^{-11} via an enhancement of $(m_\tau^2/m_\mu^2) \sim 100$ [1]. This is because of the existence of loop diagrams whose magnitude is proportional to the τ-lepton mass in such models.

Furthermore, the recent experimental hints of non-vanishing neutrino masses and mixing might suggest additional LFV effects in SUSY models. If the observed neutrino mixing and the slepton mixing are related, then the $\mu^+ \to e^+\gamma$ branching ratio is predicted. Fig.2 shows the predictions of $\mu^+ \to e^+\gamma$ decay for a large angle MSW solution (left) and a small angle MSW solution (right) for the solar neutrino deficit problem [5].

SUSY models with R-parity breaking also tend to predict a large $\mu^+ \to e^+\gamma$ branching ratio, which is sensitive to those couplings of R-parity breaking (λ and λ') which are not constrained strongly by the limits on proton decay [6].

If the photonic diagrams dominate, the $\mu^- N \to e^- N$ conversion and $\mu^+ \to e^+e^-e^-$ decay are related to $\mu^+ \to e^+\gamma$ decay as follows: $B(\mu^- N \to e^- N)/B(\mu^+ \to e^+\gamma) \approx 1/200$ and $B(\mu^+ \to e^+e^+e^-)/B(\mu^+ \to e^+\gamma) \approx 1/150$.

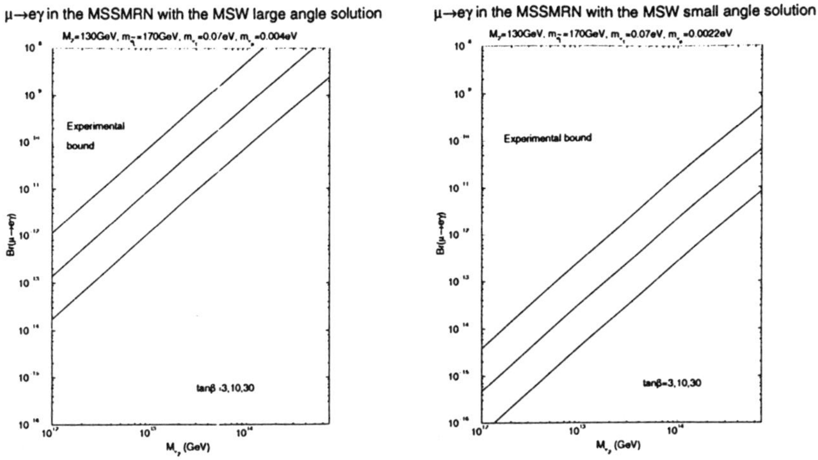

Figure 2: Predictions of $\mu^+ \to e^+\gamma$ branching ratio in MSSM with right-handed Majorana neutrino in the see-saw mechanism. The left is for a large angle MSW solution, and the right is for a small angle MSW solution.

2.1 $\mu^+ \to e^+\gamma$

The event signature of $\mu^+ \to e^+\gamma$ is a positron and a photon in coincidence, moving collinearly back-to-back with energy equal to a half of the muon mass ($m_\mu/2 = 52.8$ MeV). One of the backgrounds is from radiative muon decay, $\mu^+ \to e^+\nu\overline{\nu}\gamma$, where e^+ and photon are emitted back-to-back with the two neutrinos carrying off little energy. The other background is an accidental coincidence of an e^+ in a normal muon decay, $\mu^+ \to e^+\nu\overline{\nu}$, accompanied by a high energy photon. The sources of the latter might be either $\mu^+ \to e^+\nu\overline{\nu}\gamma$ decay, or external bremsstrahlung or annihilation-in-flight of e^+s from normal muon decay.

The most recent search was done by the MEGA collaboration at Los Alamos National Laboratory (LANL). A 90 % C.L. upper limit of $B(\mu^+ \to e^+\gamma) < 1.2 \times 10^{-11}$ has been reported from MEGA [7]. A new experiment on $\mu^+ \to e^+\gamma$ with a sensitivity of 10^{-14} is being considered at PSI where a surface muon beam of a few $\times 10^8$/sec is available [8]. The options are either a solenoid spectrometer or ring-focus spectrometer for the e^+ detection, and either a liquid Xe calorimeter or a scintillating crystals for the photon detection. For a future experiment at the 50-GeV proton synchrotron (50-GeV PS) at the KEK/JAERI project (previously JHF) [9], the use of polarized muons for a search for $\mu^+ \to e^+\gamma$ is considered for further improvement in the background rejection [10].

2.2 $\mu^- - e^-$ Coherent Conversion in Nuclei

The event signature of $\mu^- - e^-$ coherent conversion in nuclei is a mono-energetic single electron of $(m_\mu - B_\mu)$ MeV emitted from muon capture (where m_μ and B_μ are the muon mass and the binding energy of the $1s$ muonic atom respectively). Major backgrounds are muon decay in orbit from a muonic atom (in which the e^- endpoint energy is the same as the energy of the signal), radiative pion and muon capture and cosmic rays. It does not suffer from accidental backgrounds, but the beam purity and the efficiency of cosmic-ray vetoing are crucial.

The ongoing experiment is SINDRUM II at PSI. The 1993 run with a Ti target gave a 90 % C.L. upper limit of $B(\mu^- Ti \to e^- Ti) < 6.1 \times 10^{-13}$, Also for Pb, it gave $B(\mu^- Pb \to e^- Pb) < 4.6 \times 10^{-11}$. For the next stage of the experiment, the "pion-muon converter" (PMC) which consists of a straight superconducting solenoid magnet of 8.5 m in length, designed to reduce beam pion contamination, is being prepared. With this PMC, a sensitivity of 2×10^{-14} is the goal of upcoming runs [11]. There is a new proposal E940 at BNL-AGS called the MECO (Muon Electron COnversion) experiment which aims to search for $\mu^- Al \to e^- Al$ at better than 10^{-16} sensitivity [12]. The setup consists of a graded high-field superconducting solenoid for muon capture, a curved transport solenoid system which selects momentum and sign of charged particles, and a detector with good energy resolution ($\sigma_{RMS} < 300$ keV) which observes only the 105-MeV signal electrons in a second axially graded solenoidal field. A pulsed proton beam of about 1 MHz with pulse length of 50 nsec is used to minimize beam-associated backgrounds, in particular from pions.

3 A New Muon Source with High Intensity

A highly intense source of secondary beams[a] such as pions, muons and neutrinos is being considered for the 50-GeV PS. A new idea on a source as intense as $10^{11} - 10^{12}$ particles/sec have appeared recently in conjunction with $\mu^+\mu^-$

[a] This project is called as PRISM, where PRISM stands for Phase Rotated Intense Secondary Muon beam.

collider R&D [13]. The idea consists of (1) solenoid capture, (2) phase rotation, and (3) muon ionization cooling. Such an approach could be adopted to construct a unique low-energy muon source with high intensity, high brightness, and low momentum spread [14]. The working group for the 50-GeV PS has been formed to study its possibility. A schematic layout of the preliminary design is shown in Fig.3.

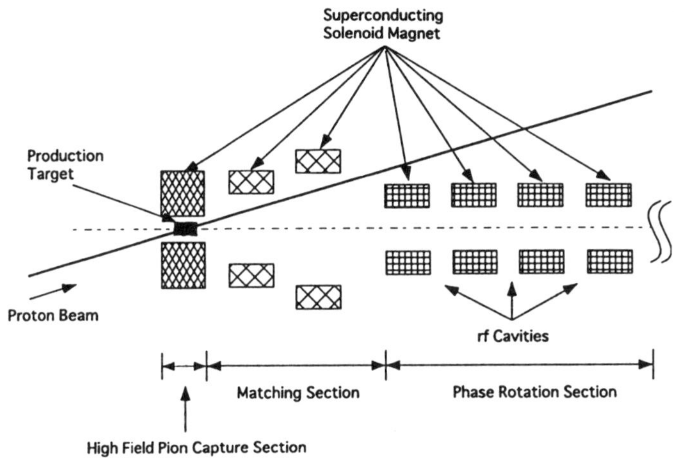

Figure 3: Schematic layout of a new high-intensity source of secondary beams at the 50-GeV PS. It consists of the high field capture section, and the RF section for phase rotation, which might be followed by the ionization cooling section (not shown).

A cooled muon beam with narrow momentum spread would be crucial to carry out LFV searches with stopped muons, since they allow higher muon stopping efficiency with less background and a thinner muon-stopping target (or even a gas target). Each of the muon LFV searches requires specific beam time structure. For example, the searches for $\mu^+ \to e^+\gamma$ and $\mu^+ \to e^+e^-e^+$ need a high repetition rate of a few 10 MHz to 100 MHz, whereas the experiments for $\mu - e$ conversion requires a slower repetition of a few 100 kHz with high beam extinction between beam pulses.

4 Conclusion

We are confident that the potential of particle physics frontier with low energy muons, in particular, a search for lepton flavor violation, is superb to uncover physics beyond the Standard Model. At the same time, extensive R&D to construct a new high-intensity muon source have just begun. With all together, the future of muon physics is undoubtfully magnificent.

Acknowledgments

The author would like to thank all the colleagues in the PRISM project and Prof. Y. Okada for discussions. This work is supported by Grant-in-Aid by the Ministry of Education, Science, Sports and Culture of Japan.

1. L.J. Hall, V.A. Kostelecky and S. Raby, *Nucl. Phys.* B **267**, 415 (1986); R. Barbieri and L. Hall, *Phys. Lett.* B **338**, 212 (1994); R. Barbieri, L. Hall and A. Strumia, *Nucl. Phys.* B **445**, 219 (1995).
2. J. Hisano, T. Moroi, K. Tobe and M. Yamaguchi, *Phys. Lett.* B **391**, 341 (1997); Erratum *ibid.* **B397**, 357 (1997).
3. J. Hisano, D. Nomura, Y. Okada, Y. Shimizu and M. Tanaka, KEK-TH-575, UT-815, OU-HET-294, hep-ph/9805367.
4. J. Hisano, D. Nomura, T. Yanagida, KEK Preprint KEK-TH-548, hep-ph/9711348.
5. J. Hisano and D. Nomura, PRD **59**, 116005 (1999).
6. M. Chaichian and K.Huitu, Phys. Lett. **B384** (1996) 157: K. Huitu *et al.*, *Phys. Lett.* B **430**, 355 (1997).
7. M.L. Brooks *et al.*, preprint hep-ex/9905013.
8. A Letter of Intent for an experiment at PSI, "Search for the decay $\mu^+ \to e^+\gamma$ with a branching ratio sensitivity of 10^{-14}", April, 1998.
9. The JAERI-KEK Joint Project Proposal, 1999, unpublished.
10. Y. Kuno and Y. Okada, *Phys. Rev. Lett.* **77**, 434 (1996): Y. Kuno, A. Maki and Y. Okada, *Phys. Rev.* D **55**, 2517 (1997).
11. A. van der Schaaf and H.C. Walter, private communications.
12. W. Molzon, private communications; M. Bachman, *et al.*, "A Search for $\mu^- N \to e^- N$ with Sensitivity Below 10^{-16}", AGS-940 (1997).
13. The muon collider collaboration, "$\mu^+\mu^-$ Collider – A Feasibility Study", BNL-52503, Fermi Lab-Conf.-96/092, LBNL-38946, 1996.
14. see, *Proceedings of the Workshop on High Intensity Secondary Beam with Phase Rotation*. Kyoto, September 21-22, 1998, unpublished.

CONSTRAINTS OF A PARITY-CONSERVING/TIME-REVERSAL-NON-CONSERVING INTERACTION

WILLEM T.H. VAN OERS

TRIUMF, 4004 Wesbrook Mall, Vancouver, B.C. Canada, V6T 2A3

and

Department of Physics, University of Manitoba

E-mail: vanoers@triumf.ca

Time-Reversal-Invariance non-conservation has for the first time been unequivocally demonstrated in a direct measurement at CPLEAR. One then can ask the question: What about tests of time-reversal-invariance in systems other than the kaon system? Tests of time-reversal-invariance can be distinguished as belonging to two classes: the first one deals with time-reversal-invariance-non-conserving (T-odd)/parity violating (P-odd) interactions, while the second one deals with T-odd/P-even interactions (assuming CPT conservation this implies C-conjugation non-conservation). Limits on a T-odd/P-odd interaction follow from measurements of the electric dipole moment of the neutron ($< 8 \times 10^{-26}$ e.cm [95% C.L.]). It provides a limit on a T-odd/P-odd pion-nucleon coupling constant which is less than 10^{-4} times the weak interaction strength. Experimental limits on a T-odd/P-even interaction are much less stringent. Following the standard approach of describing the nucleon-nucleon interaction in terms of meson exchanges, it can be shown that only charged ρ-meson exchange and A_1-meson exchange can lead to a T-odd/P-even interaction. The better constraints stem from measurements of the electric dipole moment of the neutron and from measurements of charge-symmetry breaking in neutron-proton elastic scattering. The latter experiments were executed at TRIUMF (497 and 347 MeV) and at IUCF (183 MeV). All other experiments, like detailed balance experiments, polarization - analyzing power difference determinations, and five-fold correlation experiments with polarized incident nucleons and aligned nuclear targets, have been shown to be at least an order of magnitude less sensitive. Is there room for further experimentation?

Time-reversal-invariance non-conservation has for the first time been unequivocally demonstrated in a direct measurement at CPLEAR. [1] The experiment measured directly the difference in the transition probabilities P($\overline{K}^0 \rightarrow K^0$) and P($K^0 \rightarrow \overline{K}^0$). A non-zero value of this difference gives model independent evidence of time-reversal-invariance non-conservation. It is not necessary to assume that direct CP violation in the decay of K^0 and $\overline{K}^0$ can be neglected. The result obtained for A_T with

$$A_T = \frac{R(\overline{K}^0 \rightarrow K^0) - R(K^0 \rightarrow \overline{K}^0)}{R(\overline{K}^0 \rightarrow K^0) + R(K^0 \rightarrow \overline{K}^0)}$$
$$= (6.6 \pm 1.3_{\text{stat.}} \pm 1.0_{\text{syst.}}).10^{-3}$$

is in good agreement with the measure of CP violation in neutral kaon decay. Starting with CPT conservation and the well established non-conservation of CP in kaon decays, time-reversal-invariance should also be broken. The CPLEAR measurement is the first direct confirmation of that. The question that one then can ask is: what about time-reversal-invariance non-conservation in systems other than the kaon system?

Tests of time-reversal-invariance can be distinguished as belonging to two classes: the first one deals with time-reversal-invariance-odd (T-odd)/parity violating (P-odd) interactions, while the second one deals with T-odd/P-even interactions (assuming CPT invariance this implies C-conjugation non-conservation). However, it should be noted that constraints on these two classes of interactions are not independent since the effects due to T-odd/P-odd interactions may also be produced by T-odd/P-even interactions in conjunction with standard model parity violating radiative corrections. The latter can occur at the 10^{-7} level and consequently presents a limit on the constraint of T-odd/P-even interactions, that can be derived from experiments. Limits on a T-odd/P-odd interaction follow from measurements of the electric dipole moment of the neutron (which currently stands at $< 8 \times 10^{-26}$ e.cm [95% C.L.]). It provides a limit on a T-odd/P-odd pion-nucleon coupling constant which is less than 10^{-4} times the weak interaction strength. Measurements of atomic electric dipole moments of ^{129}Xe and ^{199}Hg ($< 8 \times 10^{-28}$ e.cm [95% C.L.]) give similar constraints. [see ref. 2]

Experimental limits on a T-odd/P-even interaction are much less stringent. Following the standard approach of describing the nucleon-nucleon interaction in terms of meson exchanges, it can be shown that only charged ρ-meson exchange and A_1-meson exchange can lead to a T-odd/P-even interaction.[3] The better constraints stem from measurements of the electric dipole moment of the neutron and from measurements of charge symmetry breaking in neutron-proton elastic scattering. All other experiments, like gamma decay experiments [4], detailed balance experiments, polarization - analyzing power difference measurements, and five-fold correlation experiments with polarized incident nucleons and aligned nuclear targets, have been shown to be at least an order of magnitude less sensitive. Haxton, Hoering, and Musolf [2] have deduced constraints on a T-odd/P-even interaction from nucleon, nuclear and atomic electric dipole moments with the better constraint coming from the electric dipole moment of the neutron. In terms of a ratio to the strong ρ-meson nucleon coupling constant, they deduced for the T-odd/P-even ρ-meson nucleon coupling: $|\overline{g}_\rho| < 0.53 \times 10^{-3} \times |f_\pi^{\mathrm{DDH}}/f_\pi^{\mathrm{meas.}}|$. Here one should note that the ratio of the theoretical to the measured value of f_π may be as large as 15! [5]

In the standard model a T-odd/P-even nucleon-nucleon interaction can hardly be accommodated. It requires C-conjugation non-conservation, which cannot be introduced at the first generation quark level. It can neither be introduced into the gluon self-interaction. Consequently one needs to consider C-conjugation non-conservation between quarks of different generations and/or between interacting fields. [6]

The nucleon-nucleon scattering matrix, assuming conservation of angular momentum, parity, time-reversal-invariance, and isospin, consists of five complex amplitudes. If isospin is broken (charge symmetry breaking in the neutron-proton system), the neutron-proton scattering matrix contains a sixth complex amplitude. If in addition, one no longer assumes time-reversal-invariance there are two additional amplitudes; one of these is simultaneously timereversal-invariance violating and charge symmetry breaking. The proton-proton system may only contain one additional, sixth, time-reversal-invariance non-conserving amplitude. In a partial wave decomposition the four lowest, parity conserving, transition amplitudes in which time-reversal-invariance violation may occur are ${}^3S_1 \leftrightarrow {}^3D_1$, ${}^1P_1 \leftrightarrow {}^3P_1$, ${}^1D_2 \leftrightarrow {}^3D_2$, ${}^3P_2 \leftrightarrow {}^3F_2$. Of these only the latter one is allowed for identical nucleons because of the Pauli exclusion principle. Thus time-reversal-invariance violating effects, if these exist, are strongly suppressed in the proton-proton and neutron-neutron systems. Null tests of time-reversal-invariance do not exist in the nucleon-nucleon system. In fact, polarization - analyzing power difference determinations are based on two independent measurements with their own systematic errors in polarimeter analyzing power and beam polarization calibrations. The best calibration standards to date carry uncertainties of a few parts in 10^3. Writing the polarization - analyzing power difference as $\epsilon(1 - \mathrm{D})$, with ϵ the normalized spin-flip differential cross sections difference and with D the depolarization parameter, one can show that the difference determinations were made in an angular region where D is close to one, causing insensitivity to time-reversal-invariance non-conservation.

Charge symmetry breaking (CSB) in neutron-proton scattering manifests itself as a non-zero difference of the neutron (A_n) and proton (A_p) analyzing powers. The three precision experiments performed to date (at TRIUMF at 477 MeV [7] and at 347 MeV [8], and at IUCF at 183 MeV [9]) have unquestionably shown that charge symmetry is broken and that the results for $\Delta A = A_n - A_p$, at the zero-crossing angle of the average analyzing power, are very well reproduced by meson exchange model calculations (see Fig. 1). A T-odd/P-even interaction contains a piece which is simultaneously charge symmetry breaking. Thus, Simonius [10] deduced an upper limit on a T-odd/P-even CSB interaction from a comparison of the experimental results with the

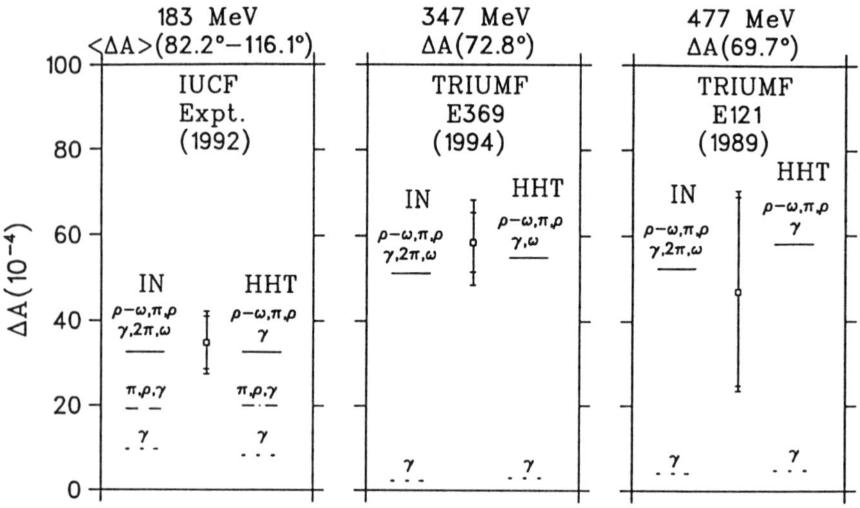

Figure 1: Experimental results of ΔA at the zero-crossing angle at incident neutron energies of 183, 347, and 477 MeV compared with theoretical predictions of Iqbal and Niskanen and Holzenkamp, Holinde, and Thomas. The inner error bars present the statistical uncertainties; the outer error bars have the systematic uncertainties included (added in quadrature). For further details see Ref.8.

theoretical predictions for the above mentioned three CSB experiments. The upper limit so derived is $|\overline{g}_\rho| < 6.7 \times 10^{-3}$ [95% C.L.]. This is therefore comparable to the upper limit deduced from the electric dipole moment of the neutron, taking the present experimental limit of f_π, and is considerably lower than the limits inferred from direct tests of a T-odd/P-even interaction. But as remarked above, it is inconceivable in the standard model to account for such an interaction. Nevertheless, there is a need to clarify the experimental constraint on a T-odd/P-even interaction by providing a BETTER experimental limit.

Such a better experimental constraint may be provided by an improved upper limit on the electric dipole moment of the neutron. In fact a new measurement with a sensitivity of 4×10^{-28} e.cm has been proposed at the Los Alamos Neutron Science Center. [11] But performing an improved n–p elastic scattering CSB experiment appears to be a very attractive alternative. One can calculate with a great deal of confidence the contributions to CSB due to one-photon exchange (the neutron magnetic moment interacting with the current of the proton) and due to the $n-p$ mass difference affecting charged one-pion and rho-meson exchange. Furthermore, one can select an energy where the $\rho^0-\omega$ meson mixing contribution changes sign at the same angle where the average of the analyzing powers A_n and A_p changes sign and therefore does not contribute. This occurs at an incident neutron energy of 320 MeV and is caused by the particular interplay of the $n-p$ phase shifts and the form of the spin/isospin operator for the $\rho^0-\omega$ mixing term. At this energy also the one-photon contribution changes sign at the same angle. The contribution due to two-pion exchange with an intermediate Δ is expected to be no more than one tenth of the overall CSB effect, essentially representing the magnitude of the theoretical uncertainty (see Fig.2). [12] It has been shown that simultaneous $\gamma-\pi$ exchanges can only contribute through second order processes and can therefore be neglected. [13] Also the effects of inelasticity are neglegibly small at 320 MeV. It appears therefore well within reach to reduce the theoretical uncertainty in the comparison between experiment and theory. Subtracting the calculated difference in the neutron and proton analyzing powers from the measured difference permits establishing an upper limit on a T-odd/P-even/CSB interaction.

The second TRIUMF experiment measuring CSB in $n-p$ elastic scattering at 347 MeV obtained the result

$$\begin{aligned}\Delta A &= A_n - A_p \\ &= (59 \pm 7[\text{stat.}] \pm 7[\text{syst.}] \pm 2[\text{syst.}]) \times 10^{-4}\end{aligned}$$

at the zero-crossing angle of the average of A_n and A_p. In the experiment polarized neutrons were scattered from unpolarized protons and *vice versa.* The polarized (or unpolarized) neutron beam was obtained using the (p,n) reaction with a 369 MeV polarized (or unpolarized) proton beam incident on a 0.20 m long LD_2 target. At the energies of the TRIUMF experiments one makes use of the large sideways-to-sideways polarization transfer coefficient r_t at 9° in the lab. The only difference in obtaining the unpolarized and polarized neutron beams was the turning off of the pumping laser light in the optically pumped polarized ion source (OPPIS). The polarized proton target was of the frozen spin type with butanol beads as target material. The same target after depolarization was used as the unpolarized proton target. Great care was taken that the two interleaved phases of the experiment were performed with identical beam and target parameters except for the polarization states. Scattered neutrons and recoiling protons were detected in coincidence in the c.m. angular range 53.4° to 86.9° in two left-right symmetric detector systems. Rather than measuring A_n and A_p directly (which would be troubled by lacking polarization calibration standards of the required precision), the zero-crossings

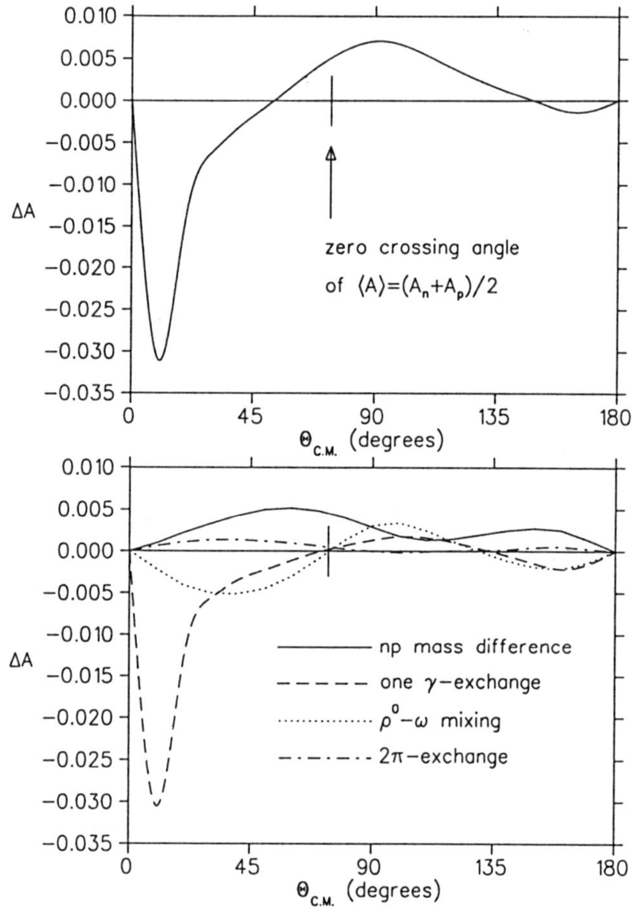

Figure 2: Angular distributions of ΔA and of the different contributions to ΔA at an incident neutron energy of 320 MeV. Note that the $\rho^0 - \omega$ mixing contribution passes through zero at the same angle as the average of A_n and A_p.

of A_n and A_p were determined in an interleaved procedure. The difference $A_n - A_p$ followed by multiplying the difference in the zero-crossing angles by the average slope of the analyzing powers (the experiment measured the slope of A_p at the zero-crossing angle). The execution of the experiment depended on a great deal of simultaneous monitoring and control measurements (see Fig. 3 for a schematic view of the experiment).

Both the statistical and systematic errors obtained in the experiment can be considerably improved upon. With the OPPIS developments that have taken place in the intervening years and with the addition of a biased Na-ionizer cell, it will be possible to obtain up to 50 μA of beam with a polarization of 80% incident on the neutron production LD_2 target (a factor of 50 increase in neutron beam intensity). A 342 MeV proton beam incident on a 0.20 m long LD_2 target would present a heat load of 500 W. LH_2 targets allowing such heat loads have been developed for electron scattering experiments. However, reducing the LD_2 target thickness to 0.05 m would give a better defined neutron energy spectrum reducing the uncertainty in the apparent difference in energies of the A_n and A_p measurements, which contributes to the systematic error. The polarized proton target should again be of the frozen spin type; choosing a target material with improved ratio of free protons to bound protons and a lower magnetic holding field (< 0.2 T) would greatly improve on the systematic error by reducing the uncertainty in the background subtraction. It appears entirely feasible to reduce the statistical and systematic errors by a factor three to four. Such an experiment would constitute a measurement of CSB in $n-p$ elastic scattering of unprecedented precision of great value on its own and simultaneously providing the best upper limit on a T-odd/P-even interaction.

True null tests of time-reversal-invariance do not exist for spin 1/2 particles scattered from spin 1/2 particles. However, null tests exist as transmission measurements for spin 1/2 particles interacting with aligned nuclear targets. [14] In such tests one measures the total cross section asymmetry $A_{y,xz}$ of vector polarized spin 1/2 particles interacting with an aligned nuclear target (e.g. a tensor polarized spin 1 deuteron target). Huffman *et al* [15] have extracted the five-fold correlation parameter $A_{y,xz}$ by observing polarized neutron transmission through nuclear spin-aligned ^{165}Ho. This resulted in an upper limit for $|\overline{g}_\rho|$ of 5.9×10^{-2} even though the measured value of $A_{y,xz}$ was $(8.6 \pm 7.7) \times 10^{-6}$. It is only the valence proton in ^{165}Ho which contributes to $A_{y,xz}$. Storage rings provide a completely different environment, with the advantages outweighing the disadvantages possibly, for high precision tests of fundamental symmetries like testing time-reversal-invariance in the GeV range. At COSY a proton-deuteron transmission experiment has been proposed. For this test of a T-odd/P-even interaction the COSY ring will serve as acceler-

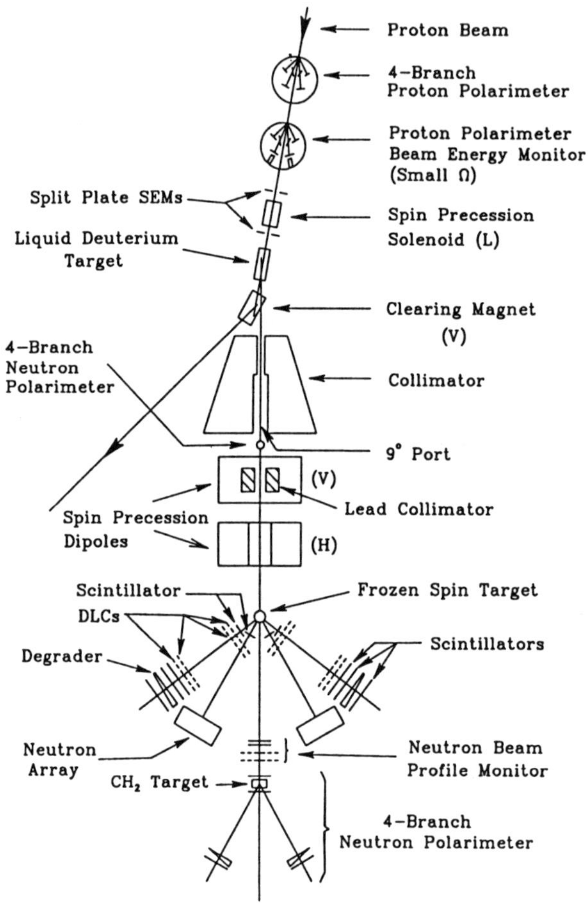

Figure 3: Schematic view of the TRIUMF CSB experiments.

ator, forward spectrometer, and detector. [16] An accuracy of 10^{-6} in $A_{y,xz}$ has been planned for the experiment, but the result on $|g_\rho|$ will depend on how inelasticity is accounted for. It remains to be seen if the experiment will reach a sensitivity comparable to the present $n-p$ CSB experiments.

References

1. A. Angelopoulos *et al.* (CPLEAR Collaboration), Phys. Lett. **B444**, 43 (1998).
2. W.C. Haxton, A. Hoering, and M.J. Musolf, Phys. Rev. **D50**, 3422 (1994).
3. M. Simonius, Phys. Lett. **58B**, 147 (1975).
4. For a recent review see: F. Boehm in Symmetries and Fundamental Interactions in Nuclei, ed. W.C. Haxton and E.M. Henley (World Scientific, Singapore, 1997).
5. S.A. Page *et al.*, Phys. Rev. **C35**, 1119 (1987); M. Bini *et al.*, Phys. Rev. **C38**, 1195 (1988).
6. M. Simonius, in Intersections between Particle and Nuclear Physics - 1984, ed. R.E. Mischke, AIP Conference Proceedings No 123, p.1115.
7. R. Abegg *et al.*, Phys. Rev. **D39**, 2464 (1989).
8. J. Zhao *et al.*, Phys. Rev. **C57**, 2126 (1998).
9. S.E. Vigdor *et al.*, Phys. Rev. **C46**, 410 (1992).
10. M. Simonius, Phys. Rev. Lett. **78**, 4161 (1997).
11. LANSCE proposal, spokespersons M.J.Cooper and S.K. Lamoreaux.
12. J. Iqbal, private communication (1999).
13. J.L. Friar and S.A. Coon, Phys. Rev. **C53**, 588 (1996).
14. H.E. Conzett, Phys. Rev. **C48**, 423 (1993).
15. P.R. Huffman *et al.*, Phys. Rev. **C55**, 2684 (1997).
16. COSY proposal, spokespersons P.D. Eversheim and F. Hinterberger.

CONSERVED VECTOR CURRENT TEST/SECOND CLASS CURRENT SEARCH IN THE A = 8 ISOTRIPLET

K. A. SNOVER, J. P. S. van SCHAGEN, D. W. STORM AND M. P. KELLY
Dept. Physics, NPL Box 354290,
University of Washington,
Seattle, WA 98195, USA
E-mail: snover@npl.washington.edu

A new experiment is underway to make a precision determination of the isovector M1 and E2 decays of the 2^+ isobaric analog resonance doublet in ^{8}Be for the purpose of an improved CVC/SCC test in the A = 8 isotriplet. Assuming CVC holds, our current sensitivity for SCC is at the level of 5% of weak magnetism.

1 Summary of results

The A = 8 and 12 isospin triplets present a unique opportunity for precision experiments to search for Second Class Currents (SCC), which produce recoil-order contributions to nuclear β-decay. As a consequence of G-parity symmetry, SCC are absent (forbidden) in the Standard Model, except for a small 'effective' SCC contribution due primarily to the up–down quark mass difference. After an interesting history spanning several earlier decades in which large SCC were eventually ruled out, current experiments are achieving much higher precision. The best sensitivity to date has been achieved in the A = 12 system by the Osaka group [1], who recently reported a small positive effect at the level of about 6% of the weak magnetism contribution given by CVC (Conserved Vector Current theory). This exciting result appears to be larger than the contribution expected due to the up–down quark mass difference [2], and is thus not understood.

It is important to perform precision searches of this sort in other systems, for several reasons. Because the observed effect in the A = 12 system is small and has a substantial uncertainty, observation of a positive effect in another system would provide independent confirmation of nonzero SCC in measurements with very different systematic errors. Secondly, measurements in several different systems are necessary if one is to understand the relative importance of nucleon vs. nuclear (many body) effects [3]. The A = 8 isotriplet offers the opportunity for a CVC/SCC search with high precision. With this goal in mind, we are remeasuring the γ-decay of the ^{8}Be 2^+ isobaric analog resonance doublet, in order to determine the weak magnetism plus second forbidden vector contributions to the β-decay of ^{8}Li and ^{8}B expected from CVC. These results will then be compared to the measured ^{8}Li – ^{8}B $\beta - \alpha$ angular correlation in order to determine the SCC contribution. The results presented here constitute a progress report in an ongoing study.

Our new measurement of the ^{8}Be γ-decay is being carried out at the Nuclear Physics Laboratory, University of Washington. We use the ^{4}He(α,γ)^{8}Be reaction, with an α-beam of $\sim$34 MeV from the Tandem-Linac and a cylindrical 9.5cm diameter gas cell target of He at 1 atm pressure. The γ-rays are detected in 3 large NaI spectrometers placed at angles of 40°, 90° and 140° relative to the beam axis. The accelerator beam energy is monitored frequently by elastic scattering at $\pm 15°$ from a thin carbon foil mounted in a separate chamber just downstream of the 90° analyzing magnet. The ^{4}He(α,γ)^{8}Be measurements are carried out with a short gas cell with Kapton windows which are viewed by the γ detectors; the window background is subtracted using runs in which the helium in the cell is replaced by hydrogen at a pressure adjusted to produce the same beam energy loss as with helium. In this manner, detailed excitation curves over the E_x = 16.62 and 16.92 MeV isospin-mixed resonance doublet have been measured from E_α = 33 to 35 MeV.

A lot is known about this resonance doublet, including the resonance energies, widths, and isospin mixing coefficients. As a result, the measured excitation curves may be fit with an R-matrix expression in which all parameters are fixed except for four: the isoscalar and isovector M1 and E2 decay strengths. One additional fit parameter is included corresponding to the beam energy calibration. The simplicity of the reaction spin sequence results in a γ-ray angular distribution that depends only on a single parameter, the E2/M1 mixing ratio. Thus simultaneous measurements at 3 angles overdetermine the mixing ratio at every E_α.

In addition, one must know the target thickness and detector response functions and efficiencies. The latter are determined at E_γ = 15.1 MeV from a comparison of the p-γ coincidence and singles yield from the ^{10}B(^{3}He,p-$\gamma_{15.1}$)^{12}C reaction, and also from the ^{12}C(p,γ)^{13}N reaction at the lowest T = 3/2 resonance. In addition the response function and efficiency are measured at E_γ = 8.3 MeV using the ^{13}C(^{3}He,p)^{15}N reaction and the response function is measured at E_γ=13.1 MeV using the ^{15}N(p,γ)^{16}O reaction. The computer code GEANT is used as a guide for interpolating the efficiencies as a function of γ-ray energy. Because the final state is broad and consists of more than one level, a precision spectrum shape measurement is also necessary to determine the final-state distribution. Such a measurement has been carried out with a long gas cell with shielded windows and a single NaI spectrometer at 90° to the beam axis.

The current goal of this work is to obtain the isovector radiative width $\Gamma_\gamma(M1)$ to a precision of $\pm 3\%$ and the isovector E2/M1 mixing ratio δ_1 to

± 0.013. These quantities enter in the CVC/SCC test in the following manner: the difference in a_2 coefficients (the coefficients of $cos^2(\theta_{\beta\alpha})$) for the $\beta-\alpha$ correlation in ^{8}Li and ^{8}B determines the quantity $\delta^-(E_\beta)M_n/E_\beta$. Now $\delta^-(E_\beta)M_n/E_\beta$ may have a weak dependence on E_β and a complete analysis must take this into account. However, for the purpose of the present discussion, we write the following approximate relation appropriate for energy averaged quantities: $\delta^-(E_\beta)M_n/E_\beta = b[1-\delta_1]/Ac + d_{II}/Ac$ where $b = AM_n[6\Gamma_\gamma(M1)/\alpha E_\gamma^3]^{1/2}$ is the weak magnetism contribution, δ_1 represents the second order CVC contribution due to the analog E2 γ-decay, A=8 is the nucleon number and M_n is the nucleon mass, c is the allowed Gamow-Teller matrix element, α is the fine structure constant and d_{II} is the SCC contribution.

There are three independent measurements of the ^{8}Li - ^{8}B $\beta-\alpha$ correlation. The results are $\delta^-(E_\beta)M_n/E_\beta = 7.0\pm0.5$ (Tribble, 1975)[4], 6.5±0.2 (McKeown, 1980)[5], and 7.1±0.2±0.2 (Beck, 1998, preliminary)[6]. The average is 6.72±0.20 (χ^2=1.5), where the error has been multiplied by $(\chi^2)^{1/2}$. In order to compute the effective value of $b[1-\delta_1]/Ac$ from the γ-decay data, one must average properly over the broad final state. In this comparison, we use the value of Γ_γ(M1) and the spectrum shape obtained in our Mark I experiment, which correspond to an energy-averaged value of b/Ac = 7.0±0.28 [7], together with our new preliminary value for δ_1 = 0.018±0.013. The result is $b[1-\delta_1]/Ac$ = 6.87±0.29. Hence our present, preliminary result for the SCC contribution, relative to the allowed Gamow-Teller term, is d_{II}/Ac = -0.16±0.35. Relative to weak magnetism, the SCC contribution is d_{II}/b = -0.02± 0.05.

It is interesting to compare our result for SCC with that obtained by the Osaka group in A = 12. They obtain, using a different notation, $2M(f_T/f_A)$ = 0.22±0.05(stat)±0.15(syst)±0.05(theor.) for the SCC contribution, which may be compared directly with our value for d_{II}/Ac quoted above. Thus the uncertainty in our result is within a factor of 2 of that obtained in A = 12. In fact, if one expresses the SCC contribution as a fraction relative to weak magnetism, then the uncertainties in the A = 8 and A = 12 results are comparable.

Our present efforts are directed toward finalizing the value and reducing the uncertainty on $b[1-\delta_1]/Ac$. We are also planning further effort to finalize the 1998 Beck value for $\delta^-(E_\beta)$.

References

1. T. Minamisono *et al.*, *Phys. Rev. Lett.* **80**, 4132 (1998).
2. H. Shiomi, *Nucl. Phys.* **A603**, 281 (1996).

3. Kuniharu Kubodera, Jean Delorme and Mannque Rho, *Phys. Rev. Lett.* **38**, 321 (1977).
4. R. E. Tribble and G. T. Garvey, *Phys. Rev. C* **12**, 967 (1975).
5. R. D. McKeown, G. T. Garvey, and C. A. Gagliardi, *Phys. Rev. C* **22**, 738 (1980).
6. M. Beck *et al.*, *Proc. of the 6th Conf. on Intersections of Particle and Nuclear Physics* , (to be published).
7. L. De Braeckeleer *et al.*, *Phys. Rev. C* **51**, 2778 (1995).

Ultra Cold Neutron Production
with
Medium Energy Proton Accelerater

Yasuhiro Masuda
Neutron Science Laboratory, KEK
1-1 Oho, Tsukuba-shi, 305-0801 Japan

Abstract: UCN production in a medium-energy-proton beamline is proposed. UCNs are produced upon neutron-phonon collision in superfluid helium. Produced UCNs are vertically extracted from the superfluid helium to an experimental chamber for neutron β decay asymmetry or EDM experiment through a strong magnetic field of solenoids, which works as a siphon. UCNs are almost completely polarized upon passage through the magnetic siphon.

Neutrons whose energy is lower than the potential which is formed by a neutron-scattering-length density of constituent nuclei in matter, are called Ultra Cold Neutrons (UCN). A typical value of the potential is 200 neV for certain metals. UCN is completely reflected from the surface of matter. As the result, UCN is confined in a bottle. This property of UCN is quite useful for neutron electric-dipole-moment (EDM), life-time and β-decay-asymmetry measurements etc. In the EDM experiment, neutron confinement is crucial, since the EDM signal is proportional to the time spent in electro-magnetic field. The systematic effect due to a neutron's motion in an electric field, an effect of the relativity is greatly reduced upon averaging neutron's motions in a bottle. In the life time measurement, the normalization for the initial number of decaying neutrons which is a dominant source of experimental errors in beam experiments is not required for confined-neutron experiments. In the β-decay-asymmetry experiment, the present error of the asymmetry results from polarization uncertainty. If we use UCN, the polarization uncertainty is greatly improved. The UCN energy at 300 neV corresponds to a neutron magnetic potential of $\sim$ 5 T, therefore almost 100% neutron polarization is realized by applying such a strong magnetic field.

Here, we propose a new method to obtain intense UCN in a spallation neutron source. In 1977, Golub and Pendlebury proposed UCN production via neutron-phonon collision in superfluid helium.[1] The neutron dispersion curve concerning energy and momentum intersects with the phonon-dispersion curve of superfluid ^{4}He at a neutron wave-length of 9 Å. At the intersection, almost all neutron's energy and momentum can be transferred to a phonon excitation. If the kinetic energy after the collision becomes lower than the potential of the superfluid-helium bottle, neutrons are confined. If we place the superfluid-helium bottle in a reactor, a tremendous number of UCN is obtained. For example, a UCN density of more than $10^5/cm^3$ is expected in a thermal flux of 10^{14} n/cm^{2}s. The UCN production was experimentally confirmed by ILL [2] and KEK [3]. However, we should remove formidable γ-heating of 1 W/g from the liquid helium at 0.5 K. It is almost impossible with present cryogenic technology. In 1987, Yoshiki et al. suggested a possibility to use a spallation-neutron source, since the spallation source has advantages in freedom of geometry choice and controllability of source power to keep the γ heating below the limit of the cryogenic technology.[4] But, γ heating was still too high to place the helium cryostat in their proposal.

We developed the idea of the spallation source, since it has an additional advantage in neutron/gamma ratio. A 5-MW spallation source is believed to correspond to the 60-MW reactor at ILL,[5] which means γ heating is much smaller in the spallation source. Here, we assume γ intensity is proportional to the source powers. We estimated the thermal neutron flux of a spallation source, which comprises a thermal moderator of 1-m-diameter spherical heavy-water and a lead-target of 20-cm-length and 10-cm-diameter. 10-cm-thick water surrounds the thermal moderator. The target is placed in the center of the thermal moderator, where spallation neutrons are produced by a medium-energy-proton beam. For the proton beam of 600-MeV and 2-mA, the neutron-production rate becomes 1.3×10^{17} n/s. Neutrons are moderated and diffuse in the heavy water. A 30-cm length is required for the neutron moderation down to thermal energies. It takes 53 μs for the moderation. The neutron diffusion time is 3.5 ms for a 0.5-m length, which is the distance from the source center to the wall of heavy water vessel. The neutron flux is expressed in terms of the product of the neutron density and velocity. If we assume the thermal neutron density in the moderator is obtained using the diffusion time as a neutron-accumulation time and the neutron production rate, we obtain an average thermal flux of 10^{14} n/cm^{2}s. We confirmed this rough estimation by a Monte-Carlo simulation. We used the LCS code of Los Alamos for the simulation. The result shows that a 1.2-MW spallation-source in this geometry realizes the same thermal flux as ILL reactor, but the γ production is much smaller. To further decrease of γ heating, we can place a bismuth shield surrounding the spallation target. However, we can not shield γ rays from neutron radiative capture in the helium-cryostat material. If we use stainless steel, which is commonly used as the cryostat material, the neutron-capture γ-ray intensity becomes too large in the high-neutron flux. Zirconium has a very small neutron-capture cross-section. We tested the cold brittleness of zirconium metal at a temperature of 4 K[6] and found it to be a suitable material for the low temperature cryostat.

We propose to build a superfluid-helium UCN-source, which is schematically shown in Fig. 1. A 16-cm-diameter zirconium-cylinder filled with helium at 0.5 K is inserted vertically into the 56-cm-diameter and 80-cm-height zirconium-cylinder filled with heavy water at 20 K which works as a cold neutron moderator. The cold neutron moderator is placed in a thermal moderator, which is heavy water at 300 K in a 100-cm-diameter and 130-cm-height tank surrounded by a 10-cm-thick water. The lead target mentioned above is placed in the heavy-water tank below the cold moderator. It is surrounded by a 10-cm-thick bismuth for γ shielding. We estimated the cold neutron flux and the heat deposit using the LCS code.[7,8] The cold-neutron flux was 0.4×10^{12} n/cm^{2}s in the liquid helium for a proton beam of 600 MeV and 20 μA. Here, the cold-neutron temperature is assumed to be 20 K. The heat deposit is less than 1 W in the liquid helium and 30 W in the 20-K heavy-water. Such amount of heat load can be removed with the present cryogenic technology. The cold neutron flux of 0.4×10^{12} n/cm^{2}s in the liquid helium results in UCN-production rate of 700 UCN/cm^{3}s. If we assume a UCN-storage time of 300 s in the liquid-helium bottle, UCN density becomes 2.1×10^5 UCN/cm^3. The values are extremely large compared with the maximum UCN density of 87 UCN/cm^3 which is obtained at the ILL turbine UCN source.

USCs are extracted from the liquid-helium bottle to an experimental apparatus for β decay or EDM experiment through a magnetic siphon, where a strong magnetic field of 6 T is applied in compensation for gravity. Since neutrons of one spin state can go upward through the magnetic siphon, UCNs are completely polarized after passage thorough the 6 T field. In the vertical extraction, liquid helium is kept in the bottle without wall, therefore, UCNs are transported to the experimental apparatus without serious loss upon transmission through the wall. 100% polarized neutrons are quite useful for β decay asymmetry experiment, since the present experimental accuracy is limited by the neutron polarization error. The neutron β decay asymmetry and life time fix the coupling constants of the nuclear weak interaction, G_V and G_A without nuclear uncertainty. The experimental value of G_V obtained from super allowed Fermi transitions contradicts the unitarity of Kobayashi-Maskawa matrix. Since the neutron life time is obtained with enough accuracy, the asymmetry of UCN β decay with relative error of 10^{-3} will show a clear answer for this problem. Polarized UCN is also useful for EDM experiment. A CP violation theory, which explains the baryon asymmetry in the universe expect a neutron EDM of 10^{-26} to 10^{-27}. Intensive polarized UCN is required for the improved EDM experiment.

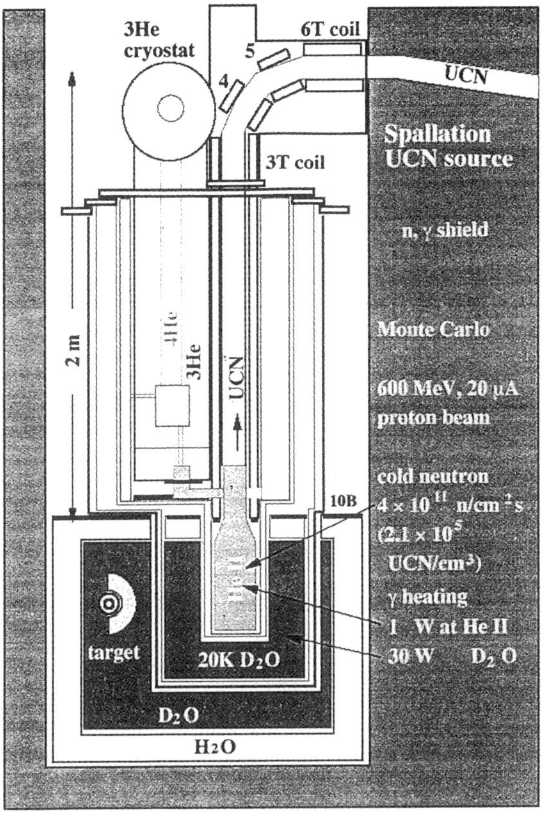

Fig. 1 UCN production in superfluid helium.

[1] R. Golub and J.M. Pendlebury, Phys. Lett. 50A(1977)337.

[2] R. Golub, C. Jewel, P. Ageron, W. Manpe, B. Heckle and A. Kilvington, Z. Phys. B51(1983)187.

[3] H. Yoshiki, K. Sakai, M. Ogura, T. Kawai, Y. Masuda, T. Nakajima, T. Takayama, S. Tanaka, and Y. Yamaguchi, Phys. Rev. Lett. 68(1992)1323.

[4] H. Yoshiki, S. Ishimoto and M. Utsuro, Z. Phys. B67(1987)161.

[5] D. Filges et al. Proc. ICANS-IV, KENS Report II(1981)249.

[6] Y. Masuda, K. Hosoyama, H. Nakai and K. Morimoto, to be published.

[7] Y. Masuda, to be published in Proc. International Workshop on "Elementaly Particle Physics with Neutrons", Oct., 1998, Grenoble.

[8] Y. Kiyanagi and T. Kamiyama, private comunication.

Solar and Atmospheric Neutrino Oscillations —Super-Kamiokande Results—

Y.Suzuki
(for the Super-Kamiokande Collaboration)
Kamioka Observatory, Institute for Cosmic Ray Research, The University of Tokyo, Higashi-Mozumi, Kamioka, Gifu 506-1205, Japan

The strong evidence for oscillation[1] of atmospheric neutrinos has been obtained by Super-Kamiokande. The data is updated to 736 effective days. A conclusive result comes from the distant dependent suppression of atmospheric muon neutrinos—the observation of the up/down asymmetry in the zenith angle distribution as well as the small value of $R=(\mu+e)_{data}/(\mu+e)_{MC}$. The data is consistent with the neutrino oscillation of $\nu_\mu \rightarrow \nu_\tau$ or ν_s. The allowed region in the oscillation parameter space is $\sin^2 2\theta > 0.9$ and $1.5\times 10^{-3}\mathrm{eV}^2 < \Delta m^2 < 6\times 10^{-3}\mathrm{eV}^2$. The supported evidence was also obtained by the zenith angle distribution of the upward through going muons and the ratio of the upward stopping to the upward through going muons.

1 detector

Super-Kamiokande, located at Kamioka Observatory[2], Institute for Cosmic Ray Research, the University of Tokyo, is a 50,000 tons imaging water Cherenkov detector which is placed underground, 1,000m water equivalent and at 137.32 degree east longitude and 36.43 degree north latitude. The inner detector consists of 32 ktons of water, viewed by 11,146 50 cm diameter PMTs. The outside of the inner detector is surrounded by an outer detector, ~2.7m thick water layers, which serves as an active shield against incoming radiations, like γ-rays and neutrons. The active part of the anti-detector is viewed by out-facing 1,881 20 cm diameter PMTs attached with the wave length shifter plates. The details on the detector will be found in ref[3]

1.1 Operation

Super-Kamiokande has been operated since April, 1996. The cumulative live-time is nearly 90%. Most of the dead time is due to the time for calibrations by the electron LINAC[4] and other methods. The number of dead inner PMTs, as of June, 1998 (about 2 years after the beginning of the experiment), is about 110 (~1%) for the inner detector and that for the outer detector is about 150.

2 Atmospheric Neutrinos [5,6,7]

Atmospheric neutrinos [8,9] are produced by the decay of the particles produced by the interaction of the primary cosmic rays on the earth's atmosphere. The peak energy of the neutrinos which produces the observable neutrino events in the detector is around 1 GeV, taking into account the neutrino cross sections.

The earth's magnetic field determines the minimum energy of the primary cosmic rays with which they are able to reach the earth's atmosphere. The geomagnetic cut-off mostly effects on the flux of neutrinos below a few hundred MeV. The magnitudes of this effect is evaluated to be a few percent at the low energy neutrinos and negligible for the high energy neutrinos above a few hundred MeV. This geo-magnetic effect works so as to increase those neutrinos coming up from the bottom of the detector at the Super-Kamiokande site, which is opposite to the effect of the neutrino oscillations.

The geo-magnetic effect causes a so called east-west effect. The geomagnetic field suppresses the primary particles entering from east than from west, due to the positive charge of the most cosmic rays. As a consequence neutrinos coming from the east has a deficit. In order to observe this effect, we selected those events entering the detector almost horizontally and look at the azimuthal angle distributions. The effect of the neutrino oscillation in this data sample is negligible, since the distance from the source to the detector is almost same for the horizontal going neutrinos. This effect is a good check on the treatment of the geo-magnetic field in our Monte Carlo Calculations. The results shown in the paper [10] clearly demonstrate that the east-west effect has been observed and agreed with the expectation very well.

2.1 Reduction of the data (contained events)

For the data analysis, we have used single ring events. Events are then classified into electron-like and muon-like events. The fully contained events are further divided into sub-GeV and multi-GeV samples with the discriminating visible energy of 1.33 GeV. The final data sample of sub-GeV and multi-GeV are listed in table 1.

2.2 R

If there is no oscillations, the double ratio R=$(e/\mu)_{data}/(e/\mu)_{MC}$ (e: electron like events; μ: μ-like events) is unity and the deviation from the unity is an evidence for neutrino oscillations. Since you can cancel most of the systematic errors by taking the ratio, the systematic uncertainty will become small (the absolute flux of the atmospheric neutrinos has an uncertainty of about 25%,

	Data	Monte Carlo
sub-GeV sample		
single-ring	3224	3765.3
e-like	1607	1502.7
μ-like	2262	2459.7
multi-ring	1271	1631.0
total	4495	5396.3
multi-GeV sample		
FC events		
single-ring	687	771.0
e-like	386	351.9
μ-like	301	419.1
multi-ring	737	935.4
total	1421	1706.4
PC events		
total(=μ-like)	374	508.9

Table 1: The sub-GeV, multi-GeV and partially contained (PC) event samples observed in 45 ktonyr exposure of Super-Kamiokande. The data are compared with the Monte Carlo prediction of Ref.[9].

mainly due to the uncertainties of the primarily cosmic ray flux, 15% and the cross sections, 15%).

From the number shown in table 1, the double ratios for sub and multi-GeV data are obtained to be $R = 0.668^{+0.024}_{-0.023} \pm 0.052(sys.)$ (sub-GeV: E^{vis}_{total} <1.33GeV) and $R_{FC+PC} = 0.655^{+0.053}_{-0.049} \pm 0.095(sys.)$ (multi-GeV: E^{vis}_{total} >1.33GeV). The total systematic errors are 7.8% and 11.8% for sub-GeV and multi-GeV samples. The major contributions are from charged and neutral current cross sections (3 to 4 %) and single ring and 2 ring separations, 3% for sub-GeV sample and 6% for multi-GeV sample. The systematic error of the flux ratio is estimated to be less than 5 %. The smaller values which apart from unity by 5.7σ for sub-GeV sample and 3.7σ for multi-GeV sample is an evidence for neutrino oscillation.

2.3 Zenith angle distribution

By an simple geometrical argument, the zenith angle distribution of the atmospheric neutrinos, if there is no oscillation, should be symmetry.

In figure 1, we have shown the zenith angle distributions for e-like and

μ-like events. You can clearly see that the suppression of the upward going muons (fig. 1 (a) and (d)).

The systematic uncertainty in the zenith angle distribution (specifically for the case of μ-like events) is mostly coming from flux calculation (less than 1~2%), 1km rock above Super-Kamiokande (1.5% only for multi-GeV events), up/down asymmetry of the energy scale(0.2~0.7%) and non-neutrino background (less than 2%). They are not the source of the observed distortion of the distribution. The geo-magnetic effect must be considered carefully, but this is already discussed in sub-section 2 and turn out to be not important for the zenith angle distributions.. Possible candidates for an exotic phenomena to produce asymmetric distribution is discussed by reference [11]. All the possible explanation have failed to explain the asymmetry, and therefore neutrino oscillation is the only cause of the up/down asymmetry of the atmospheric neutrinos.

2.4 Oscillation analysis

The χ^2 is defined for the contained data samples in order to test the oscillation and to obtain the allowed parameter regions [12]

$$\chi^2(sin^2 2\theta, \Delta m^2)\Sigma\frac{(N_{data} - N_{exp'd})^2}{\sigma^2} + \Sigma(\frac{\alpha}{\sigma_i})^2,$$

where $N_{exp'd}=N_{exp'd}(\sin^2 2\theta,\ \Delta m^2,\ \alpha_1,\ \alpha_2,...)$. α_1, α_2,... are the parameters used for the fitting. The minimum χ^2 is obtained for $\Delta m^2= 3.5\times10^{-3}$ eV ^{2}, $\sin^2 2\theta=1.0$ with $\chi^2=62.1/67_{d.o.f.}$.

The χ^2 for no oscillation is 175. The statistical significance of the oscillation, $\Delta\chi^2=133$ is enormous.

The allowed region based on the χ^2 calculations is shown in figure 2

For the consistency check, the allowed regions are obtained separately by using only R and the zenith angle distributions. They overlap very reasonable way.

If the oscillation mode is $\nu_\mu \rightarrow \nu_\tau$, then the event rates of the neutral current interaction and the electron neutrino interaction should be same as the case for no oscillation. But the neutral current interaction rate should decrease for the case of the $\nu_\mu \rightarrow \nu_s$ oscillations. In Super-Kamiokande, the neutral current interactions are identified by π^0 production. Taking the ratio, $(\pi^0/e)_{data}/(\pi^0/e)_{MC}$, we obtain 0.93 $\pm$0.07$\pm$0.19, which is consistent with both hypothesis, $\nu_\mu \rightarrow \nu_\tau$ and $\nu_\mu \rightarrow \nu_s$ and we cannot discriminate those at this time. The systematic errors of 20% are mostly due to the uncertainty of the neutral current cross sections and we hope this will be measured by

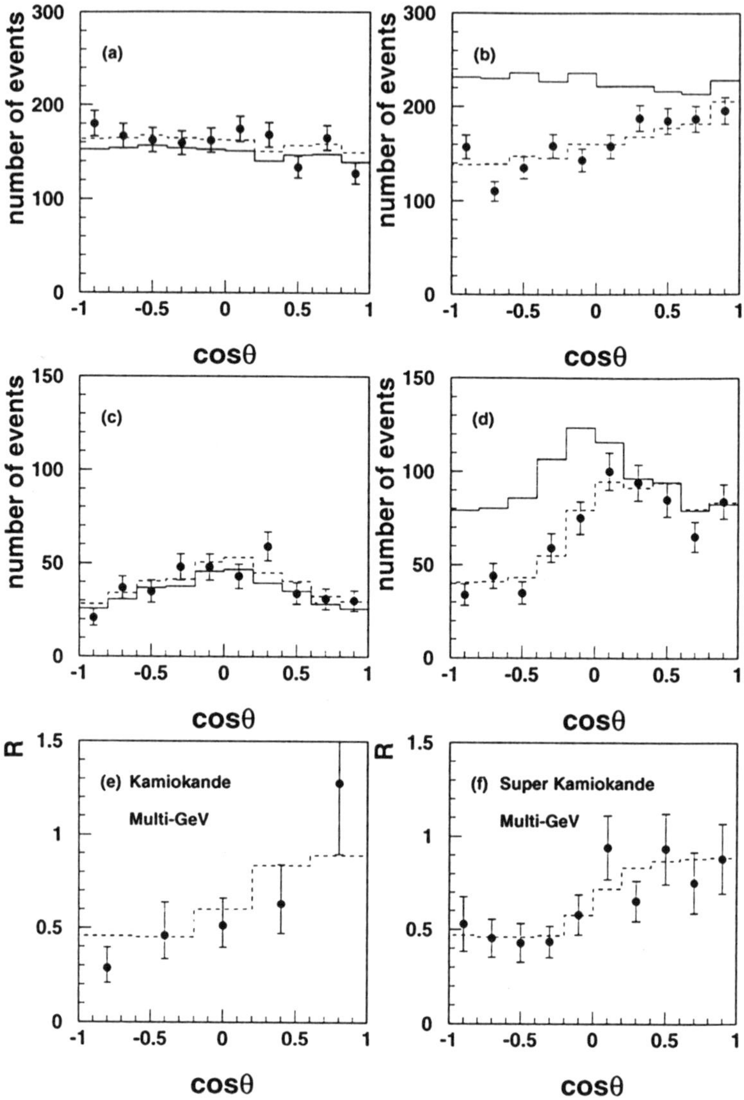

Figure 1: The zenith angle distribution of the sub-GeV and multi-GeV data samples: (a) sub-GeV e-like events, (b) sub-GeV μ-like events, (c) multi-GeV e-like events, and (d) multi-GeV μ-like events. The ratio, R for the multi-GeV events by (e) Kamiokande and (f) Super-Kamiokande are also shown.

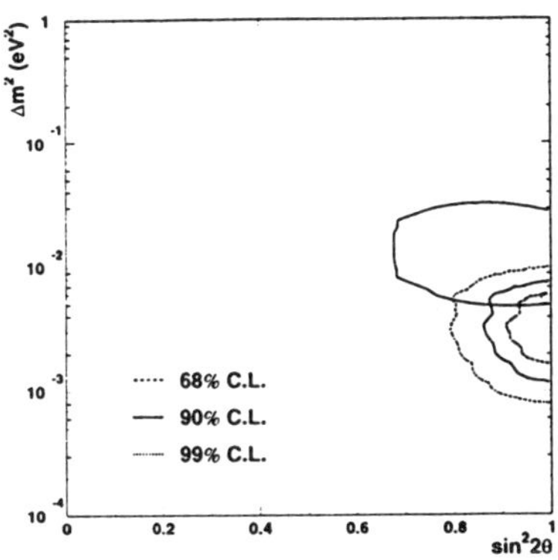

Figure 2: The allowed region obtained by the contained events for 68%, 90% and 99% C.L.. Also shown is a 90% C.L. contour for the Kamiokande Result.

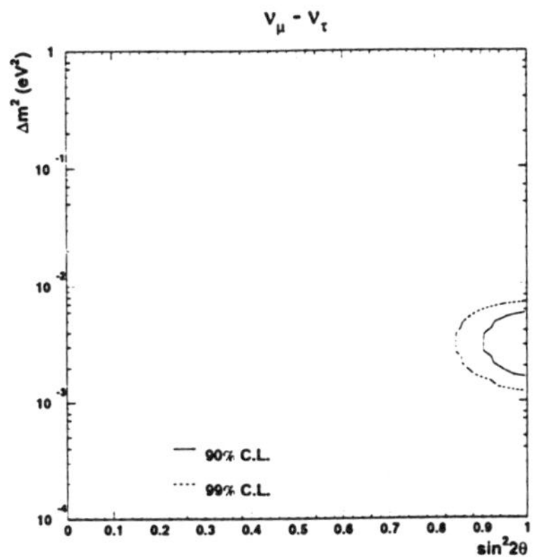

Figure 3: The allowed region obtained by all the Super-Kamiokande data , the contained events, the upward through going events and the upward going stopping events.

the front detectors of the k2k long base line experiments which has started in March 1999.

2.5 Upward through going muons

The upward going muons are produced by the interaction of the high energy neutrinos in the rocks beneath the Super-Kamiokande detector and most of them penetrate the detector (upward through going muons) and some fraction of them stop within the detector (upward stopping muons). The averaged energy for the upward through going muons is ~100 GeV.

The oscillation signature of the upward muons also comes from the zenith angle distribution. Suppose that the mass difference is 5×10^{-3}eV^2, then the horizontally coming neutrinos traveling about 1,000 km has no oscillation effect while the upward going neutrinos, traversing about 13,000 km, has the oscillation effect. We expect to observe the distortion of the zenith angle distribution. The absolute flux also expected to be suppressed. But the uncertainty of the flux is large (~ 20%) and therefore they are not the good quantity to study oscillations.

From the 534 days of data, Super-Kamiokande observed 610 upward through going muon events. By the χ^2 test for the shape we obtained 18.3 for no oscillation (9 d.o.f.) and 7.3 for the case of $\Delta\ m^2$=2.5$\times 10^{-3}$ eV^2, $\sin^2 2\theta$=1.

The allowed parameter region obtained by the through going muons are consistent with that obtained by the contained events. The upward going stopping muons are the other sample to give a supplemental information for the neutrino oscillation. The ratio of the numbers of stopping muon to the through-going muon events are sensitive to the oscillations. The mean energy of neutrinos which produce upward stopping muons are around 10 GeV and therefore effected more by the oscillation of the current oscillation parameters interested. The ratio observed is (0.22$\pm$0.023$\pm$0.014)/(0.39$\pm$0.05) is lower than unity, which indicates the neutrino oscillation. The allowed region by the ratio is also consistent with the region obtained by the contained events. All of the allowed region suggest same parameter region.

2.6 Combined analysis and conclusion of the atmospheric neutrino oscillation

Since all of the data samples—contained, upward through going muons, and upward stopping muons—indicate the same parameter region, then we are able to make a single χ^2 by using all the data. We then have obtained the allowed parameter regions by the combined data set. The result is shown in figure 3, which shows that Super-Kamiokande obtained the oscillation parameter regions of $\sin^2 2\theta > 0.9$ and $1.5\times\ 10^{-3}\mathrm{eV}^2 < \Delta m^2 < 6\times\ 10^{-3}\mathrm{eV}^2$ for the $\nu_\mu \to \nu_\tau$ mode.

1. J.Maki,M.Nakagawa,S.Sakata,Prog.Theor.Phys.,28,870-880(1962).
2. http://www-sk.icrr.u-tokyo.ac.jp/index.html
3. Y.Koshio, Ph.D. thesis, ICRR-Report-426-98-22; H.Ishino, Ph.D. thesis, ICRR-Report-446-99-5; K.Okumura, Ph.D. thesis, ICRR-Report-450-99-8.
4. M.Nakahata, Nucl. Instr. and Meth. A421,113-129 (1999).
5. Y.Fukuda et al., Phys. Lett. B433, 9 (1998).
6. Y.Fukuda et al., Phys. Lett., B436, 33-41 (1998).
7. Y.Fukuda et al., Phys. Rev. Lett., 82, 2644-2648 (1999).
8. G. Bar et al., Phys. Rev. D39,3532(1989); V. Agrawal et al., Phys. Rev. D53, 1313 (1996); T. K. Gaisser and T. Stanev, Proc. 24th Int. Cosmic Ray Conf. (Rome) Vol. 1, 694 (1995).
9. Honda et al.,Phys. Lett. B248,193(1990); Honda et al., Phys. Rev. D52, 4985 (1995).
10. Y.Fukuda et al. submitted to Phys. Rev. Lett.
11. P.Lipari and M. Lusignoli, hep-ph/9901350.
12. Y.Fukuda et al.,Phys.Rev.Lett.,81,1562-1567(1998).

SOLAR NEUTRINO DETECTORS AND (^{3}He,t) CHARGE-EXCHANGE REACTIONS

M. FUJIWARA,[1,2] H. AKIMUNE,[3] I. DAITO,[4] H. EJIRI,[1] H. FUJIMURA,[1] Y. FUJITA,[5] C.D. GOODMAN,[6] K. HARA,[1] M.N. HARAKEH,[7] F. IHARA,[1] J. JÄNECKE,[8] T. KAWABATA,[9] R.S. RAGHAVAN,[10] K. SCHWARZ,[1] M. TANAKA,[11] A.M. VAN DEN BERG,[7] T. YAMANAKA[1] M. YOSOI,[1] and R.G.T. ZEGERS[2,7]

[1] *Research Center for Nuclear Physics, Osaka University, Mihogaoka 10-1, Ibaraki, Osaka 657-0047, Japan*
[2] *Advanced Science Research Center, JAERI, Tokai, Ibaraki, 319-1195 Japan*
[3] *Konan University, Higashinada, Kobe 658, Japan*
[4] *Nagoya University, Department of Physics, 464 Nagoya, Japan*
[5] *Department of Physics, Osaka University, 560-0043, Japan*
[6] *Indiana University Cyclotron Facility, Bloomington, Indiana, USA*
[7] *Kernfysisch Versneller Instituut, 9747 AA Groningen, The Netherlands*
[8] *Department of Physics, University of Michigan, Ann Arbor, MI 48109-1120, USA*
[9] *Department of Physics, Kyoto University, Kyoto 606-8502, Japan*
[10] *Bell Laboratories, Lucent Technologies, Murray Hill, New Jersey 07974, U.S.A.*
[11] *Kobe Tokiwa Jr. College, Nagata 653, Japan*

Gamow-Teller strengths from ^{176}Yb and ^{160}Gd have been studied via the ^{176}Yb(^{3}He,t)^{176}Lu and ^{160}Gd(^{3}He,t)^{160}Tb reactions at 450 MeV. With the spectrometer "Grand Raiden", high resolution (^{3}He,t) spectra of $\Delta E \leq 150$ keV have been obtained at 0°. The excitation strengths of the two individual 1^+ states at 338.9 keV and 194.5 keV in ^{176}Lu have been deduced. Reliable solar neutrino absorption rates of the *pp* and other solar processes have been estimated from the measured B(GT) values for the two states in ^{176}Lu and the low-lying states in ^{160}Tb. The elements, Yb or Gd, are promissing for the solar neutrino detector.

1 Introduction

Recent results of solar neutrino measurements from super-Kamiokande [1] and from Gallex [2] and Sage [3] are inconsistent with the standard solar model [4] to explain the burning process based on the *pp* chain and CNO cycle in the sun. The solar neutrino flux on the earth is found to be greatly reduced, and a key ingredient concerning the neutrino disappearance seems be necessary for interpretation. If the Kamiokande results are taken as a measure of the ^{8}B neutrinos and the Gallex and Sage results taken as a measure of the integrated flux including *pp*–, ^{7}Be–, and ^{8}B-neutrinos, the low integrated flux measured by

Gallex and Sage implies a negligible flux of ^{7}Be neutrinos. This is inconsistent with the energy generation cycle in the sun, since the element, ^{8}B, is created in a minor branch (0.02%) after the ^{3}He$(\alpha,\gamma)^7$Be reaction (14%). This problem has led to the speculation that the electron neutrinos from the ^{7}Be decay are converted via neutrino oscillation in their long flight path of 1.5×10^{11} m from the center of the sun to the detector on earth [5,6,7]. This idea is strongly supported by the result of atmospheric neutrino measurements at Kamioka, which suggests the evidence for the oscillation between ν_μ and ν_τ [8]. A natural conclusion of this observation is that the electron neutrino mass should be smaller than 10^{-2} eV. At present, the most important is a direct measurement of the electron neutrino flux from the sun at the energy corresponding to the ^{7}Be neutrinos, which have never been separately detected so far. If we can reliably detect the ^{7}Be neutrinos, major difficulties tó understand the long-standing solar-neutrino problem will be solved.

Raghavan[9] suggests that it is possible to construct a low-threshold, real time, flavor specific scheme for detecting solar neutrinos based on the Gamow-Teller transitions for the (ν,e^-) process on ^{176}Yb and ^{160}Gd. An unique idea is that absorption events of *pp* and ^{7}Be solar neutrinos can be identified by a delayed coincidence between the prompt electron from the GT transition followed by the γ-decay of the excited state, and that the background events are greatly reduced by a factor of more than 10^6 because of the coincidence measurement. A project for constructing a new solar-neutrino detector LENS (Low Energy Neutrino Spectrometer) [10] using Yb- or Gd-loaded scintillation-counters is now in progress at Gran Sasso. This LENS project will make a real-time spectroscopy of solar *pp* and ^{7}Be neutrinos possible, for the first time.

To determine the sensitivity for *pp* and ^{7}Be neutrinos, we need to know the B(GT) values for the 1^+ states excited by the ^{176}Yb$(\nu,e^-)^{176}$Lu and ^{160}Gd$(\nu,e^-)^{160}$Tb reactions, independently. We therefore measured a high resolution (^{3}He,t) spectra at zero degrees using a ^{3}He beam at 450 MeV to complement the (p,n) measurements. The (^{3}He,t) reaction at 450 MeV has already been demonstrated to be useful in getting the B(GT) values. Since the full Fermi strength B(F)=N–Z is exhausted by the IAS, the B(GT) values can be relatively precisely obtained from the singles (^{3}He,t) spectra on ^{176}Yb and ^{160}Gd by using the relative ratio of the excitation strengths of 1^+ states and the IAS [11].

2 Experiment

The (^{3}He,t) experiment was performed using the spectrometer Grand Raiden [12] at RCNP, which was set at $\theta = 0°$ with the solid angle of $\Delta\Omega = 1.6$ msr. Both the incident ^{3}He^{++} beam and the reaction products entered into the spectrometer

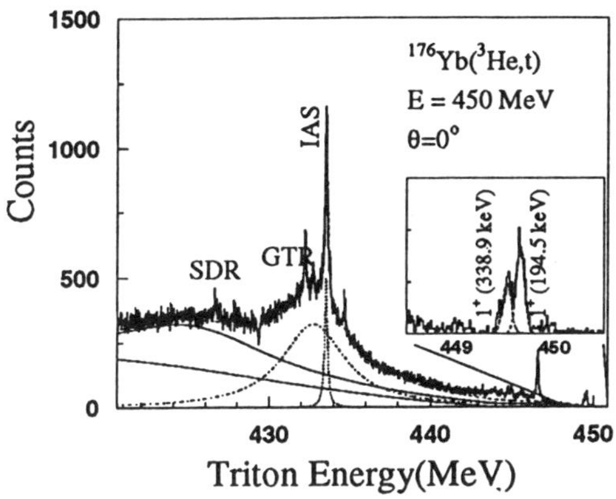

Figure 1: Triton-particle energy spectrum from the ^{176}Yb(^{3}He,t) reaction at 0° taken with a ^{3}He beam of 150 MeV/A. A sharp IAS peak and two broad bumps corresponding to a Gamow-Teller resonance and a spin dipole resonance are observed. The global structure of the spectrum are fitted by assuming an underlying continuum due to quasi-free scattering. Discrete states in the low excitation energy are displayed in the expanded spectrum.

set at 0°. The metallic targets used are ^{176}Yb (96.7 % enrichment, 3.3 mg/cm^2) and ^{160}Gd (97.8 % enrichment, 3.0 mg/cm^2). The ^{3}He^{++} beam at 450 MeV was passed through the aperture of the entrance slit of the spectrometer, bent inward, and stopped at the internal Faraday cup in the first dipole (D1) magnet because of its lower magnetic rigidity. Triton particles were analyzed by the spectrometer and were detected in the focal-plane detector system consisting of two multi–wire–drift–chambers (MWDC's) and two ΔE plastic scintillation counters, of which signals were used for particle identification as well as for triggering particle events. The MWDC's gave information on the arrival positions at the focal plane and on the scattered angles of tritons.

An extremely good particle-identification with the focal-plane detector of the spectrograph was required to avoid the huge background due to the ^{3}He^{++} → ^{3}He$^+$ atomic charge-exchange process [13], since the Q-values for the ^{176}Yb, ^{160}Gd(^{3}He,t)^{176}Lu, ^{160}Tb reactions leading to the lowest 1^+ states is small as 0.3 MeV. For this purpose, we added a plastic scintillation counter with a thickness of 1 mm in front of the two ΔE counters of 10 mm thickness for particle identification. This front counter was thin in order to avoid tritons induced by reactions in the scintillator itself. Triton yields due to the incoming ^{3}He$^+$ particles could greatly reduced in the previous experiment. We could obtain a nice spectrum, being free from the ^{3}He^{++} → ^{3}He$^+$ atomic charge-exchange process.

To achieve a high energy resolution around 100 keV, we installed an energy defining slit in the beam transport system from the K=400 MeV ring cyclotron to the target position. The image defining slit was set at the first beam-focusing position of the beam-line. The technique of dispersion matching between the beam-line and the spectrometer was employed for a high resolution needed for the measurement.

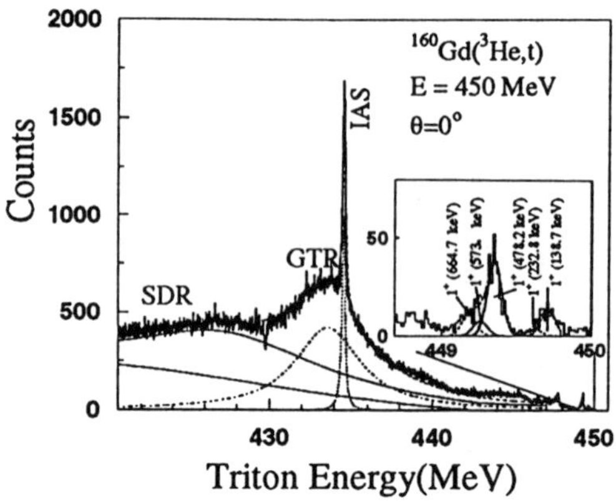

Figure 2: Triton-particle energy spectrum from the ^{160}Gd(^{3}He,t) reaction at 0° taken with a ^{3}He beam of 150 MeV/A. A sharp IAS peak and two broad bumps corresponding to a Gamow-Teller resonance and a spin dipole resonance are observed. The global structure of the spectrum are fitted by assuming an underlying continuum due to quasi-free scattering. Discrete states in the low excitation energy are displayed in the expanded spectrum.

An overall energy resolution was $\Delta E = 100{\sim}140$ keV. This resolution was enough to obtain the cross section values for the low-lying Gamow-Teller states in both ^{160}Gd and ^{176}Yb.

To confirm the validity of the particle identification in the present experiment, we made an additional experiment where the ^{176}Yb target was backed with a thin mylar foil (0.086 mg/cm^2) to reduce the huge $^3He^{++} \rightarrow {}^3He^+$ yields. In this run, the high resolution measurement was not possible because of the achromatic mode beam transportation. However, we could confirmed that the summed yields for the 1^+ states at 195 keV and 339 keV has the same relative intensity in the 0° ^{176}Yb(^{3}He,t) spectrum taken in the dispersion-matching high-resolution mode.

A complementary measurement to obtain the reliable B(GT) values in the (^{3}He,t) analyses was made using the ^{164}Dy(^{3}He,t) reaction leading to the ground 1^+ state in ^{164}Ho. From the log *ft* value of 4.6 (Ref.[14]), the B(GT) value to this state can be deduced to be 0.293±0.006, and was used to calibrate the B(GT) values for the 1^+ states observed in the 450 MeV (^{3}He,t) reactions.

3 Discussion

Fig. 1 and 2 shows the (^{3}He,t) spectra on ^{176}Yb and ^{160}Gd at θ=0°. The overall shape of these spectra is very similar to the (p,n) spectra measured at 120 MeV and 150 MeV at IUCF[15]. The transition strengths of the low-lying 1^+ state in ^{171}Lu and ^{160}Tb are found to be very small, consistent with the (p,n) results. We deduced the transition strengths to the low-lying states by fitting the peaks with a Gaussian shape (see Fig. 1 and 2). The transition strengths to the excited states were compared with those to the IAS.

The excitation strengths to GT states and IAS are expressed by

$$\left(\frac{d\sigma}{d\Omega_F}\right)(q\approx0) = \left(\frac{\mu}{\pi\hbar^2}\right)^2\left(\frac{k_f}{k_i}\right)N_\tau^D|J_\tau|^2B(F), \tag{1}$$

$$\left(\frac{d\sigma}{d\Omega_{GT}}\right)(q\approx0) = \left(\frac{\mu}{\pi\hbar^2}\right)^2\left(\frac{k_f}{k_i}\right)N_{\sigma\tau}^D|J_{\sigma\tau}|^2B(GT), \tag{2}$$

where the volume integrals of the central part of the effective interaction are J_τ and $J_{\sigma\tau}$, respectively. The distortion effects are factorized by using N_τ^D, and $N_{\sigma\tau}^D$. Using the observed strengths to the ground 1^+ state and the IAS in the ^{164}Dy(^{3}He,t)^{164}Ho reaction at the same bombarding energy of 450 MeV, we calibrated the ratio of the interaction strengths including the distortion effects as

$$\frac{N_{\sigma\tau}^D|J_{\sigma\tau}|^2}{N_\tau^D|J_\tau|^2} = \frac{(\frac{d\sigma}{d\Omega})_{GT}B(GT)}{(\frac{d\sigma}{d\Omega})_{IAS}B(F)} = 7.44\pm0.79, \tag{3}$$

where we used B(GT)=0.293±0.006, B(F)=N-Z=32, and $(\frac{d\sigma}{d\Omega})_{GT}/(\frac{d\sigma}{d\Omega})_{IAS}$ is 0.068±0.007 for the transition from ^{164}Dy to ^{164}Ho, which is obtained in the ^{164}Dy(^{3}He,t)^{164}Ho reaction. The value of 7.44±0.79 for the (^{3}He,t) reaction at 150 MeV/A locates between the (p,n) values of 4.76 and 8.46 at 120 MeV and 160 MeV, respectively[16].

The present calibration method is applicable since the distortion effect and the ratio of the effective interactions $|J_{\sigma\tau}/J_\tau|$ are expected to be the same as those for the ^{160}Gd, ^{176}Yb(^{3}He,t) reactions. Empirically, the volume integrals of the effective interaction and the distortion effect are not different among target nuclei in the same mass region.

The B(GT) values for the 1^+ states in ^{160}Tb and ^{176}Lu were obtained using the equation;

$$B(GT) = B(F)\frac{1}{7.44}\frac{(\frac{d\sigma}{d\Omega})_{GT}}{(\frac{d\sigma}{d\Omega})_{IAS}}, \tag{4}$$

where B(F) is (N-Z), and the ratio of $(\frac{d\sigma}{d\Omega})_{GT}/(\frac{d\sigma}{d\Omega})_{IAS}$ is obtained from the (^{3}He,t) spectra at 0°. The deduced B(GT) values are tabulated in Table 1.

Using the standard solar model (SSM) flux by Bahcall[4], the capture rates of the solar neutrinos are obtained, which are listed in Table 2 and in Table 3.

If the SSM flux is assumed, the total solar neutrino capture rate amounts to ~300 counts/year with 20 tons of Yb/Gd materials. If a delayed-coincidence technique with γ-rays following neutrino absorption is well applied, it is easy

Table 1: Gamow-Teller strengths B(GT) from the $^{176}_{70}$Yb(^{3}He,t)$^{176}_{71}$Lu and $^{160}_{64}$Gd(^{3}He,t)$^{160}_{65}$Tb reactions at θ=0°. The excitation energies and the Fermi strengths are also given.

$^{176}_{70}$Yb(^{3}He,t)$^{176}_{71}$Tb		
J^π	E_x(keV)	B(GT) or B(F)
1^+_1	194.5	0.20±0.04
1^+_2	338.9	0.11±0.02
0^+_{IAS}	16026 ± 6	36
$^{160}_{64}$Gd(^{3}He,t)$^{160}_{65}$Tb		
J^π	E_x(keV)	B(GT) or B(F)
1^+_1	138.7	0.054±0.009
1^+_2	232.8	0.014±0.002
1^+_3	478.2	0.16 ±0.03
1^+_4	573.0	0.021±0.004
1^+_5	664.7	0.031±0.005
0^+_{IAS}	15019 ± 6	32

Table 2: Absorption rate (per year) of solar neutrinos from the ground state of ^{176}Yb to the two states at 194.5 keV and 338.9 keV in ^{176}Lu for a detector with natural Yb (12.73 %) with 20 tons.

neutrino type	E_ν (keV)	number of neutrino absorption
pp	$\leq$ 420 keV	172.9
^{7}Be	383 keV, 862 keV	179.0
^{8}B	$\leq$14940 keV	5.3
pep	1442 keV	9.7

Table 3: Absorption rate (per year) of solar neutrinos for natural Gd(21.86% ^{160}Gd) with 20 tons.

neutrino type	E_ν (keV)	number of neutrino absorption
pp	$\leq$ 420 keV	151.0
^{7}Be	383 keV, 862 keV	135.0
^{8}B	$\leq$14940 keV	6.2
pep	1442 keV	10.8

to overcome a common background problem due to the α-decay radioactive isotopes.

4 Summary

We report the experimental results on the Gamow-Teller transitions from ^{176}Yb and ^{160}Gd using (^{3}He,t) reaction at $E(^3\text{He})$=450 MeV. High resolution measurement enables us to resolve the low-lying 1^+ states in the residual nuclei. By taking the ratios of the excitation strengths for the IAS and the Gamow-Teller state, we deduce the B(GT) values for low-lying 1^+ states after calibrating the reaction mechanism. The solar neutrino absorption ratios are given for Yb and Gd in this work. The present results will serve to estimate the scale of the possible new solar neutrino detector.

The large-scale online solar-neutrino detector with a Gd or Yb doped scintillation counter is now proposed under the name of LENS (Low Energy Neutrino Spectroscopy). In this program, flavor-specific, source-specified, and real-time spectroscopy of the pp, pep, ^{7}Be, ^{8}B and CNO electron neutrinos from the sun are detected in the online mode. Hopefully, the neutrino oscillation will be examined by detecting the individual solar neutrinos.

Acknowledgment

The authors acknowledge the RCNP cyclotron staff for their support during the experiment. The present work was supported by the Minstry of Education, Science, Sports and Culture (Monbusho) with Grant No. 09041108. The present work was also supported in part by the Japan Society for the Promotion of Science (JSPS).

References

1. Y. Fukuda *et al.*, Phys. Rev. Lett. 81 (1998) 1158.
2. W. Hampel *et al.*, *Phys. Lett.* **B388** (1996) 384.
3. V.N. Gavrin, Nucl. Phys. B **77** (1999) 20.
4. John N Bahcall, Neutrino Astrophysics, Cambridge University Press 1988.
5. S.P. Mikheyev, A.Y. Smirnov, Prog. Part. Nucl. Phys. **23** (1989) 41, L. Wolfenstein, Phys. Rev. **D17** (1978) 2369.
6. Physics of Neutrinos by M. Fukugita and T. Yanagida, pp. 1, in "Physics and Astrophysics of Neutrinos" edited by M. Fukugita and A. Suzuki (Springer-Verlag 1994).
7. W.C. Haxton, Nucl. Phys. A570, 125 (1994); and references therein.
8. Y. Fukuda *et al.*, Phys. Rev. Lett. 81 (1998) 1162.

9. R.S. Raghavan, Phys. Rev. Lett. **78** (1997) 3618.
10. Letter of Intent to LNGS on "LENS (Low Energy Neutrino Spectrometer) January 1999.
11. M. Fujiwara *et al., Nucl. Phys.* A599 (1996) 223c.
12. M. Fujiwara, H. Akimune, I. Daito, H. Fujimura, Y. Fujita, K. Hatanaka, H. Ikegami, I. Katayama, K. Nagayama, N. Matsuoka, S. Morinobu, T. Noro, M. Yoshimura, H. Sakaguchi, Y. Sakemi, A. Tamii, and M. Yosoi, Nucl. Instrum. Method Phys. Res., Sec A **422** (1999) 484–488.
13. K. Dennis, H. Akimune, G.P.A. Berg, S. Chang, B. Davis, M. Fujiwara, M.N. Harakeh, J. Jänecke, J. Liu, K. Pham, D.A. Robert, and E.J. Stephenson, Phys. Rev. A **50** (1994) 3992.
14. E.N. Shurshikov and N.V. Timofeeva, Nuclear Data Sheets **65** (1992) 365.
15. C.D. Goodman, private communication.
16. C.D. Goodman, Nucl. Phys. A577 (1994) 3c.

Double Gamow-Teller Excitation by Second-Order Born Approximation

Kazuo Muto

Graduate School of Science and Technology, Tokyo Institute of Technology
e-mail : muto@th.phys.titech.ac.jp

Forward-angle cross sections of double charge exchange reactions with heavy ions are calculated in second-order plane-wave Born approximation. The calculated results indicate that the cross sections are in the order of nb/sr, and that those reactions do not simulate Gamow-Teller transitions, even at forward angles, as single charge exchage reactions.

1 Introduction

The nuclear double beta ($\beta\beta$) decay is one of the most promising processes which reveal fundamental nature of neutrinos and weak interactions beyond the standard model of the electroweak interactions [1]. It occurs by double Gamow-Teller (DGT) transitions, though the the daughter states of $\beta\beta$ decay carry only a tiny fraction of the sum-rule DGT stregnth [2]. Double charge exchange (DCX) reactions by heavy ions are expected to excite DGT states, see Fig. 1. There has been a DCX experiment which searches for DGT excitation by ^{24}Mg(^{18}O, ^{18}Ne)^{24}Ne reaction at 76 A MeV. The observed forward-angle cross sections to low-lying states are about 1 nb/sr, which are unexpectedly smaller than simple estimations of μb [3]. Motivated by these considerations, this work aims, first, at estimating DCX reaction cross sections by second-order Born approximation in order to examine experimental feasibility, and, secondly, at clarifying the transition mechanism if the hadronic DCX reaction simulates $\beta\beta$ transitions by the weak interaction.

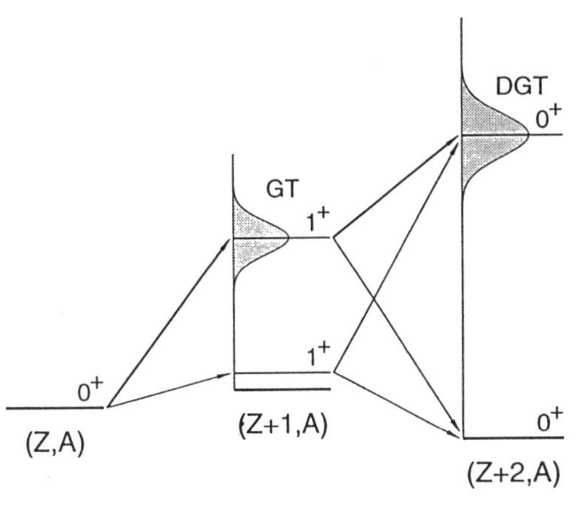

Fig. 1 Double Gamow-Teller excitation. Sequential GT transitions lead, through the GT giant resonance, mainly to the DGT giant resonance, which is predicted to appear just above the double IAS and carry almost the entire strength of the sum-rule value [2].

2 DCX Cross Sections at 0 Degree

2.1 Second-order Born approximation

Differential cross sections of DCX reaction is expressed in second-order Born approximation [4] as

$$\begin{aligned}\frac{d\sigma}{d\Omega} &= \frac{m_\beta m_\alpha}{(2\pi\hbar^2)^2} \,|\langle \beta \,|\, V\,G\,V \,|\, \alpha \rangle|^2 \\ &= \frac{m_\beta m_\alpha}{(2\pi\hbar^2)^2} \left| \sum_\gamma \int d\boldsymbol{r}_\beta \int d\boldsymbol{r}_\alpha \, M_{\beta\gamma}(\boldsymbol{r}_\beta)\, G_\gamma(\boldsymbol{r}_\beta, \boldsymbol{r}_\alpha)\, M_{\gamma\alpha}(\boldsymbol{r}_\alpha) \right|^2 , \end{aligned} \tag{1}$$

with the sum over intermediate states. α (β, γ) represents the initial (final, intermediate) state of the reaction, m_α (m_β, m_γ) the reduced mass, V the *NN* interaction, and G_γ is Green function which describes propagation in the intermediate state γ. Nuclear transition amplitudes are

$$M_{\beta\gamma}(\boldsymbol{r}_\beta) = \langle \beta \,|\, V \,|\, \gamma \rangle_{\boldsymbol{r}_\beta} \phi(\boldsymbol{k}_\beta, \boldsymbol{r}_\beta), \qquad M_{\gamma\alpha}(\boldsymbol{r}_\alpha) = \langle \gamma \,|\, V \,|\, \alpha \rangle_{\boldsymbol{r}_\alpha} \phi(\boldsymbol{k}_\alpha, \boldsymbol{r}_\alpha). \tag{2}$$

In the ^{24}Mg(^{18}O, ^{18}Ne)^{24}Ne case, nuclear matrix elements of *NN* interaction are

$$\begin{aligned} \langle \beta \,|\, V \,|\, \gamma \rangle_{\boldsymbol{r}_\beta} &= \langle {}^{18}\mathrm{Ne}, {}^{24}\mathrm{Ne} \,|\, V \,|\, {}^{18}\mathrm{F}, {}^{24}\mathrm{Na} \rangle_{\boldsymbol{r}_\beta} \\ \langle \gamma \,|\, V \,|\, \alpha \rangle_{\boldsymbol{r}_\alpha} &= \langle {}^{18}\mathrm{F}, {}^{24}\mathrm{Na} \,|\, V \,|\, {}^{18}\mathrm{O}, {}^{24}\mathrm{Mg} \rangle_{\boldsymbol{r}_\alpha} \end{aligned} \tag{3}$$

in which integrations are performed for internal coordinates of the colliding nuclei, leaving the difference between the center-of-mass coordinates. We assume the plane-wave approximation for the incoming and outgoing wave fucntions

$$\phi(\boldsymbol{k}_\beta, \boldsymbol{r}_\beta) = \exp(-i\boldsymbol{k}_\beta \cdot \boldsymbol{r}_\beta), \qquad \phi(\boldsymbol{k}_\alpha, \boldsymbol{r}_\alpha) = \exp(i\boldsymbol{k}_\alpha \cdot \boldsymbol{r}_\alpha), \tag{4}$$

accordingly for the Green function as well

$$G_\gamma(\boldsymbol{r}_\beta, \boldsymbol{r}_\alpha) = \frac{m_\gamma}{2\pi\hbar^2} \frac{\exp(ik_\gamma |\boldsymbol{r}_\beta - \boldsymbol{r}_\alpha|)}{|\boldsymbol{r}_\beta - \boldsymbol{r}_\alpha|}, \tag{5}$$

and decompose them into partial waves.

2.2 A simple estimation

Single charge exchange reactions, such as (p, n) and (^{3}He, t), have been employed to evaluate GT strengths, by making use of the proportionality between the forward-angle cross sections and B(GT). When momentum is not transfered to the target nucleus, the predominant component of *NN* interaction is $\boldsymbol{\tau}\cdot\boldsymbol{\tau}\,\boldsymbol{\sigma}\cdot\boldsymbol{\sigma}$, and it induces GT transitions of the target and projectile nuclei. If the projectile is scattered at 0 degree at both vertices in DCX, the cross section is reduced to

$$\frac{d\sigma}{d\Omega} = \frac{m_\beta m_\alpha {m_\gamma}^2}{(\pi\hbar^2)^4} \frac{v(0)^4}{R^2} B(\mathrm{DGT})^{\mathrm{Projectile}} B(\mathrm{DGT})^{\mathrm{Target}} \tag{6}$$

where $v(q)$ is the Fourier transform of the potential $V_{\tau\sigma}(r)$ in the spin-isospin channel. The Green function has also been simplified, $G \approx m_\gamma/(2\pi\hbar^2 R)$, with R being the nuclear radius.

This extreme case of DCX reaction may simulate the two-neutrino mode of $\beta\beta$ decay, see Fig. 2.. The latter is the second-order process of the weak interaction. The spin-isospin operator $\tau\sigma$ induces nuclear GT transitions at each weak vertex with zero-momentum transfer. However, the $2\nu\beta\beta$ nuclear matrix element involves the energy denominator, which is indispensable for an accurate evaluation of the nuclear matrix element. On the other hand, the strong DCX process does not well take into account energies of the nuclear intermediate states.

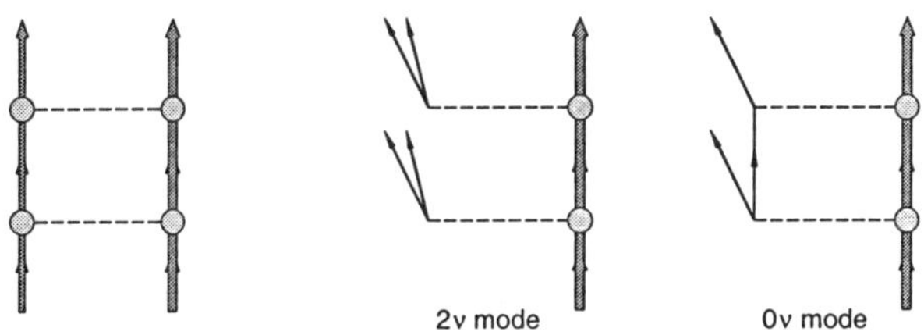

Fig. 2. Diagrams of second order strong/weak processes. Left: DCX by heavy ions, Right: $\beta\beta$ decays of two-neutrino and neutrinoless modes. Dashed lines denote mesons in the strong process, and weak bosons W in the weak processes. In the $2\nu\beta\beta$ mode, two pairs of electron and antineutrino are emitted, while in the $0\nu\beta\beta$ mode two electrons are emitted from the nucleus and a virtual Majorana neutrino propagates in the nucleus.

The top panel of Fig. 3 shows the calculated DGT strengths (^{24}Mg $\rightarrow$ ^{24}Ne) for 0^+ states of the the residual nucleus ^{24}Ne. The ground state carries a sizable B(DGT) value. This is because the target nucleus ^{24}Mg has no neutron excess, and therefore there appears no giant DGT resonance. As the neutron excess is increased, the DGT strength tends to shift to a higher excitation region, leading to a coherent giant DGT resonance, as illustrated in Fig. 1.

2.3 Finite Momentum Transfer

The assumption of zero-momentum transfer at both vertices is too crude for DCX. With higher probabilities, the projectile is scattered at finite angles with, accordingly, finite momentum transfer. When the transfered momenta at both vertices are compensated, namely, the sum of the momenta transfered to the target nucleus vanishes, the ejectile appears at forward angles.

Fig. 4 shows momentum-transfer dependence of B(DGT) to the ground state of ^{24}Ne. The B(DGT) shows a very rapid decrease with momentum transfer q, about three orders of magnitude from $q = 0$ to $q = 1.0$ fm^{-1}. Secondly, the $L = 0$ transfer dominates the transition, and higher partial waves give negligibly small contributions.

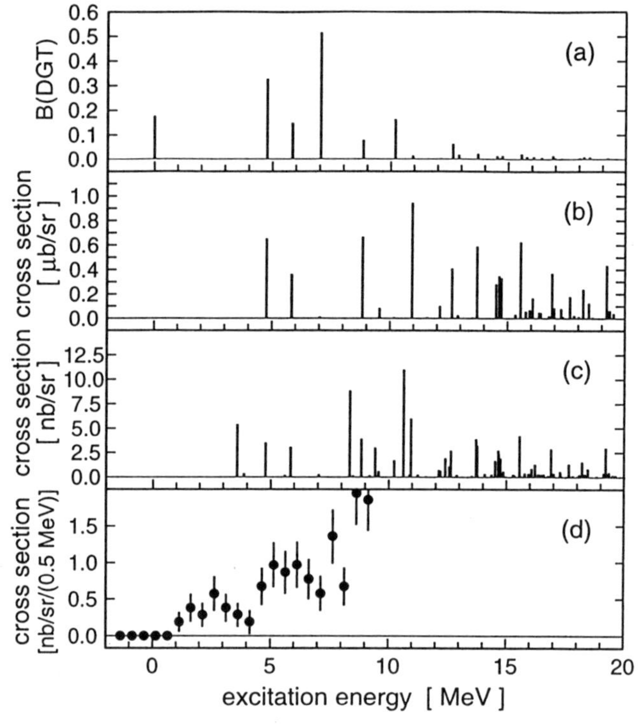

Fig. 3. DCX ^{24}Mg(^{18}O, ^{18}Ne)^{24}Ne at $E(^{18}\mathrm{O}) = 76$ A MeV. (a) DGT strengths of the target nucleus: $^{24}\mathrm{Mg}_{g.s.} \rightarrow {}^{24}\mathrm{Ne}(0^+)$, (b) PWBA cross sections at 0 degree, (c) the same as (b) but with damping effects. (d) experimental cross sections [3]. Calculations were made with shell-model wave functions [5] and M3Y interaction [6].

PWBA cross sections at 0 degree calculated with the momentum-transfer dependence in Fig. 4 are shown in the second panel (b) of Fig. 3. The cross sections are as small as $\approx$ 1 μb/sr, reflecting the rapid decrease of B(DGT) with momentum transfer. It is also noted that B(DGT) in Fig. 3(a) is redistributed. Both observations come from finite momentum transfer at both strong vertices and from different behaviors of dependence among final states of the target nucleus. For example, the ground state of ^{24}Ne, which carries a large B(DGT), has small cross sections when momentum transfer is taken into account. One can see a tendency that cross sections of higher-lying final states are enhanced to a larger extent.

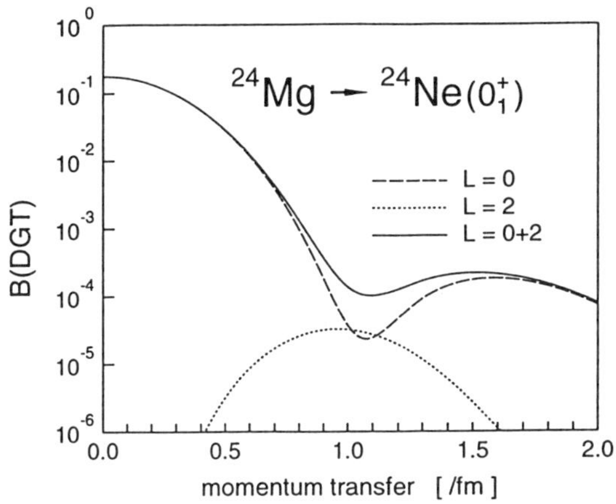

Fig. 4 Reduced DGT transition probability as a function of momentum transfer calculated for $^{24}\mathrm{Mg}_{g.s.} \rightarrow {}^{24}\mathrm{Ne}_{g.s.}$. Contributions of $L = 0$ and $L = 2$ transfers are determined by the spin-isospin component $V_{\tau\sigma}(r)$ of the NN interaction, together with the nuclear wave functions.

2.4 Distortion Effects

The relative-motion wave functions have so far been assumed to be plane waves, which obviously do not describe the behavior of the actual wave functions. Distorted waves can be calculated, for example, with optical potentials. However, the most important effect of distortion for the very rapidly oscillating wave functions would be damping due to flow of flux to other open channels, when the two nuclei overlap with each other.

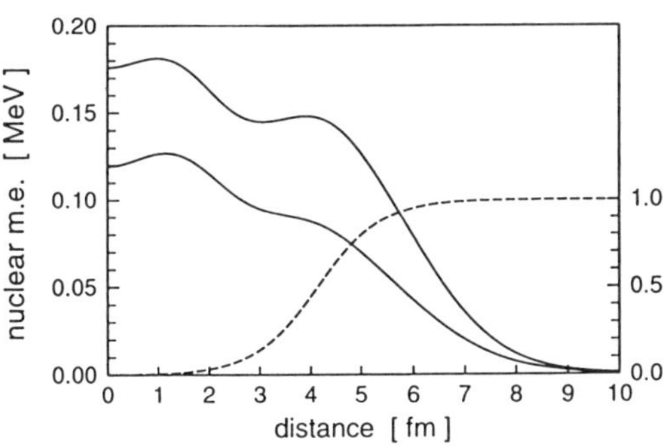

Fig. 5. Calculated nuclear transition amplitudes in DCX reaction ^{24}Mg(^{18}O, ^{18}Ne)^{24}Ne as function of distance between two nuclei, for the second order transition via the lowest 1^+ states of intermediate nuclei. The dashed line is a function which is used in the PWBA calculation, in order to take into account effects of damping of wave functions in the nuclear interior.

Here, the damping effect is taken into account in a simple way by multiplying a real function, indicated by a dashed line in Fig. 5, which decreases rapidly in the nuclear interior. In Fig. 5 are also shown nuclear matrix elements in DCX reaction ^{24}Mg(^{18}O, ^{18}Ne)^{24}Ne for the second order transition via the lowest 1^+ states of intermediate nuclei,

$$\langle\, {}^{18}\mathrm{Ne}(0^+_{g.s.}),\ {}^{24}\mathrm{Ne}(0^+_{g.s.})\,|\,V\,|\,{}^{18}\mathrm{F}(1^+_1),\ {}^{24}\mathrm{Na}(1^+_1)\,\rangle_{r_\beta}$$
$$\langle\, {}^{18}\mathrm{F}(1^+_1),\ {}^{24}\mathrm{Na}(1^+_1)\,|\,V\,|\,{}^{18}\mathrm{O}(0^+_{g.s.}),\ {}^{24}\mathrm{Mg}(0^+_{g.s.})\,\rangle_{r_\alpha}.$$

The forward-angle cross sections, as given in Fig. 3(c), are reduced by distortion by a factor of about hundred, ≈ 10 nb/sr for strongest transitions. The predicted peak at 3.6 MeV corresponds to excitations to the first 2^+ states in both target and projectile nuclei. The calculated cross sections are comparable to those of the experimental DCX reaction, shown in Fig. 3(d).

3 Summary

Cross sections of DCX reaction at 0 degree are calculated in second order PWBA with distortion effects. It is shown that the cross sections are in the order of nb/sr, due to combined effects of finite momentum transfer at each vertex, interference of nuclear transition amplitudes through various nuclear intermediate states, and distortion of the relative wave functions.

In the context of $\beta\beta$ decay, DCX reactions mediated by the strong interaction simulate neither the two-neutrino decay because of finite momentum transfer, nor the neutrinoless decay since spin-parities of nuclear intermediate states are determined by the $V_{\tau\sigma}$ component of *NN* interactions.

References

[1] K. Muto and H.V. Klapdor, in *Neutrinos*, (Springer, 1988).
[2] K. Muto, Phys. Lett. **277B** (1991) 13.
[3] J. Blomgren *et al.*, Phys. Lett. **362B** (1995) 34, and references therein.
[4] N. Austern, *Direct Nuclear Reaction Theories*, (Wiley-Interscience, 1970).
[5] B.H. Wildenthal, Prog. Part. Nucl. Phys. **11** (1984) 5.
[6] W.G. Love, in *The (p,n) Reaction and the Nucleon-Nucleon Force*, (Plenum, 1980).

SOME NEW RESULTS ON $\beta\beta$ DECAY: NEMO AND OTHERS

A.S. BARABASH
Institute of Theoretical and Experimental Physics, B.Cheremushkinskaya 25, 117259 Moscow, Russia

Last results and perspectives for the future of NEMO and DBA experiments and a new positive result for the $2\nu\beta\beta$ decay of ^{100}Mo to the first 0^+ excited state in ^{100}Ru are presented. Possible time dependence of weak interaction constant is discussed.

1 Introduction

A brief review of the experiments in which author is involved is done in this paper. The structure of the paper is the following: final results from NEMO-2 and the present status of NEMO-3 detector are presented in Sec. 2; results with liquid argon ionization chamber (DBA experiment) are given in Sec. 3; results of studying $2\nu\beta\beta$ decay of ^{100}Mo to the first 0^+ excited state in ^{100}Ru are presented in Sec. 4; possible time variation of weak interaction constant is discussed in Sec. 5.

2 NEMO experiment

The NEMO collaboration is building the NEMO-3 [1] detector for double beta decay experiments which will be capable of studying $\beta\beta 0\nu$ decays of ^{100}Mo and other nuclei with a sensitivity up to $\sim 10^{25}$ y corresponding to neutrino masses of 0.1 to 0.3 eV. Two prototype detectors, NEMO-1 [2] and NEMO-2 [3], have been constructed as research and development efforts to establish reliable techniques.

2.1 NEMO-2 detector and $\beta\beta$ decay experiments with ^{100}Mo, ^{116}Cd, ^{82}Se, ^{96}Zr and ^{94}Zr

NEMO-2 [3] consists of a $1 m^3$ tracking volume filled with helium gas and 4% ethyl alcohol. (Fig. 1). Vertically bisecting the detector is the plane of the source foil (1m × 1m). The tracking portion of the detector is made of open Geiger cells with octagonal cross sections defined by 100μm nickel wires.

On each side of the source there are 10 planes of 32 cells which alternate between vertical and horizontal orientations. The cells provide three-dimensional tracking of charged particles by recording the drift time and two plasma propagation times in each cell.

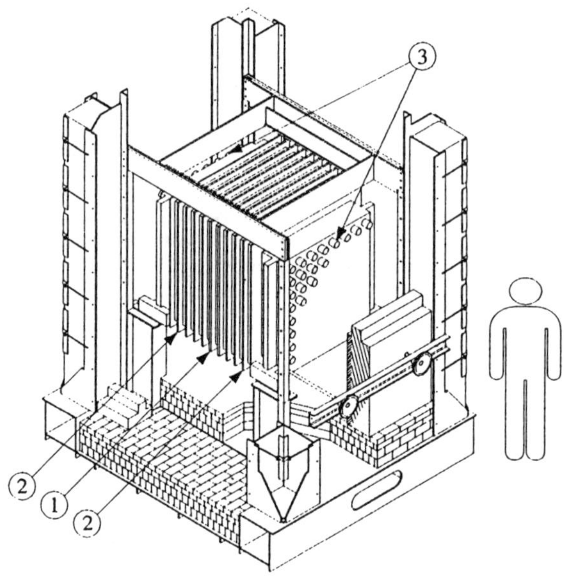

Figure 1: The NEMO-2 detector. (1) Central frame with the metallic foil. (2) Tracking device of 10 frames with 2 × 32 Geiger cells each. (3) Scintillator array. (The shielding is not shown.)

A calorimeter made of scintillators covers two opposing, vertical sides of the tracking volume. Two configurations of the calorimeter have been implemented. The first one consisted of 2 planes of 64 scintillators (12 cm×12 cm×2.25 cm) associated with "standard" photomultiplier tubes (PMTs). This configuration was used in the experiment with ^{100}Mo. The next configuration has included 2 planes of 25 scintillators (19 cm×19 cm×10 cm) with PMTs made of low radioactive glass. The tracking volume and scintillators are surrounded by a lead (5 cm) and iron (20 cm) shield.

The performance and operating parameters are as follows. The threshold for the scintillators is set at 50 keV, the energy resolution (FWHM) is 18% at 1 MeV and the time resolution is 275 ps for a 1 MeV electron (550 ps at 0.2 MeV).

NEMO-2 detector was operated in the Frejus Underground Laboratory (4800 m w.e.) from 1991 to 1997. During this period $\beta\beta$ decay prosesses in ^{100}Mo, ^{116}Cd, ^{82}Se, ^{96}Zr and ^{94}Zr were investigated. Half-life values for $2\nu\beta\beta$ decay and half- life limits on $0\nu\beta\beta$, $0\nu\beta\beta\chi^0$, and the $0\nu\beta\beta$ transition to the 2^+ excited state have been extracted from the data and are presented in Table 1.

Table 1: Main results obtained with the NEMO-2 detector.

Isotope	$T_{1/2}(2\nu)$, 10^{19} y	$T_{1/2}$, 10^{21} y (90% CL)		
		0ν	$0\nu\chi^0$	$0\nu(0^+ - 2^+)$
^{100}Mo [4]	$0.95 \pm 0.04(stat) \pm 0.09(syst)$	> 6.4	> 0.5	> 0.8
^{116}Cd [5]	$3.75 \pm 0.35(stat) \pm 0.21(syst)$	> 5.0	> 1.2	> 0.6
^{82}Se [6]	$8.3 \pm 1.0(stat) \pm 0.7(syst)$	> 9.5	> 2.4	> 2.8
^{96}Zr [7]	$2.1^{+0.8}_{-0.4}(stat) \pm 0.2(syst)$	> 1.0	> 0.35	> 0.39
^{94}Zr [7]	> 0.011	> 0.019	> 0.0023	-

2.2 NEMO-3 detector

In contrast to the ^{76}Ge experiments, the NEMO experiments [1] use tracking detectors which are not only able to measure the full energy released, but other parameters of the process such as the single electron energy, the angle between electrons, the coordinates of events, etc. The optimal operating parameters of the detector were investigated with the prototype NEMO-2 [3,4,5,6,7]. Currently the NEMO-3 detector is under construction and will be able to accommodate up to 10 kg of various double beta decay candidates (^{100}Mo, ^{116}Cd, ^{82}Se, ^{130}Te, ^{96}Zr, ^{150}Nd, etc). The lifetime limits after 5 years of measurement will be at the level 10^{25}y for $0\nu\beta\beta$ decay with $< m_\nu > \sim (0.1 - 0.3)$ eV, and $\sim 10^{23}$ y for $0\nu\beta\beta\chi^0$ decay ($< g_{ee} > \sim 10^{-5}$), and finally $\sim 10^{22}$ y for $2\nu\beta\beta$ decay.

A general view of the detector's cylindrically symmetric geometry is shown in Fig. 2. The detector consists of a tracking volume filled with helium gas, a thin source foil divides the tracking volume vertically into two concentric cylinders with a calorimeter at the inner and outer walls. The tracking system consists of 6180 Geiger cells 2.7 m long which are parallel to the detector's vertical axis. Energy and time-of-flight measurements are performed by the plastic scintillators covering the two concentric surfaces discussed above and their associated end caps. The total number of low radioactive photomultipliers will be 1940. A magnetic field (~ 30 Gauss) will be used to reject backgrounds connected with pair creation and incoming electrons. A shield consisting of 20 cm of iron and 20 cm of paraffin will protect the detector from external radioactivity. 15 sectors over 20 are already built. The placement of completed sectors on the frame in Frejus laboratory started in December 1998. This final

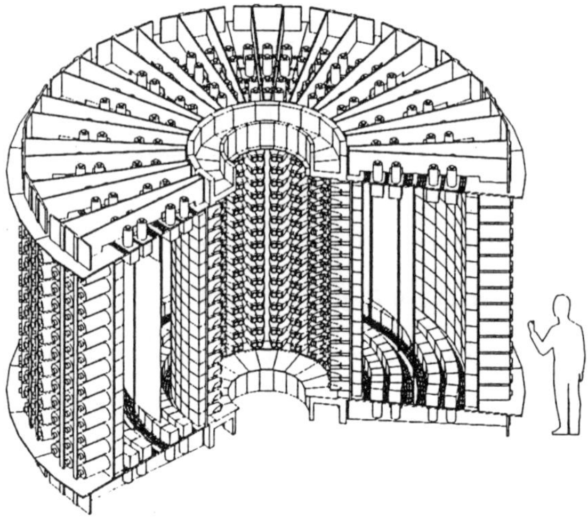

Figure 2: Schematic view of the NEMO3 detector.

stage of the construction is expected to continue until the beginning of 2000. Presently, it is planned to start operating with 7 kg of ^{100}Mo, 1 kg of ^{82}Se and 0.5 kg of ^{116}Cd, with some sectors filled with foils especially designed to check background.

3 Investigation of $\beta\beta$ decay of ^{100}Mo with liquid argon ionization chamber (DBA experiment)

The experiment is located in the Gran Sasso Undeground Laboratory (3500 m w.e.). The experimental setup consists of a liquid Ar ionization chamber placed in a lead passive shielding, gas system and electronics. The registration part of the chamber is composed of identical measuring sections. Each section consists of two paired flat ionization chambers with screening grids and common cathode. The sensitive volume diameter is 30 cm and its height is 56 cm. The chamber contains 14 cathodes, 15 anodes and 28 screening grids. The grid-anode distance is 5.5 mm and grid-cathode one is 14.5 mm. The chamber cathodes are made of molybdenum foil of 50 mg/cm^2 thick. During last three runs (202 h, 238 h and 313 h) 4 cathodes made of enriched molybdenum (98.35% of ^{100}Mo) were used. The total mass of ^{100}Mo under investigation was 138.7 g. The chamber has been assembled with four cathodes containing enriched ^{100}Mo and ten cathodes with natural Mo. The enriched cathodes have been put in the center of the detector so that the natural Mo and ^{100}Mo alternated each other. This configuration is used to provide background subtraction for two-neutrino double beta decay measurement. More detailed description of the chamber and experiment one can find in [8]. Here we present the results which were extracted from 753 h data.

0ν-decay. Only two events in enriched foils and 4 events in natural ones in energy interval (2.7–3) MeV were detected. Taking into account the efficiency (23.4%) we can obtain the limit on 0ν-decay of ^{100}Mo, $T_{1/2} > 6.7(3.5) \cdot 10^{21}$ y at 68%(90%) CL.

$0\nu\chi^0$-decay. The energy interval (2.3–3.0) MeV was investigated. 214 events in enriched and 188 events in natural foils were found. Using the efficiency value of 5.7% the following limit was obtained, $T_{1/2} > 1.2 \cdot 10^{20}$ y at 90% CL.

2ν-decay. The (1.6–3) MeV energy region was investigated because of quite high background at low energy. 1417 and 1292 events were detected in 4 enriched and 4 natural foils, respectively. The difference is (125 ± 52) events. If all 152 events are connected with 2ν-decay of ^{100}Mo then the half-life is equal to $\sim 8.5 \cdot 10^{18}$ y which is in argeement with the last NEMO result [4].

At present the total mass of ^{100}Mo in our detector is increased up to ~300 g. This quantity is optimal to investigate the $2\nu\beta\beta$ mode with "differential method" when the electrode system is assembled with alternating cathods made of natural Mo and ^{100}Mo. In addition we plan to improve the passive shielding (to add antineutron shielding and 10 cm of copper.) As a result the half-life for the $2\nu\beta\beta$ decay of ^{100}Mo will be measured with accuracy of ~10%, sensitivity to 0ν- and $0\nu\chi^0$-modes will be increased to the level of $\sim (1-2)\cdot 10^{23}$ y and $\sim (1-2) \cdot 10^{22}$ y, respectively. In the future the mass of ^{100}Mo can be increased up to ~500 g.

4 $2\nu\beta\beta$ decay of ^{100}Mo to the first 0^+ excited state in ^{100}Ru

The experimental work [10] has been performed in the Modane Underground Laboratory (depth of 4800 m w.e.). A set of ^{100}Mo enriched metal powder samples were measured by using for low background HPGe detectors with volumes of 100, 120, 380 and 400 cm^3. Data of 17 measurements were analysed and the sum spectrum were obtained. The product $P = \sum m_i \eta_i t_i \varepsilon_i$ is equal 154.85 kg·h for the γ-line at 539.53 keV. Here summing up are made on 17 measurements, m_i is the mass of a sample, η_i is its enrichment, t_i is the counting time and ε_i is the detection efficiency. The small peaks were found in places of interest (Fig. 3a). They are equal to (86 ± 25) events at 539.53 keV and (67 ± 23) events at 590.76 keV. To increase the effect one can add up two peaks. The peak at 539.53 keV with surrounding channels were shifted

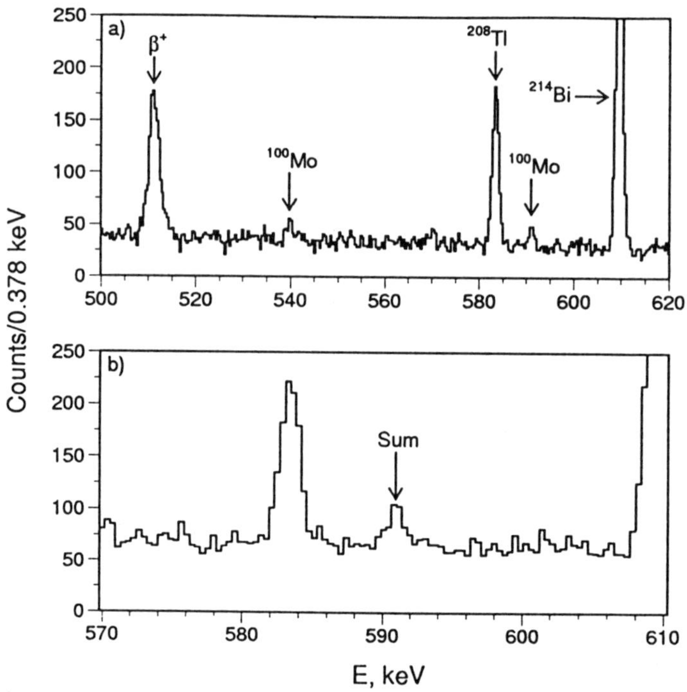

Figure 3: Total γ-ray spectrum in the energy range corresponding to the $2\nu\beta\beta$ decay of ^{100}Mo to the first 0^+ excited state (a). The sum of two peaks, 539.53-keV and 590.76-keV, with surrounding channels is shown lower (b).

to 590.73 keV and summed up. The result is shown in Fig. 3b. The total peak is estimated as 153 ± 35 counts. The calculated value for half-life is $9.3^{+2.8}_{-1.7} \cdot 10^{20}$ y. The systematic error is estimated as 15%. If the previous data[9] is used additively one can produce the estimate with greater statistics, $T_{1/2} = 7.6^{+1.6}_{-1.1} \cdot 10^{20}$ y. The indefinity of 15% connected with systematical errors in the efficiencies calculated keeps.

5 Does the weak interaction constant depend on time?

The comparison[11] of the probability of $2\nu\beta\beta$ decay in direct (counter) and geochemical experiments has shown the following:

1. A discrepancy exists between the values of the half-life of ^{82}Se which were obtained in modern counter experiments and in geochemical measurements.
2. The ^{96}Zr data show the same tendency as the ^{82}Se data – the present-day value of $T_{1/2}$ is less than the values obtained in geochemical measurements.
3. Geochemical measurements on young samples give lower values of $T_{1/2}(^{130}\mathrm{Te})$ as compared to measurements on old samples.

These discrepancies can all be explained (at least partially) by a change in weak interaction constant G_F with time. If this is indeed the case, then this will have the most serious consequences for modern physics and astrophysics. But, this is why it is necessary to confirm (or refute) reliably the reality of these discrepancies. This can be done only by performing new and more accurate measurements. It was proposed the following:
- precise laboratory measurements of the present-day values of the $2\nu\beta\beta$ decay half-lives of ^{82}Se, ^{96}Zr and ^{130}Te should be performed;
- new, precise measurements of the half-lives of ^{82}Se, ^{96}Zr and ^{130}Te in geochemical experiments should be performed;
- the possibility of perfoming geochemical experiments with ^{100}Mo, ^{116}Cd, ^{124}Sn, ^{110}Pd, ^{150}Nd and ^{76}Ge should be investigated, and if possible such measurements should be performed. One of the most interesting isotopes is ^{100}Mo which has high probability of decay, high abundance in the natural Mo (9.6%) and ^{100}Ru (not gas!) in the final state. In addition, the precise value of $T_{1/2}$ for ^{100}Mo extracted from direct experiment[4] is known.

Acknowledgments

This work were supported by Russian Foundation for Branch Investigations under contract 97-02-17344 and by INTAS under grant 96-0589.

References

1. NEMO Collaboration, *preprint LAL* 94-29 (1994).
2. D. Dassié *et al.*, *Nucl. Instrum. Methods* **A309**, 465 (1991).
3. R. Arnold *et al.*, *Nucl. Instrum. Methods* **A354**, 338 (1995).
4. D. Dassié *et al.*, *Phys. Rev.* **D51**, 2090 (1995).
5. R. Arnold *et al.*, *Z. Phys.* **C72**, 239 (1996).
6. R. Arnold *et al.*, *Nucl. Phys.* **A636**, 209 (1998).
7. R. Arnold *et al.*, *Nucl. Phys.* A, to be published.
8. V.D. Ashitkov *et al.*, *Nucl. Phys. B (Proc. Suppl.)***70**, 233 (1999).
9. A.S. Barabash *et al.*, *Phys. Lett.* **B345**, 408 (1995).
10. A.S. Barabash *et al.*, *Physics of Atomic Nuclei*, to be published.
11. A.S. Barabash, *JETP Lett.* **68**, 1 (1998).

Double Beta Decays of ^{100}Mo by ELEGANT V at Oto Cosmo Observatory

N. Kudomi[a], H. Ejiri[a], K. Fushimi[b], K. Hayashi[a], R. Hazama[c], T. Kishimoto[c], K. Kume[a], H. Kuramoto[a], T. Matsuoka[c], H. Ohsumi[c], K. Takahisa[a], Y. Tsujimoto[a], and S. Yoshida[a]

a)RCNP, Osaka Univ., Ibaraki, Osaka 567, Japan

b)Facul. of Integ. Arts and Sci., The Univ. of Tokushima, Tokushima, 770, Japan

c)Dept. of Phys., Osaka Univ., Toyonaka, Osaka 560, Japan

Exclusive measurements of neutrino-less double beta decays($0\nu\beta\beta$) of ^{100}Mo were made by means of ELEGANT V. ELEGANT V was moved to Oto Cosmo Observatory from Kamioka underground lab. The present status of the double beta decay experiment with ELEGANT V is presented. The data at Oto lab., being combined with the data at Kamioka, gives stringent limits on half-lives for $0\nu\beta\beta$ and $\langle m_\nu \rangle < 1.9$eV for the Majorana neutrino mass.

Double beta decays are of current interest from both astroparticle and nuclear physics view points [1,2,3]. The neutrino-less double beta decays($0\nu\beta\beta$), which violate the lepton number conservation law, provide one with very sensitive tests for the Majorana neutrino mass, the right handed weak currents and so on(Fig. 1B-E). Two neutrino double beta decay($2\nu\beta\beta$) (Fig. 1A) give directly the nuclear matrix element[4]. This is used to verify nuclear structure calculations and to select appropriate spin-isospin interactions $H_{\tau\sigma}$ to be used for evaluating the nuclear matrix element $M^{0\nu}$ for $0\nu\beta\beta$. Since $\beta\beta$ transition rates depend largely on the nuclear spin-isospin responses(matrix elements), it is important to study $\beta\beta$ decays on several nuclei in order to extract universal values of (limits on) physics quantities beyond the standard theory.

The $\beta\beta$ processes for electron(β^-) decays are schematically illustrated in Fig. 1. The transition rate for $2\nu\beta\beta$(Fig. 1A) is simply written in terms of the nuclear matrix element $M^{2\nu}$ as

$$[T^{2\nu}_{1/2}]^{-1} = G^{2\nu}|M^{2\nu}|^2, \tag{1}$$

where $T^{2\nu}_{1/2}$ is the half-life and $G^{2\nu}$ is the phase space factor. Thus $M^{2\nu}$ is derived experimentally from the observed $2\nu\beta\beta$ rate. The transition rate for $0\nu\beta\beta$ associated with the neutrino exchange (Fig. 1B) is written in terms of the neutrino mass term $\langle m_\nu\rangle$ and the RHC terms of $\langle\lambda\rangle$ and $\langle\eta\rangle$ as [1,2,3],

$$\begin{aligned}[T^{0\nu}_{1/2}(\nu)]^{-1} = G^{0\nu}|M^{0\nu}|^2(\langle m_\nu\rangle^2 + C_{\lambda\lambda}\langle\lambda\rangle^2 + C_{\eta\eta}\langle\eta\rangle^2 \\ +C_{m\lambda}\langle m_\nu\rangle\langle\lambda\rangle + C_{\lambda\eta}\langle\lambda\rangle\langle\eta\rangle + C_{m\eta}\langle m_\nu\rangle\langle\eta\rangle),\end{aligned} \tag{2}$$

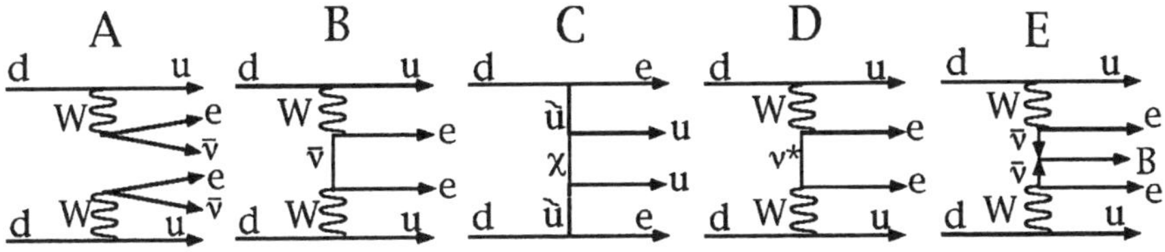

Figure 1: Schematic diagrams of $2\nu\beta\beta$ and $0\nu\beta\beta$. A: $2\nu\beta\beta$, B: $0\nu\beta\beta(\nu)$, C: $0\nu\beta\beta(\chi)$, D: $0\nu\beta\beta(\nu^*)$, E: $0\nu\beta\beta$B,

where $G^{0\nu}$ and $M^{0\nu}$ are the phase space factor and the nuclear matrix element for the mass term. C_{ij}, with i, j being m, λ and η, are the nuclear responses in units of $M^{0\nu}$. These $0\nu\beta\beta$ transition rates include both the nuclear matrix elements (form factors) relevant to nuclear structures and the physics quantities relevant to particle physics.

So far, $0\nu\beta\beta$ decay rates have been studied on several nuclei. Unique features of the present work of ^{100}Mo with ELEGANT V[6] are as follows.

i) Energy and angular correlations of two β rays and γ rays are measured by ELEGANT V. Thus limits on the $0\nu\beta\beta$ processes for individual terms of $\langle m_\nu \rangle$, $\langle \lambda \rangle$, $\langle \eta \rangle$ and others can be obtained.

ii) ^{100}Mo has large phase space factors for $0\nu\beta\beta$ and $2\nu\beta\beta$ because of the large Q values. Furthermore, the $2\nu\beta\beta$ matrix element is known to be large from the previous work[5].

iii) ELEGANT V has been improved so that the background rate has been reduced to the level of 0.1counts/day/kg at the $0\nu\beta\beta$ region. Origins of the background events are well investigated by the β(e) and γ correlation studies in the off-line analysis. Thus the correction for them is possible.

The obtained limits on $\langle m_\nu \rangle$, $\langle \lambda \rangle$, $\langle \eta \rangle$, $\langle g_B \rangle$ and others, depending somewhat on the matrix elements used, are most stringent for ^{100}Mo. They are same orders of magnitudes as the values derived from other nuclei. Thus the present data with the unique points as given above, together with other data, may give stringent limits on the relevant values beyond the standard theory.

ELEGANT V consists of three drift chambers of DC-A, DC-B and DC-C for measuring two β trajectories, plastic scintillator arrays(PL) for β ray energies and arrival times, and NaI scintillator(NaI) arrays for γ and X rays[6]. They are schematically shown in Fig. 2. The total weight of ^{100}Mo is 171gr. The central drift chamber(DC-C) was also used for detecting delayed α-rays following ^{210}Po produced by the β decay of ^{214}Bi in the source region. PL's are calibrated by γ-rays of 511keV and 1275keV from the standard source of ^{22}Na by fitting the shape of the Compton spectrum[7], and NaI's are calibrated

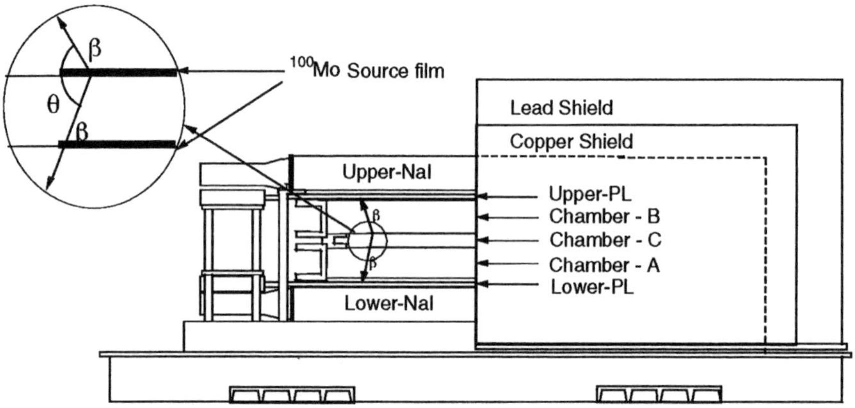

Figure 2: Schematic view of ELEGANT V [6]. DC-A, DC-B and DC-C are lower, upper and central drift chambers.

by fitting the photo peaks of γ-rays from the ^{22}Na. The detection efficiencies of DC's are checked by the β-ray from ^{90}Sr passing through DC's.

At present, the $\beta\beta$ decays of ^{100}Mo are measured at Oto Cosmo Observatory with 1200m.w.e.. The sum energy($E_1 + E_2$) spectrum is obtained by selecting events with several conditions with live time for the 3639hrs. The high energy part of the spectrum is shown in Fig. 3. The major part of the spectrum is the $2\nu\beta\beta$ component.

Background contributions from the natural radioactive contaminations were estimated as follows. The ^{214}Bi with Q_β = 3.28MeV and ^{208}Tl with Q_β = 4.99MeV are two major isotopes, which may give background events in the $0\nu\beta\beta$ energy window of $E_1 + E_2 \sim$ 3MeV. These isotopes are decay products of ^{238}U- and ^{232}Th-chain isotopes contained in the source film and detector element. ^{214}Bi is produced also from Rn contained in the air around the source and the detector element.

As the first step, the total amounts of the ^{214}Bi and ^{208}Tl were evaluated from the single electron/β-ray event rates in coincidence with γ-rays characteristic of the decays of ^{214}Bi and ^{208}Tl. Here the γ-rays were measured by the NaI detector array of ELEGANT V. The electron energy, time and trajectory, which were measured by the DC's and PL, were used to identify the locations of the ^{214}Bi and ^{208}Tl isotopes in ELEGANT V. The obtained ^{214}Bi and ^{208}Tl contents are shown in Table 1.

On the basis of the estimated ^{214}Bi and ^{208}Tl contents, Monte Carlo calculations were carried out to evaluate the BG(fake) event rates caused by these

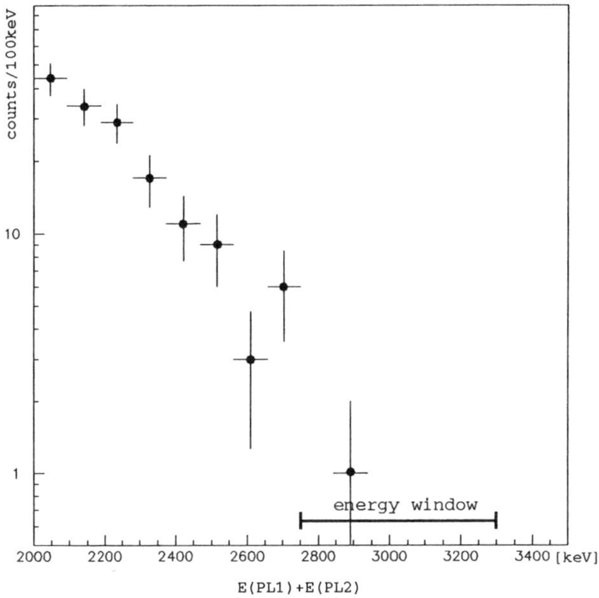

Figure 3: The sum energy spectrum($E_1 + E_2$) for the two electron events.

Table 1: The ^{214}Bi and ^{208}Tl contents, which contribute to the $0\nu\beta\beta$ energy window, evaluated in Oto measurement together with those in Kamioka measurement[8]. Locations of source, DC-gas, and PL indicate the ^{100}Mo source film, the gas inside DC's and the gas around the source, and mainly the air around PL, respectively. The contents of ^{214}Bi and ^{208}Tl in source in Oto was referred the evaluation in Oto, because the same source films were used in Oto and Kamioka.

Origin	Location	Amount		
		Oto	Kamioka	
^{214}Bi	Source	$(8.3 \pm 1.7) \times 10^{-3}$	$(8.3 \pm 1.7) \times 10^{-3}$	Bq/kg
	DC-gas	$(2.2 \pm 0.5) \times 10^{-2}$	$(2.0 \pm 0.2) \times 10^{-2}$	Bq/m^3
	PL	$(6.5 \pm 1.5) \times 10^{-2}$	(2.80 ± 0.14)	Bq/m^3
^{208}Tl	Source	$(2.1 \pm 0.7) \times 10^{-5}$	$(2.1 \pm 0.7) \times 10^{-5}$	Bq/kg

Table 2: Limits with 68%CL on the half-lives for the $0\nu\beta\beta$ decay of ^{100}Mo at Oto lab., at the Kamioka lab. and those obtained by combining the two limits. For each decay mode, the energy window, the obtained yield, the expected $2\nu\beta\beta$+BG yield and the detection efficiency(ϵ) are given.

(Oto(Preliminary), Live Time=3639hrs)

mode		E window (MeV)	Yield (counts)	$2\nu\beta\beta$+BG (counts)	ϵ	$T_{1/2}$ (10^{21}y)
$0^+ \to 0^+$	$\langle m_\nu \rangle$ term	2.7-3.3	1	1.1	0.15	>25
	$\langle \lambda \rangle$ term	2.7-3.3	1	1.1	0.10	>18
		2.7-3.3	0*	0.4*	0.063*	>17*
	$\langle \eta \rangle$ term	2.7-3.3	1	1.1	0.14	>23
$0^+ \to 2^+$	$\langle \lambda \rangle$ term	2.1-2.8	0	<0.8	0.035	>9.2
Majoron	emission	2.5-3.0	11	7.8	0.034	>1.8

(Kamioka[8], Live Time=7333hrs)

mode		E window (MeV)	Yield (counts)	$2\nu\beta\beta$+BG (counts)	ϵ	$T_{1/2}$ (10^{21}y)
$0^+ \to 0^+$	$\langle m_\nu \rangle$ term	2.8-3.3	3	4.1	0.19	>52
	$\langle \lambda \rangle$ term	2.8-3.3	3	4.1	0.11	>31
		2.8-3.3	0*	0.6*	0.075*	>39*
	$\langle \eta \rangle$ term	2.8-3.3	3	4.1	0.18	>51
$0^+ \to 2^+$	$\langle \lambda \rangle$ term	2.1-2.8	0	0.23	0.048	>14
Majoron	emission	2.5-3.0	26	28.	0.045	>5.4

(Combine)

mode		$T_{1/2}$ (10^{21}y)	physics quantity
$0^+ \to 0^+$	$\langle m_\nu \rangle$ term	67	$\langle m_\nu \rangle$ <1.9eV
	$\langle \lambda \rangle$ term	41	
	$\langle \lambda \rangle$ term	56*	$\langle \lambda \rangle < 3.1 \times 10^{-6}$
	$\langle \eta \rangle$ term	64	$\langle \eta \rangle < 2.2 \times 10^{-8}$
$0^+ \to 2^+$	$\langle \lambda \rangle$ term	34	
Majoron	emission	4.8	$\langle g_B \rangle < 7.7 \times 10^{-5}$

* Values derived by additional energy and angular windows for the $\langle \lambda \rangle$ term $0\nu\beta\beta$ process.

isotopes, which may survive after the $\beta\beta$ selections mentioned above. The major BG $\beta\beta$ events come from the β and conversion-electrons, and the single β or the Compton-electron passing through the source film. The observed spectrum is reproduced by the sum of the estimated BG spectrum and the $2\nu\beta\beta$ spectrum with the previously known half-life of 1.15×10^{19}y [5].

There are no excess $0\nu\beta\beta$ counts beyond the statistical fluctuations of the $2\nu\beta\beta$+BG events. In order to obtain upper limits on the number of the counts, the detection efficiencies for individual processes of the $\langle m_\nu \rangle$, $\langle \lambda \rangle$ and $\langle \eta \rangle$ terms were evaluated by Monte Carlo simulations, as shown in Table 2. The limits with the 68%CL on the half-lives are obtained from the number of the observed counts and the number of the estimated $2\nu\beta\beta$+BG counts. The standard method of the likelihood analysis was used [10]. The results of the measurement at Oto together with those of the previous work at Kamioka are summarized in Table 2.

The combination of these two measurement is made with the statistical approach presented in ref. [11]. The results are given in Table 2.

Using the nuclear matrix elements given in ref. [9], the half-life limits for the $0^+ \rightarrow 0^+$ $0\nu\beta\beta$ and $0^+ \rightarrow 0^+$ $0\nu\beta\beta$B processes leads to the upper limits of $\langle m_\nu \rangle < 1.9$eV, $\langle \lambda \rangle < 3.1 \times 10^{-6}$, $\langle \eta \rangle < 2.2 \times 10^{-8}$, and $\langle g_B \rangle < 7.7 \times 10^{-5}$.

The present work measures the energy and angular correlations of two β rays, which are important for the study of individual double beta processes. It is also important to study the double beta decay of several nuclei, because the values of $\langle m_\nu \rangle$, $\langle \lambda \rangle$, $\langle \eta \rangle$, $\langle g_B \rangle$ and others depend somewhat on the matrix elements which are calculated with some approximations.

Acknowledgments

The authors thank Oto village and Nishi-Yoshino village for their kind support and for the Oto lab. construction. The present work is supported by the Grant-in-Aid of Scientific Research and Ministry of Education, Science and Culture, Japan. K.F., R.H., N.K., K.K., H.M. and K.N. were supported by JSPS Research Fellowships.

References

1. W.C. Haxton and G.J. Stephenson, *Jr. Prog. in Part. Nucl. Phys.* **12**, 409(1984).
2. M. Doi, T. Kotani and E. Takasugi, *Prog. Theor. Phys. Suppl.* **83**, 1(1985).
3. H. Ejiri, Int. J. Mod. Phys. **E 6**, 1(1997)
4. H. Ejiri and H. Toki, *Joun. of Phys. Soc. of Japan,* **65**, 7(1996).
5. H. Ejiri, et al., *Phys. Lett.* **B 258**, 17(1991); H. Ejiri, et al., *J. Phys. G. Nucl. Part. Phys.* **17**, S155(1991).
6. H. Ejiri, et al., *Nucl. Instr. and Meth.* **A 302**, 304(1991). K. Nagata, et al. *Nucl. Instr. and Meth.* **A 362**, 261(1995).
7. N. Kudomi, *Nucl. Instr. and Meth.* **A** (1999) to be published.
8. H. Ejiri, et al., Nucl. Phys. A611, 85(1996).
9. T. Tomoda, *Rep. Prog. Part. Phys.* **54**, 53(1991).
10. Particle Data Group, Phys. Rev. **D 50**, (1994)1173.
11. O. Helene, Nucl. Instr. and Meth in Phys. Res. **A390**, 383(1997)

On the effective meson-nucleon Lagrangian for neutrino-less double β decay from R-parity violating SUSY model

K. Suzuki and H. Toki
RCNP, Osaka University, Osaka 567-0047, Japan
E-mail: ksuzuki@rcnp.osaka-u.ac.jp

We study the effective meson-nucleon Lagrangian which describes the neutrino-less double β decay in the R-parity violating SUSY. The low energy effective Lagrangian which induces $dd \to uu + 2e^-$ transition generally consists of the nucleon-nucleon contact interaction, pion-nucleon Yukawa type and the pion-pion contact parts. Making use of the soft pion theorem, we find that the large contribution comes from the pion-nucleon term in close analogy with the $\Delta I = 1/2$ enhancement of the non-leptonic weak transitions of hyperons.

1 Introduction

There exist several possible mechanisms to generate the neutrinoless double β decay ($0\nu\beta\beta$) beyond the Standard Model. Most popular process is induced by the exchange of the heavy Majorana neutrino. Another one originates from the R-parity violating supersymmetric extension of the standard model (SUSY) [1], in which the $0\nu\beta\beta$ is mediated by the superparticles and the lepton number conservation is violated. Thus, observation of the $0\nu\beta\beta$ decays in nuclei can provide constraints on the physics beyond the Standard Model. Here, we stress the R-parity violating SUSY, where R-parity is +1 for the normal particles and -1 for superparticles. Conservation of the R-parity is not ensured by the gauge invariance. If the R-parity is conserved, for example, the lightest superparticle could be a good candidate of the dark matter. Therefore, the experimental test for the R-parity conservation is of great interest.

To make theoretical predictions for the life time of the $0\nu\beta\beta$ decay width to be compared with the experiment, we need the following steps. First, we construct the low energy effective Lagrangian for quarks and gluons by integrating out the superparticle degrees of freedom within R-parity violating SUSY. Then, the effective Lagrangian at the hadronic level in terms of baryons and mesons are constructed by taking into account the quark-gluon structure of the hadrons. Finally, we apply the resulting hadronic Lagrangian to the nuclear many body system, and evaluate the transition matrix elements in nuclei, ${}^A_Z X \to {}^A_{Z+2} Y + 2e^-$. In this work, we try to construct a possible effective Lagrangian for the hadrons. We shall deal with the pion-nucleon dynamics and the $0\nu\beta\beta$ interaction consistently based on the current algebra beyond the naive factorization approximation, taking analogy with the pionic decay of hyperons in the standard weak theory. In the following sections, we show the basic steps to construct the effective Lagrangian in the R-parity violating SUSY. We remark, however, the procedure presented here is rather general and can be applied to any mode beyond the Standard Model.

2 Construction of the effective Lagrangian

The explicit R-parity violating term of the SUSY potential[2] is written as,

$$W_{\not{R}p} = \lambda_{ijk} L_i L_j \bar{E}_k + \lambda'_{ijk} L_i Q_j \bar{D}_k + \lambda''_{ijk} \bar{U}_i \bar{U}_j \bar{D}_k \ , \tag{1}$$

where we use the standard notations. From this superpotential, possible diagrams which contribute to the $0\nu\beta\beta$ transition are considered in ref.[1,2]. Typical one is shown in Fig.1. The low energy quark-lepton Lagrangian after integrating out the superfields, in which the $0\nu\beta\beta$ decay is caused by the subprocess $dd \to uu + 2e^-$, is given by

$$\mathcal{L}_{ql} = \frac{G_F^2}{2m_p} e(1+\gamma^5)e^c \left[\eta_{ps} \mathcal{O}_{ps} - \frac{1}{4}\eta_T \mathcal{O}_T \right] \ , \tag{2}$$

where the operators are

$$\mathcal{O}_{ps} = \bar{u}\,(1+\gamma^5)\,d\,\bar{u}(1+\gamma^5)\,d \tag{3}$$

$$\mathcal{O}_T = \bar{u}\,\sigma^{\mu\nu}(1+\gamma^5)\,d\,\bar{u}\,\sigma_{\mu\nu}(1+\gamma^5)\,d\,. \tag{4}$$

Here, the coupling constants η are expressed in terms of the R-parity violating coupling constant λ'_{111}, other SUSY parameters and the unification scale[2,3].

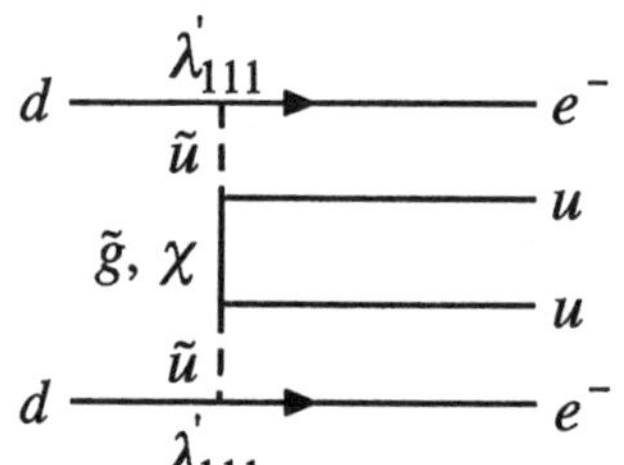

Figure 1: Typical diagram for $0\nu\beta\beta$ decay in R-parity violating SUSY

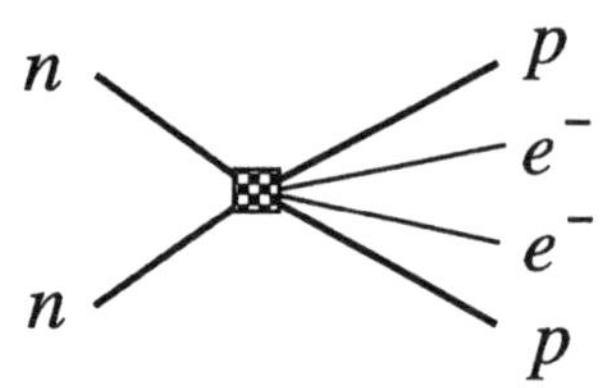

Figure 2: Diagram for $\mathcal{L}_{2N}$

To describe the decays in nuclei, we try to construct the effective Lagrangian for the pion-nucleon system, which can be generally written as

$$\mathcal{L}_{Had} = \mathcal{L}_{2N} + \mathcal{L}_{N\pi} + \mathcal{L}_{2\pi} \tag{5}$$

The first term represents the point interaction between nucleons and electrons shown in Fig.2, $nn \to pp + 2e^-$, which is expressed as

$$\mathcal{L}_{2N} = A_{2N} \frac{G_F^2}{2m_p} \bar{p}\Gamma_i n \, \bar{p}\Gamma_i n \, e(1+\gamma^5)e^c \tag{6}$$

where Γ_i denote the appropriate gamma matrices. Contributions of this term to nuclear decay matrix elements are evaluated in ref. [1,2]. This transition would be suppressed very much by the nucleon-nucleon hard core potential [3], although the transition matrix element does not vanish due to the finite size of nucleons.

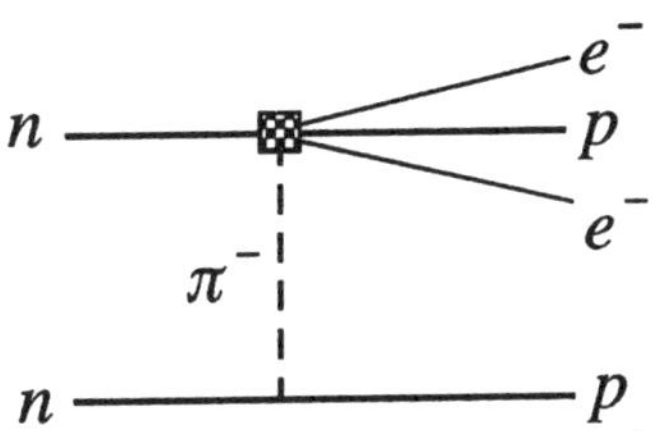

Figure 3: Diagram for $\mathcal{L}_{N\pi}$

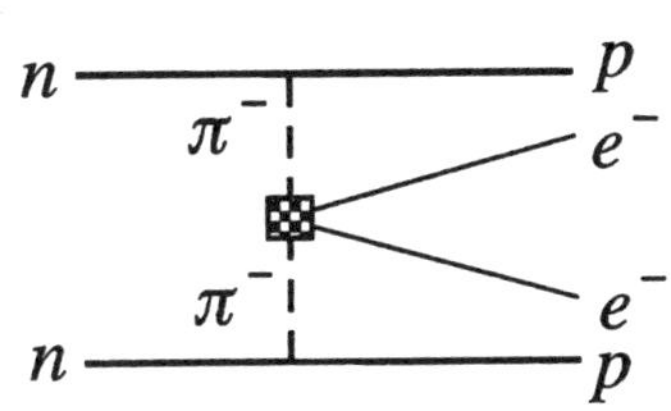

Figure 4: Diagram for pion exchange current $\mathcal{L}_{2\pi}$

Remaining terms which involve pion fields can be written by

$$\mathcal{L}_{N\pi} = \frac{G_F^2}{2m_p} m_\pi^2 \, e(1+\gamma^5)e^c \, \left[A_{N\pi} \, \bar{p} n \, \pi^- + \, B_{N\pi} \bar{p} \, i\gamma^5 n \, \pi^-\right] \tag{7}$$

$$\mathcal{L}_{2\pi} = -A_{2\pi} \frac{G_F^2}{2m_p} m_\pi^4 \, e(1+\gamma^5)e^c \, (\pi^-)^2 \tag{8}$$

These contributions are schematically shown in Fig.3 and 4, respectively. We note that the first term of eq. (7) is not considered in the previous works. In order to fix unknown coupling constants, $A_{N\pi}, B_{N\pi}, A_{2\pi}$, of the effective hadronic Lagrangian, we exploit the matching conditions;

$$\langle p \, 2e^- | \mathcal{L}_{ql} | n \, \pi^- \rangle = \langle p \, 2e^- | \mathcal{L}_{N\pi} | n \, \pi^- \rangle \tag{9}$$

$$\langle \pi^+ 2e^- | \mathcal{L}_{ql} | \pi^- \rangle = \langle \pi^+ 2e^- | \mathcal{L}_{2\pi} | \pi^- \rangle \tag{10}$$

These equations relate the coupling constants of the effective hadronic Lagrangian with $\eta_{ps,T}$ of the original Lagrangian (2). Our task here is to calculate the left hand side of eqs. (9,10) using some non-perturbative method based on QCD, and determine the values of the coupling constants, A_i, B_i.

Faessler etal. have estimated the magnitude of each contribution using the vacuum insertion (factorization) approximation, and found that the pion exchange current coming from the third term provides dominant contributions to the $0\nu\beta\beta$ decay over other processes.

Our purpose of this work is to construct the improved effective Lagrangian in more systematic way based on the chiral dynamics, instead of the crude vacuum insertion approximation adopted in ref. [3]. Let us consider the non-leptonic weak transitions of hyperons [4], $B_i \to B_f + \pi^a$, because the weak interaction Hamiltonian has a very similar operator structure with eqs. (3,4). In such a decay, one has to calculate

$$\langle B_f\, \pi^a | (\bar{u}s)_{V-A}\, (\bar{d}u)_{V-A} | B_i \rangle \ . \tag{11}$$

The vacuum insertion approximation simply provides the following result;

$$\sim \text{const.}\, \langle B_f | (\bar{u}s)_{V-A} | B_i \rangle \, \langle \pi^a | (\bar{d}u)_{V-A} | 0 \rangle \tag{12}$$

where the constant is determined by suitable Fierz rearrangements. However, this approximation gives about 1/10 of the experimental values. In addition, relative magnitudes for various hyperon decays cannot be explained by this approximation. This fact certainly suggests that the vacuum insertion is not a proper approximation scheme any more for the strong interacting pion-nucleon system. In case of the non-leptonic hyperon decay, we have a well-established technique to calculate the 4-quark matrix elements based on the chiral dynamics of QCD. Here, we apply the similar procedure to calculate the left hand side of eq. (9).

We make use of the soft pion theorem based on the current algebra which is proved to work very well for the pionic hyperon decay process of the standard weak theory. Using the reduction formula and the PCAC relation, one gets

$$\begin{aligned}
\langle p\, \pi_a | \mathcal{O}_{ps,T} | n \rangle &= \int d^4x\, e^{-iqx} (\partial^2 + m_\pi^2) \langle p | T \{ \pi^a(x), \mathcal{O}_{ps,T} \} | n \rangle \\
&= \frac{-1}{f_\pi} \left\langle p \left| \left[i \int d^3x\, A_a^0(x), \mathcal{O}_{ps,T} \right] \right| n \right\rangle \\
&\quad + \frac{iq_\mu}{f_\pi} \int d^4x\, e^{-iqx} \langle p | T \{ A_a^\mu(x), \mathcal{O}_{ps,T}(0) \} | n \rangle
\end{aligned} \tag{13}$$

One can see that the first term of eq. (13) corresponds to the so called commutator term, shown in Fig.5. The second one provides the baryon pole contribution illustrated in Fig.6, by inserting the intermediate baryon states between

$A_\mu(x)$ and $\mathcal{O}_{ps,T}(0)$.

$$\sum_i \frac{2M_N}{f_\pi}\left(\frac{G^a_{pi}W_{in}}{M_N - M_i} + \frac{W_{pi}G^a_{in}}{M_N - M_i}\right) , \tag{14}$$

where the matrix elements G_{ij}, W_{ij} are defined as,

$$\langle B_i|A^{a\mu}|B_j\rangle = G^a_{ij}\bar{u}(i)\gamma^\mu\gamma^5 u(j)$$
$$\langle B_i|\mathcal{O}_{ps,T}|B_j\rangle = h_{ij}\bar{u}(i)u(j)$$

In the present case, the intermediate states are assumed to be identified with Δ resonances. The commutator and the baryon pole correspond to the first and the second terms of the Lagrangian (7), respectively. The $\Delta I = 1/2$ enhancement of the non-leptonic weak hyperon decay is well reproduced by taking into account contributions coming from these diagrams.

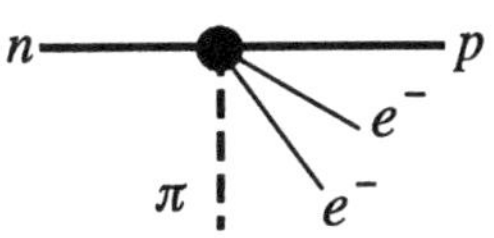

Figure 5: First term of eq. (13). The filled circle represents the non-trivial vertex which arises from the chiral dynamics and the operator (3,4)

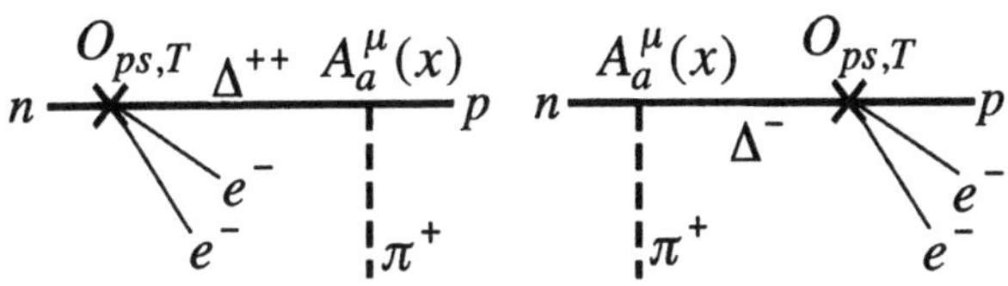

Figure 6: Second term of eq. (13). The cross denotes the insertion of the operator (3,4) and A_μ is the axial-vector current which produces the pion. Intermediate baryon states are assumed to be Δ.

3 Estimates by numerical calculations

In order to estimate the values of the matrix elements such as $\langle p|(\bar{u}d)(\bar{u}d)|n\rangle$, $\langle p|(\bar{u}\sigma^{\mu\nu}d)(\bar{u}\sigma_{\mu\nu}d)|n\rangle$, we must rely on the quark model of the nucleon and pion. Here, we simply take the non-relativistic constituent quark model with the Harmonic Oscillator wave functions, and make rough estimates of the coupling constants $\mathcal{L}_{N\pi,2\pi}$. The oscillator size parameter is chosen to reproduce the radius 0.6 fm. Since the 4-quark operator reduces to the delta function in the coordinate space in the non-relativistic limit, $\mathcal{O} \to a^\dagger_{i'}a^\dagger_{j'}\delta^{(3)}(\vec{r}_i - \vec{r}_j)a_i a_j$, the relevant integral for our purpose is given by

$$\int d^3\rho d^3\lambda\ \Psi^*(\rho,\lambda)\ \delta^{(3)}(\vec{\lambda})\ \Psi(\rho,\lambda) = 0.002\text{GeV}^3$$

where we have used the Jacobi coordinate for the quark wave function $\Psi(\rho, \lambda)$.

The matching condition eq. (9) relates $A_{N\pi}, B_{N\pi}$ with the SUSY parameters, $\eta_{ps,T}$. Calculation of the commutator in eq. (13) term yields

$$A_{N\pi} \simeq 27\eta_{ps} + 60\eta_T \, . \tag{15}$$

Also, calculating the baryon pole term, we obtain the coupling constant $B_{N\pi}$ of eq. (7) as

$$B_{N\pi} \simeq -195\eta_{ps} - 292\eta_T \tag{16}$$

For comparison, we show results obtained by using the factorization approximation [3]

$$B_{N\pi}^{\text{fact.}} \simeq 96\eta_{ps} + 64\eta_T \tag{17}$$

We conclude that large contributions come from the non-factorized terms, that is, the commutator and the baryon pole. However, the complete dominance over the factorization results, which is seen in the pionic decay of the hyperon, is not realized.

In the present problem for $0\nu\beta\beta$, the situation is slightly different from the non-leptonic hyperon decay due to the Dirac structure of the 4-quark operators. Consider the weak decay of hyperons first. The standard weak Hamiltonian has the $(V - A) \otimes (V - A)$ structure. Hence, when we apply the vacuum insertion approximation to this process, we always find that the matrix element is proportional to $\langle \pi^a | \bar{\psi}\gamma^\mu\gamma^5\psi | 0 \rangle \propto f_\pi q^\mu \sim \mathcal{O}(m_\pi)$, which is very small due to the small current quark masses. This is the reason why the factorization contribution is negligibly small for the non-leptonic weak decay. On the other hand, the operators for the $0\nu\beta\beta$ decay of eqs. (3,4) have the $(S+P) \otimes (S+P)$ structure. Thus, the vacuum insertion provides a factor $\langle \pi^a | \bar{\psi}\gamma^5\psi | 0 \rangle \propto m_\pi^2/m_q \sim \mathcal{O}(1)$ [5], which is about 10 times as large as the $(V - A)$ case. Therefore, the factorization already gives substantial contributions to the $0\nu\beta\beta$ matrix elements.

On the other hand, the transition matrix element of $\pi^- \to \pi^+ \to 2e^-$ which contributes to the exchange current shown in Fig.4 is also evaluated by using the Nambu and Jona-Lasinio model [6]. In the lowest order, our NJL calculation [5] agrees with the result of ref. [3], which confirms the argument that the pion exchange current gives sizable contributions to the $0\nu\beta\beta$ matrix element.

4 Conclusions

We have studied the pion-nucleon effective Lagrangian in order to describe the $0\nu\beta\beta$ decay, starting from the R-parity violating SUSY. We generally construct the effective interaction with the pion and nucleon degrees of freedom, which is divided into three parts, two nucleon contact term, pion-nucleon Yukawa term and the two pion contact term. We have calculated the transition matrix elements using the technique based on the current algebra which provides a good description for the non-leptonic weak decay of hyperons, paying attention to the similarity of both processes. We have emphasized that the vacuum insertion is a poor approximation for the strong interacting pion-nucleon system. Due to the $(S+P)$ structure of the operators, the factorization approximation already gives substantial contribution to the double β decay matrix elements. Nevertheless, the contributions from the diagrams shown in Fig.5 and 6 are important, and their magnitudes are at least comparable with those from the vacuum insertion approximation.

Here, we concentrate on the R-parity violating SUSY, however, our treatment can be applicable to any model that generates the $0\nu\beta\beta$ decay. To calculate the $0\nu\beta\beta$ matrix elements in nuclei, all the processes arising from $\mathcal{L}_{2N}, \mathcal{L}_{N\pi}$ and $\mathcal{L}_{2\pi}$ should be consistently taken into account in any model.

Acknowledgments

K.S. would like to thank COE program at RCNP, Osaka University, which enables him to work out this subject.

References

1. J.D. Vergados, *Phys. Lett.* B **184**, 55 (1987).
2. M. Hirsch, H.V. Klapdor-Kleingrothaus, S.G. Kovalenko, *Phys. Rev.* D **53**, 1329 (1996).
3. A. Faessler, S. Kovalenko, F. Simkovic, *Phys. Rev.* D **58**, 115004 (1998) A. Faessler, S. Kovalenko, F. Simkovic, J. Schwinger, *Phys. Rev. Lett.* **78**, 183 (1997).
4. J.F. Donoghue, E. Golowich and B. Holstein, *Phys. Rep.* **131**, 319 (1986), and references therein
5. K. Suzuki and H. Toki, *Mod. Phys. Lett.* **A9**, 1059 (1994)
6. Y. Nambu and G. Jona-Lasinio, *Phys. Rev.* **122**, 345 (1961)

SINGLE AND DOUBLE BETA DECAY WITH THERMAL DETECTORS

Ettore Fiorini

Dipartimento di Fisica G.Occhialini dell' Universita' di Milano-Bicocca e Sezione di Milano dell' INFN, via Celoria 16, 20133 Milan (Italy)

Abstract

The present status of experiments on single and double beta decays where the use of cryogenic detectors is providing or is going to provide competitive results is reported and discussed. Special attention is devoted to double beta decay and to the direct measurement of the neutrino mass. Some application of cryogenic detectors to solar neutrino experiments is also reported.

1. Introduction

I would like to report here some of the very attractive experiments being carried out with cryogenic detectors on weak interactions. and their already interesting results.

I will first consider single beta decay especially in view of the measurement of the neutrino mass, of searches for rare events and of the detection of solar neutrinos. The status and experimental prospective in searches on double beta decay will conclude my talk.

2. The problem of neutrino mass

The existence of a non zero mass for the neutrino (and of course for the antineutrino) is predicted by the theories beyond the Standard Model and would have important consequences in astroparticle physics. It could in fact account, at least partially, for the existence of the ubiquitous Dark Matter.

Upper limits of 0.17 and 18 MeV have been obtained for the mass of the ν_μ and ν_τ, respectively [1] . The situation of the much more sensitive experiments to determine the mass of the electron neutrino (and antineutrino) is more complicated.

Most of these experiments are addressed to the decay :

$$^{3}H \rightarrow {}^{3}He + e^{-} + \overline{\nu}_e \qquad (1)$$

with a transition energy E_o = 18.6 keV. They are all based on the measurement of the electron momentum. Various difficulties are present: uncertainties on the experimental resolution and on E_o, which must be fitted together to the neutrino mass to the experimental distribution, and low counting rate in the high energy region of the experimental distribution. In addition no experiment has been carried out so far with pure tritium nuclei. As a consequence molecular effects play an important role and could have been responsible for the negative value for the square of the neutrino mass found in most experiments. Recent results, but also intriguing problems, come from high statistic experiments carried out with spectrometers based on adiabatic magnetic focusing by the Troisk [2] and Mainz [3] groups. In the Troisk experiment there is no evidence for a massive neutrino , but a bump appears near the end of the Kurie plot. It corresponds only to about 10^{-10} of the total rate , but its shape and intensity varies with time following a semi-annual modulation. A bump seems to appear also in the Mainz experiment, but lower statistics and shorter duration of the experiment prevent confirmation of the striking semi-annual modulation claimed by Troisk .

Cryogenic detectors with a beta active absorber allow a complementary calorimetric approach to the measurement of the neutrino mass. The Genoa [4] and Milano [5] groups are actively studying the decay:

$$^{187}Re \rightarrow {}^{187}Os + e^{-} + \overline{\nu}_e \qquad (2)$$

with the lowest transition energy (~ 2.5 keV) in nature. The Genoa approach is based on the use of absorbers of polycrystalline rhenium and Neutron Transmutation Doped (NTD) thermistors. A very important result in the field of material science has been the experimental detection [6] by this group of oscillations in the β particle spectrum due to the interaction of the emitted electron with its local crystal environment Fig.1). This Beta Environmental Fine Structure (BEFS) is analogous , but cheaper, than the well known Extended X-ray Absorption Fine Structure (EXAFS) [7,8].

In order to avoid the superconductivity of rhenium the Milan group has adopted $AlReO_4$ absorbers (where incidentally the expected BEFS is lower) and both NTD and Si:P thermistors. A ten bolometer array with ~250 μg absorbers is presently in operation. Preliminary measurements with four bolometers (fig.2) yield a value of 2450 ± 11 eV for the transition energy and of 43 ± 5 Gy for the lifetime which is very important in geochronology [9].

Even if CPT conservation prevents difference between the particle and antiparticle mass, many experiments have been carried out on the mass of the electron *neutrino* with limits less stringent than for its antiparticle. They are either based on the photon spectrum in radiative Electron Capture (EC) or on the different rate of EC to the various atomic levels of the daughter atoms , which depend obviously on the neutrino mass. This last approach has been adopted by the Genoa group [10] for the decay:

$$e^- + {}^{163}Ho \rightarrow {}^{163}Dy + \bar{\nu}_e \qquad (3)$$

The four MI, MII, NI, NII lines are resolved in this promising preliminary experiment.

Among the other "thermal" results I would like to quote the evidence against the 17 keV neutrino by the Oxford group [11] and the measurements of the decay of polarized tritium nuclei performed in Leuven [12].

3. Searches for rare beta decays

The fourth forbidden unique transition

$$^{113}Cd \rightarrow {}^{113}I + e^- + \bar{\nu}_e \qquad (4)$$

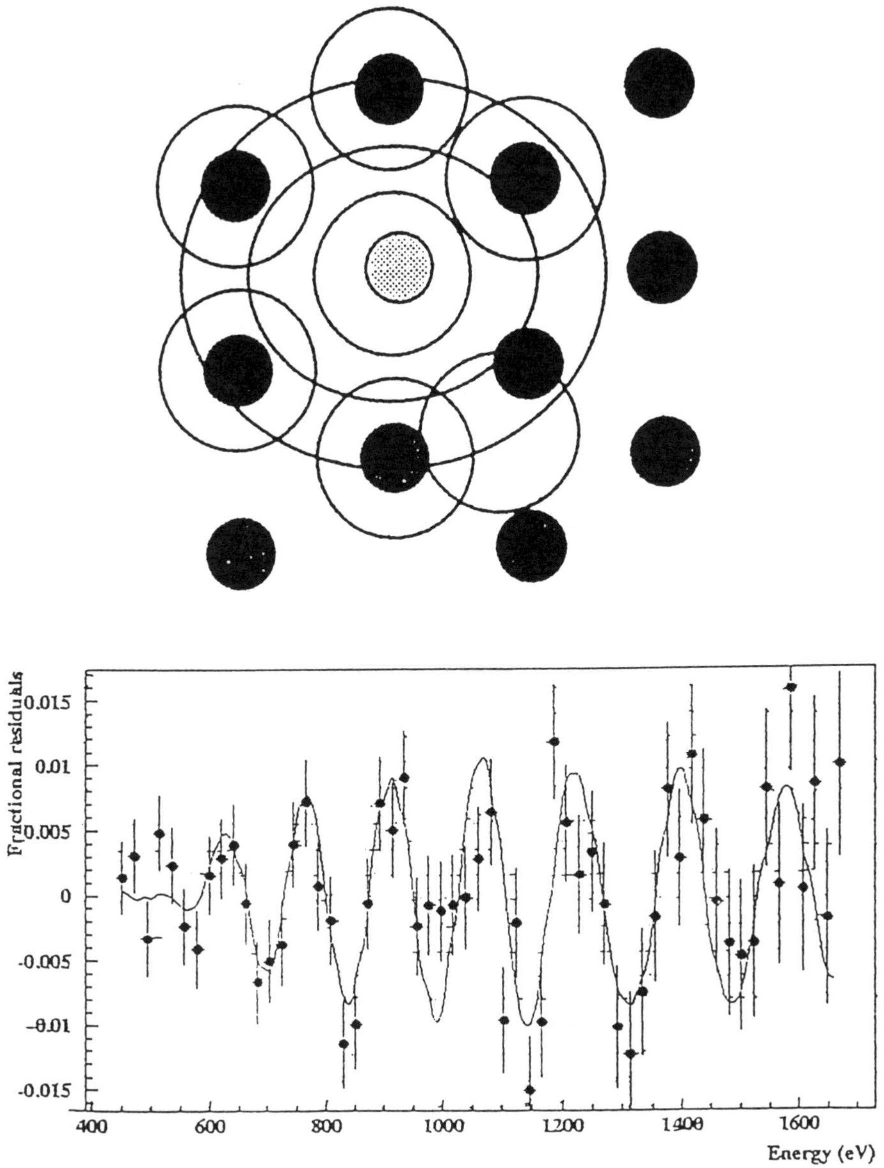

Fig.1: The BEFS effect from the Genoa group

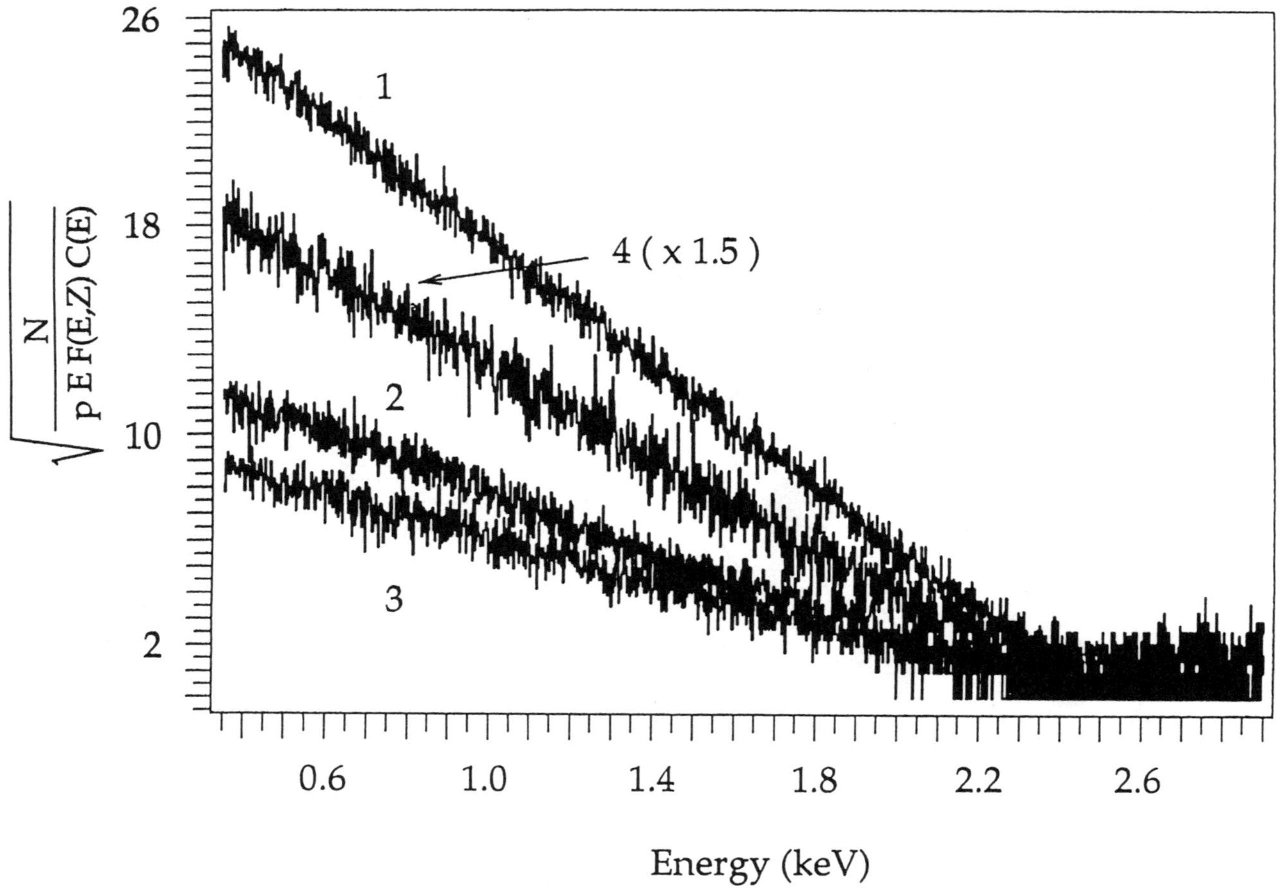

Fig. 2: Four Kurie plots of the Milano group

has been studied underground by the Milan-Kiev collaboration [13] with crystals of $CdWO_4$ of 58 and 144 grams .The values of the transition energy (318±5 keV) and of the lifetime (9 ± 1) x 10^{15} years, are the best in the literature. The spectrum of this decay has been obtained for the first time (fig.3).

The electron capture process

$$e^- + {}^{123}\mathrm{Te} \rightarrow {}^{113}\mathrm{Sb} + \bar{\nu}_n \quad (5)$$

has been investigated by the Milano group worried that the value of the lifetime quoted in the literature (1.3 x 10^{13} years) could imply a serious pile-up in their ββ experiment with large TeO_2 bolometers. Four detectors of 340 grams each where operated in coincidence and anticoincidence. It was thus found that in previous experiments the excitation K line at 27 keV of Tellurium was wrongly taken as the 30.5 keV EC line of ^{123}Te. In the coincidence spectrum both lines appeared and could be energetically separated, while in the anticoincidence one only the latter line was detected. The resulting correct lifetime was of 2.9 ± .9 x 10^{19} years, one million times larger than in the literature!

4. Cryogenic detectors and solar neutrinos

Experiments on the interactions of solar neutrinos in large arrays of thermal detectors have been proposed and look technical feasible, even if very expensive [14]. A less ambitious approach is the "thermal" detection of the radioactive nuclei produced by these interactions. The Genoa-Moscow collaboration [15] is planning an experiment to detect the reaction:

$$\nu_e + {}^7\mathrm{Li} \rightarrow {}^7\mathrm{Be} + e^- \quad (6)$$

where the produced nucleus undergoes electron capture with a lifetime of 53 days. Previous planned experiments on this reaction where based on the detection of Xrays from EC of ^{7}Be on a 320 keV excited level of ^{7}Li, which occurs however with

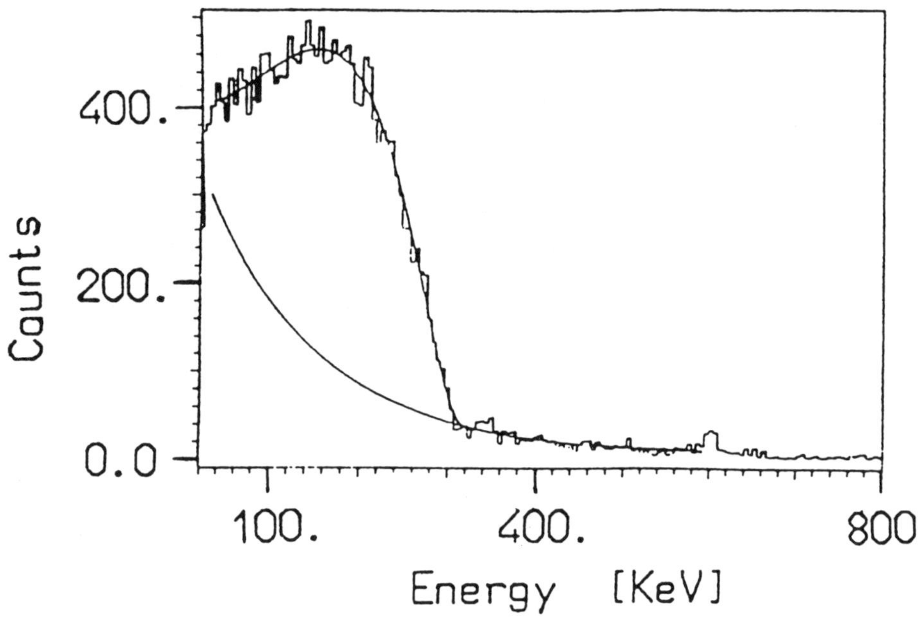

Fig.3 : The decay spectrum of ^{113}Cd

a branching ratio of 10 % only. K electron capture to the ground state (90% B.R.) would deliver an energy of 112 eV, hard to be revealed with any conventional detector. Using a bolometer with an absorber of Be and later of BeO the Genoa group was able to separate the 112 keV line from a line at ~75 eV.due to the sum of the 29 eV recoil from electron capture to the 320 keV excited state and the 57 eV signal expected from L electron capture down to the ground state.

Another thermal approach to the detection of the products of solar neutrinos is due to the Garching group who is participating to the Gallium Neutrino Observatory (GNO) experiment . The reaction in this case is:

$$\nu_e + {}^{71}\mathrm{Ga} \rightarrow {}^{71}\mathrm{Ge} + e^- \qquad (7)$$

where ^{71}Ge undergoes electron capture with a lifetime of 11.4 days and the production of L and K lines at 1.3 and 10.4 keV, respectively. The corresponding signals are presently measured with proportional counters. The Garching group [16] has however implemented a 4π bolometric detector of ^{71}Ge, with a much better resolution and 100% efficiency.

5. Double beta decay

Double beta decay is a very rare process [17,18] which could in principle occur in three channels:

$$(A,Z) \rightarrow (A,Z+2) + 2\,e^- + 2\,\bar{\nu}_e \qquad (8)$$

$$(A,Z) \rightarrow (A,Z+2) + 2\,e^- + \chi \qquad (9)$$

$$(A,Z) \rightarrow (A,Z+2) + 2\,e^- \qquad (10)$$

The two neutrino double beta decay (8) is allowed by the Standard model, while process (9), where one or more Goldstone boson named majorons are emitted, and the neutrinoless process (10) violate the conservation of the lepton number. The study of this last decay is very promising because a peak corresponding to the total transition energy would appear in the sum spectrum of the two electron energies. In addition its rate would be strongly enhanced with respect to the two neutrino one, thus allowing a very sensitive test of lepton number conservation. From the rate of neutrinoless double beta decay one could extract an *average* neutrino mass $<m_\nu>$ subject however to considerable uncertainties in the nuclear matrix calculations.

Milking experiments are based on the detection of the daughter isotopes produced by decay of large amounts of $\beta\beta$ materials stored underground. *Geochemical* experiments search on the contrary abnormal isotopic abundance of the daughter nuclei in geologically old rocks rich in the daughter nuclides. They prove the existence of $\beta\beta$ decay of ^{82}Se, ^{96}Zr, ^{128}Te, ^{130}Te and ^{238}U, but do not allow to discriminate directly among the three above mentioned channels.

Direct experiments are based on the detection of the two emitted electrons and can be divided into two categories: those where the source is different from the detector and those where the detector itself is made with a $\beta\beta$ active material.

Two neutrino $\beta\beta$ decay has been found, or at least indicated for ten nuclei: ^{48}Ca, ^{76}Ge, ^{82}Se, ^{96}Zr, ^{100}Mo (also to an excited state), ^{116}Cd, ^{128}Te, ^{130}Te, ^{150}Nd, and ^{238}U. No evidence has been on the contrary obtained for neutrinoless $\beta\beta$ decay. The most constraining limits on the lifetime and on the average neutrino mass obtained with direct methods are reported Table 1. These latter are strongly affected by uncertainties on the nuclear matrix elements.

Due to the uncertainties of the nuclear matrix elements and to the ample choice of candidate nuclei in the source=detector approach, cryogenic detectors made with $\beta\beta$ active materials look very promising for the future of these experiments. The only experiment performed so far and still running in the Gran Sasso Underground laboratory is carried out with an array of 20 crystals of TeO_2 with total mass of 6.8 kg [19]. It is by far the most massive cryogenic operating detector and, according to the nuclear matrix element calculations, already yields the most sensitive limit on $<m_\nu>$ after the Ge experiments.

Various absorbers can be considered for thermal searches on double beta decay, as show in Table 2. Tests on all them have been performed successfully with the exception of NdF_2. Crystals of this material where cooled both by the Milano and Santa Barbara group who were however unable to reach temperatures low enough to operate them as detectors. The Milano group has on the contrary constructed a CaF_2 *scintillating bolometer* where light and heat were recorded simultaneously [20] in view of reduction by coincidence of the background due to α particles.

A large experiment for double beta decay, dark matter and low energy nuclear physics searches, based on cryogenic detectors , named CUORE (for Cryogenic Underground Observatory) has been proposed by a collaboration involving the Berkeley, Florence, Leyden, Milan, Neuchatel, South Carolina and Zaragoza groups [21]. It consists in an array of 1020 crystals of 5x5x5 cm^3 of TeO_2 with a total sensitive mass of 800 kg. It will be composed of 17 towers each made by 15 "floors" of four crystals. The cost would be about 10 M$, to be doubled if enriched Tellurium will be adopted. The construction of a first tower of CUORE, named CUORICINO (Fig.4) has been approved and funded. I would like to add that CUORE is the one of the two large projects for double beta decay. The other is GENIUS with 300 conventional enriched ^{76}Ge diodes of a total mass of a ton , suspended in liquid nitrogen which also acts as a shield [22].

An array with four of the 5x5x5 TeO_2 crystals of CUORICINO has been already successfully tested [23] . Thanks to a new system of suspensions their resolution are equal or better than those of the 20 3x3x6 cm^3 array. One of the crystal has a resolution similar to that of the best Ge diodes for high energy • rays and definitely superior (4.2 keV at FWHM) for α particles.

6. Conclusions

Cryogenic detectors, even if still in their infancy as far as their application to fundamental physics is concerned, already show their potential in the field of weak interactions. Their complementary approach will be in most cases very useful for the comparison with the results with conventional detector. In some fields they can provide the only way to overcome the bottlenecks of conventional experiments.

Table I : Recent results on 0ν ββdecay

Decay	Limit on τ (years)	Limit on <m_ν> (eV)
$^{48}Ca - ^{48}Ti$	9.5×10^{21}	12-55
$^{76}Ge - ^{76}Se$	1.3×10^{25}	0.3-1.3
$^{82}Se - ^{82}Kr$	5×10^{21}	11-24
$^{100}Mo - ^{100}Ru$	5.2×10^{22}	1.4 – 264
$^{116}Cd - ^{116}Sn$	2.9×10^{22}	4 –13
$^{130}Te - ^{130}Xe$	1×10^{23}	1.8-4.5
$^{136}Xe - ^{136}Ba$	4.2×10^{23}	2-5
$^{150}Nd - ^{150}Sm$	2.1×10^{21}	4 – 16

Table 2 : "Thermal " candidates for ββ decay

Absorber	Isot.ab. (%) of ββ candidate	Transition en. (keV)
^{48}**Ca** F_2	.187	4272
^{76}**Ge** (thermal)	7.44	2039
^{116}**CdWO**$_4$	8.73	2995
^{124}**Sn** (grains)	5.79	2287
^{130}**TeO**$_2$	33.9	2528
^{150}**TeF**$_2$	5.64	3368

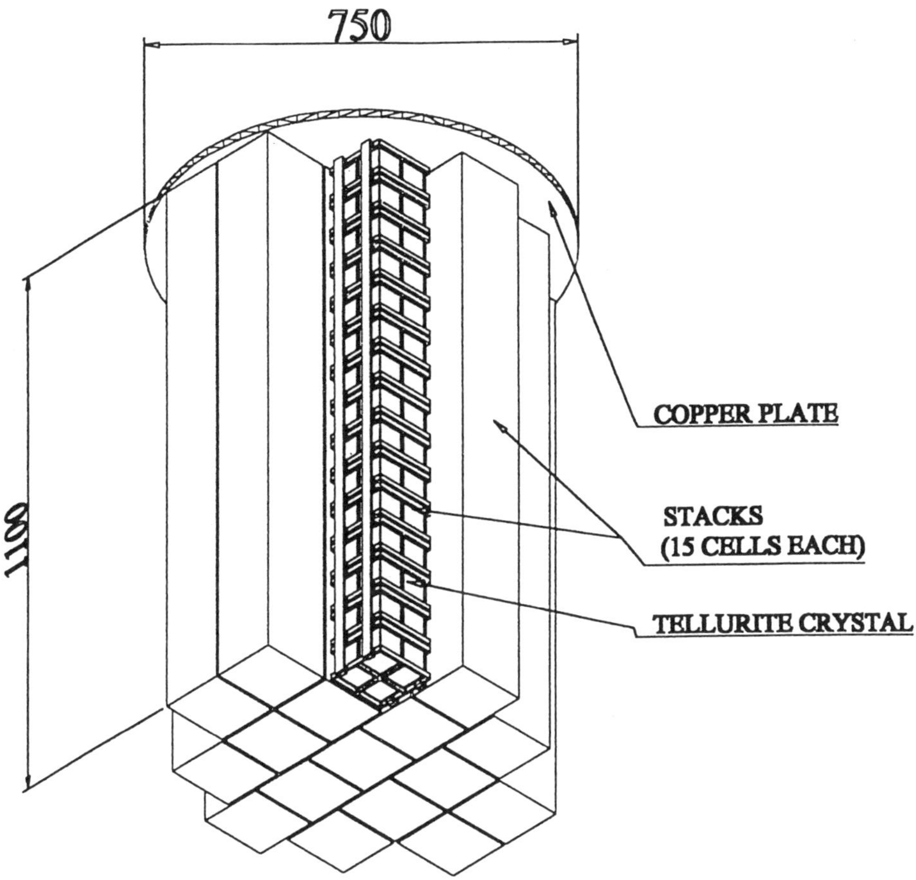

Fig.4: The CUORE proposed array and the CUORICINO tower

This paper has been funded in part by the Commission of European Communities under contract FRMX-CT98-0167.

References

[1] Review of Particle Properties , Eur. Phys.J. C 3 (1998) 1

[2] V. M. Lobashev et al: Neutrino mass and anomaly in the tritium beta-spectrum, presented to Neutrino98, Tokayama/ Japan , June 1998, in Nucl.Phys.B (Proc.Suppl) 327

[3] H. Barth et al: Results of the Mainz Neutrino Mass Experiment, presented to Neutrino98, Tokayama/ Japan , June 1998, Nucl.Phys.B (Proc.Suppl.) 321

[4] F. Gatti et al: First results of the calorimetric spectrometer for beta decay of Rhenium-187, Talk to LTD8, Dalfsen (The Netherlands) , August 16-20, 1999, NIM A (in press)

[5] A. Nucciotti et al : Neutrino mass measurement with an array of high resolution AgReO4 microcalorimentes, Tälk to LTD8, Dalfsen (The Netherlands) , August 16-20, 1999, NIM A (in press)

[6] F. Gatti et al, Nature 397 (1999) 137

[7] S.E. Koonin, Nature 354 (1991) 468

[8] G. Benedek et al, Nucl.Instrum.and Meth.A 426 (1999) 147

[9] M. F. Horan, M. I. Smoliar, and R. J. Waker, Geoch. and Cosmoch. Acta 62 (1998) 545 and references therein; J.T. Chesley and J. Ruiz, Earth and Planetary Sci.Lett. 154 (1998) 1

[10] F. Gatti et al, Phys.Lett. B 398 (1997) 415

[11] R.J. Gaitskell et al, Phys.Lett. B 370 (1996) 163

[12] P. de Moor et al, Proc. of LTD-7, 27 July- 2 August 1997, Munich / Germany, ed. by S. Cooper, page 181

[13] A. Alessandrello et al, Nucl.Phys.B (Proc.Suppl) 35 (1994) 394

[14] A. Alessandrello et al, Astrop. Phys. 3 (1995) 239, also for previous references

[15] M. Galeazzi et al, Phys.Lett. B 398 (1997) 187

[16] M. Altmann et al : A promising option for the GNO Solar Neutrino Experiment? Talk to LTD8, Dalfsen (The Netherlands) , August 16-20, 1999, NIM A (in press)

[17] A.Morales : Review of Double Beta Experiments and Comparison with Theory, Invited talk at the XVIII International Conference on Neutrino Physics and Astrophysics, Neutrino 98, June 1998 , Tokayama/ Japan, Nucl. Phys.B (Proc.Suppl) 335

[18] E. Fiorini: Double beta decay, invited paper to Neutrino Telescope, Venice, February 1999 , ed. by M. Baldo Ceolin

[19] S. Pirro: Present status of MI-BETA cryogenic experiment and preliminary results for CUORICINO , Talk to LTD8, Dalfsen (The Netherlands) , August 16-20, 1999, NIM A (in press)

[20] A. Alessandrello et al, Phys.Lett.B 420 (1998) 109

[21] E.Fiorini, Phys.Rep. 307 (1998) 309, and letter of intent to the Gran Sasso Scientific Committee

[22] L. Baudis et al, Phys.Rep. 307 (1998) 301, and private communications by L.Baudis and H. Klapdor-Kleingrothaus

[23] A. Alessandrello et al: Vibrational and Thermal Noise Reduction for Cryogenic Detectors, this conference

AXION SEARCH EXPERIMENT IN KYOTO

M. TADA, Y. KISHIMOTO, M. SHIBATA, K. KOMINATO, C. OOISHI, S. YAMADA[1], H. FUNAHASHI[1], K. YAMAMOTO[2], A. MASAIKE[3], and S. MATSUKI
(CARRACK COLLABORATION)

Nuclear Science Division, Institute for Chemical Research, Kyoto University, Uji, Kyoto 611-0011, Japan

[1] *Physics Department, School of Science, Kyoto University, Kyoto 606-0065, Japan*

[2] *Nuclear Engineering Department, School of Engineering, Kyoto University, Kyoto 606-0065, Japan*

[3] *Faculty of Engineering, Fukui University of Technology, Fukui 910, Japan*

Progress in dark matter axion search in Kyoto with the Rydberg-atom cavity detector is reported. Based on the prototype apparatus CARRACK1, a new experimental apparatus CARRACK2 was constructed to search for axions over a wide range of axion mass, 2 μeV to 30 μeV. Status of the apparatus and axion search is presented

1 Introduction

Quest for dark matter particles is one of the most challenging issues related to particle physics and cosmology. The axion, originally proposed to solve the so-called strong CP problem, is one of the well motivated and most promising candidates for the non-baryonic dark matter particles [1]: still open window of the axion mass is between 1 μeV to 1 meV [2]. From the production mechanism of axions theoretically conjectured in the early Universe [1], the axion with its mass around 10 μeV is expected to be the main component of dark matter, assuming the axion density occupy all the matter density in the Universe ($\Omega_m \sim 0.35$) [1].

Such extremely low mass axions can be searched for by converting axions into photons in a strong magnetic field via the Primakoff process. The sensitivity of recent LLNL experiment along this line of detection with a cryogenic HEMT amplifier reached the limit to exclude KSVZ-model of dark halo axions at the mass region of 3 μeV [3].

We have developed a sensitive microwave single-photon detector to search for axions with Rydberg atoms in resonant cavities [4,5,6]. In our method, photons produced by this conversion in a resonant cavity are absorbed by Rydberg atoms passed through the cavity and the Rydberg atoms thus excited to an upper state are selectively field-ionized and counted just after getting out the cavity. A prototype detector, CARRACK-1, was constructed and is now in operation to search for axions with mass around 10 μeV [6]. The progress in the development of the search experiment is here reported.

Based on this prototype apparatus CARRACK1, a large-scale system CARRACK2 was constructed to cover a wide range of axion mass from 2 to 30 μeV. This new system has several improved characteristics compared to the prototype system. The main characteristics of the new system are also presented here.

2 Characteristics of the Rydberg-atom cavity detector

2.1 *Principle of the detector*

Principle of the present experimental method [4,5] is schematically shown in Fig. 1. The dark matter axions are converted into microwave photons in a resonant microwave cavity (called conversion cavity) in a strong magnetic field. The photons are then transferred to another coupled cavity (called detection cavity) and absorbed by Rydberg atoms passed through the cavity.

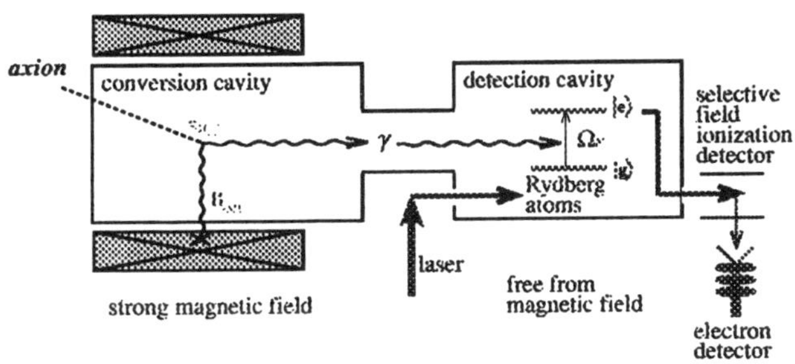

Figure 1: Principle of the CARRACK experiment to search for dark matter axions with Rydberg atoms in cooled resonant cavities.

The Rydberg atoms are initially prepared to a lower state with two-step laser excitation where the lasers have narrow frequency-width enough to excite only the lower state. The transition frequency between the lower and a upper state is adjusted to be approximately equal to the cavity resonant frequency. The atoms excited to the upper state by absorbing the axion-converted photons, are then ionized with the selective field ionization [7] just outside the detection cavity. The electrons thus produced from the ionization process are counted with the electron detector.

To reduce thermal background photons in the cavity, the whole cavities are cooled down to 10 mK range with a dilution refrigerator (Oxford-Kelvinox300).

2.2 Prototype apparatus CARRACK1

The apparatus mainly consists of 6 parts: superconducting magnets, coupled cavity, dilution refrigerator, atomic beam system, laser excitation system, and selective field ionization system. In Fig. 2-a shown is a schematic diagram of the apparatus CARRACK1.

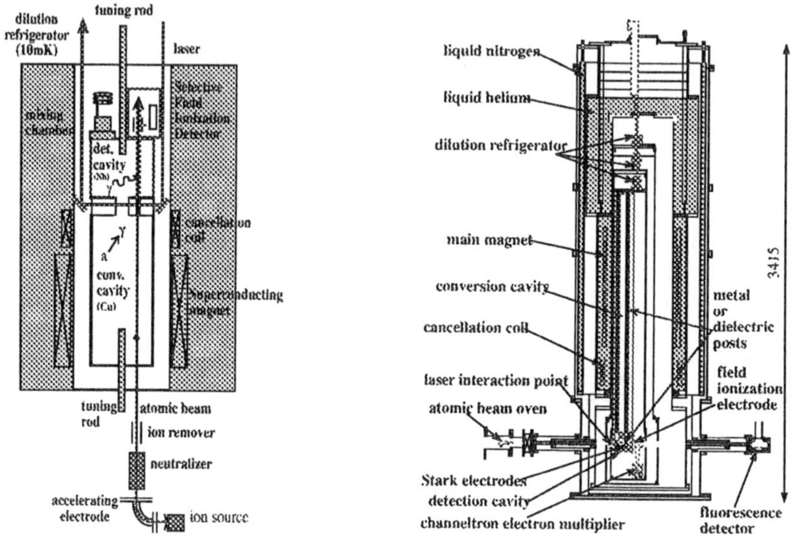

a) b)

Figure 2: Schematic view of the experimental system CARRACK to search for dark matter axions with Rydberg atoms in cooled resonant cavities: a) CARRACK1, b) CARRACK2

The magnet system consists of 2 solenoid coils: one is the main coil which, of 150 mm inner diameter and of 420 mm length, produces the magnetic flux density of 7T at the center. The other coil (cancellation coil) is placed at the upper side of the main coil to reduce the magnetic field at the region of the detection cavity to less than 0.09 T. The detection cavity is made of niobium to prevent the magnetic field from penetrating into the detection cavity with the Meissner effect. This is to avoid complicated level splitting and level shifts in the Rydberg states due to the Zeeman effect.

The Rydberg atoms are produced with a two-step laser excitation from the ground state $5s_{1/2}$ to $5p_{3/2}$ and then to $ns_{1/2}$ state of alkaline ^{85}Rb atoms. The transition from the $ns_{1/2}$ to the $np_{1/2}$ states is used to absorb the axion converted photons.

The cavity mode used is the cylindrical TM_{010} mode and the the cavity resonant frequency is tuned by a dielectric (aluminum-oxide) rod inserted along the cylindrical axis. The frequency can be tuned over 15 % for one cavity with this tuning system. The transition frequency of the Rydberg atoms must approximately coincide with the cavity resonant frequency [5,8]. The frequency is tuned coarsely by selecting a suitable lower state with the principal quantum number n and then finely by applying Stark electric field with parallel plate electrodes in the detection cavity.

The tuning rod for the conversion cavity is moved from the outside of the cryostat with a stepping motor. In order to reduce the thermal heat leak from the room temperature side to the mostly cooled cavity at 10 mK range, the rod is moved through Kevlar strings which have especially low heat conductivity and still are strong enough to move the rod. The driving mechanism of the tuning rod with Kevlar string is schematically shown in Fig. 3-a.

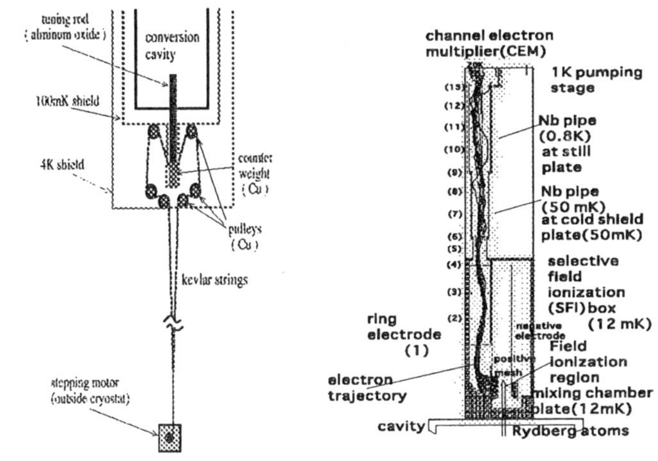

Figure 3: a) Schematic view of the rod driving mechanism to tune the cavity frequency, b) Structure of the field ionization electrodes and the channeltron electron detector with a series of ring-electrodes to transport electrons

Two atomic beam system can be used: non-thermal beam of 10 to 100 eV energy and thermal beam of $\sim$ 0.04 eV. As analyzed in detail [8], the detection efficiency of the present method is strongly dependent on the velocity of the atoms, and thus the atomic beam with suitable velocity to meet the optimum condition for the detection efficiency is utilized from these two atomic-beam system.

The whole cavity system including the selective field-ionization (SFI) detection system is cooled down to 10 mK range with a dilution refrigerator (DR). The field ionization electrodes are placed on the exit of the detection cavity which is located on the bottom plate of the mixing chamber of the dilution refrigerator. The electrons produced by the field ionization process are then transported through a series of ring electrodes to a channeltron detector located at the 1K pumping stage of the DR system. The channeltron is heated up to about 20 K with a heating coil surrounding it to get enough amplification gain even in the environment of such low temperature. This detection system has been successfully used for a long period of measuring time without any deterioration of the performance. In Fig. 3-b shown is the structure of the SFI detection system and typical example of the transported electron trajectories.

Typical cooling behavior of the cavity system is shown in Fig. 4, where the temperature of the cavity at the mixing chamber was measured by using the nuclear-spin orientation method with a ^{60}Co crystal. The lowest achieved temperature is $12 \sim 15$ mK.

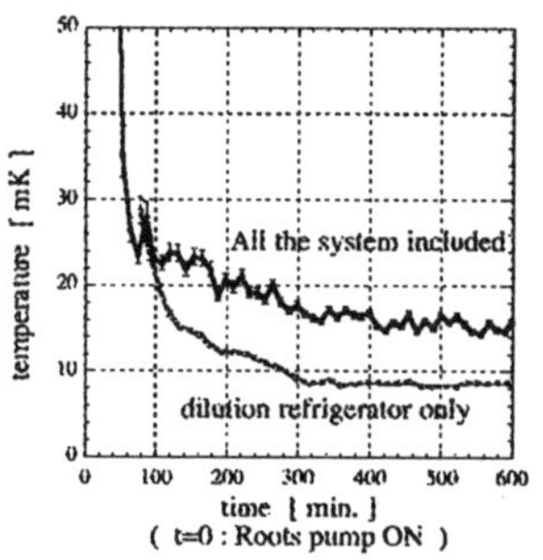

Figure 4: Cooling behavior of the cavity and SFI detection system with the operation of dilution refrigerator.

In Fig. 5-a shown is SFI counts measured by varying the frequency of the second laser. The observed peak corresponds to the excitation of $111s_{1/2}$ state with a two step laser excitation. Also shown in Fig. 5-b is the dependence of the SFI signal counts on the applied electric field at the field ionization electrodes. The lower peak (marked p-state) is from the the $111p_{1/2}$ state excited from the initially prepared lower $s_{1/2}$ state by absorbing thermal blackbody photons in the resonant cavity, while the peak at the higher field (marked s-state) is from the prepared Rydberg states $111s_{1/2}$. Note that the field ionization in the present system is of differential type, so that the total number of Rydberg states prepared initially by laser excitation is obtained by integrating all the counts from the threshold field to higher field region.

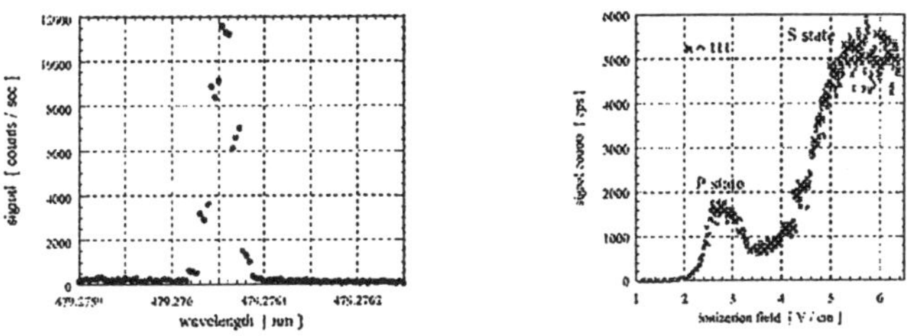

Figure 5: a) Excitation spectrum of Rb $111s_{1/2}$ state measured with the SFI detector by varying the wavelength of the second laser, b) Field ionization counts as a function of the applied electric field at the electrode.

3 New apparatus CARRACK2

Based on the prototype apparatus CARRACK1 above, a new system CARRACK2 was constructed to cover a wide range of axion mass with a large (inner-diameter of 540 mm and length of 1 m) superconducting magnet. The central magnetic flux density is 7T. Schematic diagram of the new system is shown in Fig. 2-b. Several new features have been incorporated to improve the performance compared to CARRACK1: specifically the detection cavity is located at the lower part of the cryostat, thus the atomic beam as well as the laser beam are introduced into the cryostat horizontally. This configuration makes the access easier to the cavity system from the outside, although the structure is more complicated than before. In this new system, the cavity resonant frequency is tuned with dielectric posts in both cavities which are moved in the cavity along the radial direction (post method). The detection efficiency for axions with this system is better than the rod system adopted in the prototype system.

4 Status

With the CARRACK1 system data were taken for the axion mass around 10 μeV over $\sim$ 6% range. In a preliminary analysis, the sensitivity reached the KSVZ limit of axion model. Further detailed analysis is in progress to have conclusive results on the limit.

The whole parts of the new apparatus CARRACK2 were assembled and overall test experiment is being performed. The magnetic field at the conversion cavity and the field cancellation at the detection cavity with the cancellation coil and the Nb cavity were successfully achieved as originally designed. With a new selective field ionization detector, the Rydberg states of n $\sim$ 100 were successfully observed and the selectivity to separate the lower and the

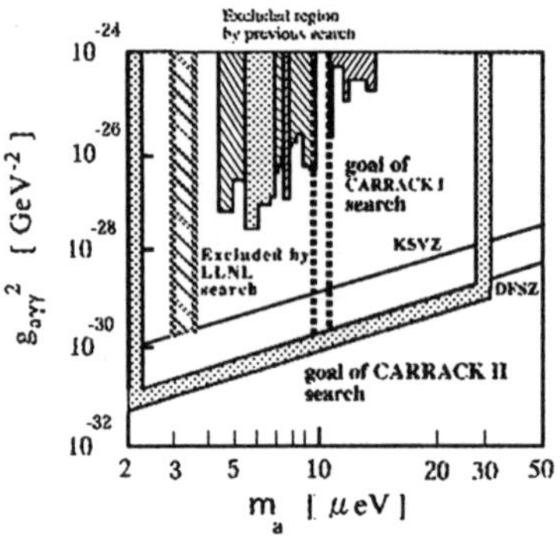

Figure 6: Goal of the CARRACK2 experiment: upper limit of the axion-photon coupling constant to be determined as a function of the axion mass.

upper states was carefully examined. The resulting selectivity was found to be satisfying, suggesting the effectiveness of the non-adiabatic method for the ionization developed here [9]. The final goal of the CARRACK2 experiment is shown in Fig. 6 where the expected upper limit of the axion-photon coupling constant $g^2_{a\gamma\gamma}$ to be achieved is plotted as a function of axion mass.

Acknowledgments

The authors would like to thank Prof. H. Ejiri for his continuous encouragement throughout this work. This research is partly supported by a Grant-in-Aid for Specially Promoted Research by the Ministry of Education, Science, Sports, and Culture, Japan under the program number 09102010.

References

1. Recent reviews include E. Kolb and M. Turner, *The Early Universe*, (Addison-Wesley, New York, 1990); M. Turner, Astro-ph/9901113.
2. Recent reviews include G. Raffelt, *Phys. Rep.* **198** 1 (1990);
3. C. Hagmann *et al*, *Phys. Rev. Lett.* **80**, 2043 (1998).
4. S. Matsuki and K. Yamamoto, *Phys. Lett.* B **263**, 523 (1991).
5. I. Ogawa, S. Matsuki and K. Yamamoto, *Phys. Rev.* D **53**, R1740 (1996).
6. M. Tada *et al.*, *Nucl. Phys.* **72B**, 164 (1999).
7. T. F. Gallagher, *Rydberg Atoms*, Cambridge Univ. Press, Cambridge, 1994.
8. K. Yamamoto and S. Matsuki, *Nucl. Phys.* **72B**, 132 (1999).
9. CARRACK collaboration, to be published.

ZERO MODES ENHANCEMENT QUANTUM MODEL OF THE YANG-MILLS VACUUM

V. GOGOHIA, H. TOKI

Research Center for Nuclear Physics (RCNP), Osaka University, Mihogaoka 10-1 Ibaraki, Osaka 567, Japan

gogohia@rcnp.osaka-u.ac.jp, toki@rcnp.osaka-u.ac.jp

Using the effective potential approach for composite operators we have formulated new zero modes enhancement quantum model of the QCD ground state. It is based on the existence and importance of such kind of the nonperturbative, topologically nontrivial excitations of the gluon field configurations (due to self-interactions of massless gluons only), which can be effectively corrctly described by the q^{-4}-type behavior of the full gluon propagator in the deep infrared domain. The ultraviolet part of the full gluon propagator is approximated by the asymptotic freedom to-leading order perturbative logarithm term of the running coupling constant. We minimize the regularized and correspondingly subtracted effective potential at a fixed scale as a function of the physically meaningful parameter. When it is zero, then perturbative phase survives only in our model. This makes it possible for the first time to establish relation between the nonperturbative scale parameter and flavorless QCD asymptotic scale parameter.

1 Introduction

Today there are no doubts left that the dynamical mechanisms of the important nonperturbative quantum phenomena such as quark confinement and dynamical (or equivalently spontaneous) chiral symmetry breaking (DCSB) are closely related to the complicated topologically nontrivial structure of the QCD vacuum [1]. On the other hand, it also becomes clear that the nonperturbative infrared (IR) dynamical singularities, closely related to the nontrivial vacuum structure, play an important role in the large distance behaviour of QCD [2]. For this reason, any correct non-perturbative model of quark confinement and DCSB necessarily turns out to be a model of the true QCD vacuum and the other way around.

Our model of the QCD true ground state is based on the existence and importance of such kind of the nonperturbative, quantum excitations of the gluon field configurations (due to self-interactions of massless gluons only, i.e, without explicit involvment of some extra (to QCD) degrees of freedom) which can be effectevely correctl described by the q^{-4} behaviour of the full gluon propagator in the IR domain (at small q^2) [3]. These excitations are also topologically nontrivial in comparison with the free gluon structure, q^{-2}. Let us consider the full gluon propagator

$$iD_{\mu\nu}(q) = \left\{T_{\mu\nu}(q)d(-q^2, a) + \xi L_{\mu\nu}(q)\right\} \frac{1}{q^2}, \tag{1}$$

where ξ is a gauge fixing parameter ($\xi = 0$, Landau gauge) and $T_{\mu\nu}(q) = g_{\mu\nu} - q_\mu q_\nu / q^2 = g_{\mu\nu} - L_{\mu\nu}(q)$. Evidently, its free (perturbative) counterpart is obtained by simply setting the full gluon form factor $d(-q^2, \xi) = 1$ in (1). The exact expression for the full gluon propagator is not known. For this reason, we approximate it by the sum of its deep IR and ultraviolet (UV) asymptotics, i. e. setting

$$d(-q^2, a) = \frac{\mu^2}{(-q^2)} + \frac{c}{\ln(-q^2/\Lambda^2_{YM})} \tag{2}$$

in the whole range. Here μ is the appropriate mass scale parameter, responsible for the nonperturbative dynamics. Evidently, the UV piece is nothing but the running coupling constant $\bar{g}^2(-q^2)$, which is the leading order solution to the renormalization group equations [4] and $c = (16\pi^2/11)$ is the inverse of the first coefficient of the β-function in the case of Yang-Mills (YM) gluon fields (pure gluodynamics). In the above mentioned Ref. [4] it is denoted as $2/b_0$. The flavorless QCD asymptotic scale parameter is denoted as Λ_{YM}. In the weak coupling limit ($-q^2 \to \infty$) precisely the second term becomes dominant (asymptotic freedom). On the other hand, we obtain the generally accepted form of the IR singular asymptotics for the full gluon propagator [4-7] (and references therein)

$$D_{\mu\nu}(q) \sim (q^2)^{-2}, \qquad q^2 \to 0, \tag{3}$$

which may be equivalently refered to as the strong coupling regime [4]. It describes the zero modes enhancement (ZME) dynamical effect in QCD at large distances. In our papers [3,8] it is shown that this behavior of the full gluon propagator in the deep IR domain can be directly related to quark confinement and DCSB. Moreover, it is in fair agreement with chiral QCD topology [9,10] as well.

2 The effective potential for composite operators

In this section we are going to analytically investigate the quantum part of the vacuum energy density which is due to q^{-4}-type nonperturbative, topologically nontrivial excitations of the gluon field configurations in the QCD vacuum complemented by the renormalization group improvements, (2) for the UV region. Let us start from the nonperturbative gluon part of the vacuum energy density which to-leading order (log-loop level $\sim \hbar$) is given by the effective potential for composite operators [11] as follows

$$V(D) = \frac{i}{2} \int \frac{d^n q}{(2\pi)^n} Tr\{\ln(D_0^{-1}D) - (D_0^{-1}D) + 1\}, \tag{4}$$

where $D(q)$ is the full gluon propagator (1) and $D_0(q)$ is its free (perturbative) counterpart. Here and below the traces over space-time and color group indices are understood. The effective potential is normalized as $V(D_0) = 0$, i. e. the free perturbative vacuum is normalized to zero. In order to evaluate the effective potential (4) we use the well-known expression, $Tr \ln(D_0^{-1}D) = 8 \times \ln det(D_0^{-1}D) = 8 \times 4 \ln\left[\frac{3}{4}d(-q^2) + \frac{1}{4}\right]$. It becomes zero (in accordance with the above mentioned normalization condition) when the full gluon form factor is replaced by its free counterpart as well as it does not explicitly depend on a gauge choice.

Going over to four ($n = 4$) dimensional Euclidean space by introducing the appropriate dimensionless momentum variable as $x = q^2/\Lambda_{YM}^2$ and after doing some algebra in (4), on account of (2), we finally obtain the following expression for the regularized vacuum energy density, $\epsilon_g = V(D)$,

$$\epsilon_g = \frac{1}{\pi^2} \Lambda_{YM}^4 I_g(x_0; b), \tag{5}$$

where

$$I_g(x_0;b) = \int_0^{x_0} dx\, x \left\{ \ln x + \frac{3b}{4x} - a + \frac{3c}{4\ln x} - \ln[3b + x + \frac{3cx}{\ln x}] \right\} = I_g^r(x_0;b) + I_g^s(x_0;b), \tag{6}$$

and

$$\begin{aligned} I_g^r(x_0;b) &= \frac{1}{2} x_0^2 \ln x_0 (\ln x_0 + 3c) - \frac{1}{4} x_0^2 + \frac{3}{4} x_0 b - I_2(x_0;b), \\ I_g^s(x_0) &= -\frac{\bar{a}}{2} x_0^2 + \frac{3c}{4} li(x_0^2) + \frac{1}{2} x_0^2 \ln\ln x_0 + \frac{1}{2} Ei(1, -2\ln x_0), \end{aligned} \tag{7}$$

with $\bar{a} = a + \ln(\ln x_0 + 3c)$. In this integral x_0 is the corresponding dimensionless UV cut-off, $x_0 = q_0^2/\Lambda_{YM}^2$ while $b = \mu^2/\Lambda_{YM}^2$ and $li(x)$ is the logarithm-integral function. $Ei(1, -2\ln x_0)$ is the exponential-integral function. It has a singularity when formally $x_0 = 1$ as well as the preceding term. The constant $a = (3/4) - 2\ln 2 = -0.6363$. In the compositions shown in (7), the corresponding integral is defined as

$$I_2(x_0; b) = \int_0^{x_0} dx\, x \ln[(3b + x)\ln x + 3cx]. \tag{8}$$

In principle, we must sum up all contributions in order to obtain total vacuum energy density (the confining quark part of the vacuum energy density is not considered here). However, approximating the full gluon propagator by the sum of its deep IR (confinement) and UV asymptotics in (2), the nonzero terms appear which precisely violates the above mentioned normalization condition of the perturbative vacuum to be zero when the parameter b goes to zero ($\mu \to 0$). These terms are collected in the expression (7) for $I_g^s(x_0)$. It also contains two terms which suffer from unphysical singularity - Landau pole - at $x_0 = 1$. This should be traced back to (2) when $-q^2 = \Lambda_{YM}^2$ since we use it in the whole range. So these terms should be subtracted from the gluon part of the vacuum energy density. We regularize the gluon contribution to the vacuum energy density by subtracting unwanted terms by means of the ghost contribution, i. e. $\epsilon_g + \epsilon_{gh} = \epsilon_g^{reg} = \pi^{-2}\Lambda_{YM}^4 I_g^r(x_0; b)$. Thus our regularization procedure is in agreement with general physical interpretation of ghosts to cancel the effects of the unphysical degrees of freedom of the gauge bosons [4].

The regularized and subtracted vacuum energy density becomes

$$\Omega_g \equiv \Omega_g(x_0; b) = \frac{1}{\Lambda_{YM}^4}\epsilon_g^{reg} = \frac{1}{2\pi^2}\Big\{x_0^2 \ln x_0 (\ln x_0 + 3c) - \frac{1}{2}x_0^2 + \frac{3}{2}x_0 b - 2I_2(x_0; b)\Big\}, \tag{9}$$

where we introduce the effective potential at a fixed scale Λ_{YM}, namely Ω_g.

The asymptotic behaviour of the effective potential (9) depends on the asymptotic properties of the integral $I_2(x_0; b)$, (8). It is almost obvious that its asymptotics at $x_0 \to 0, \infty$ (and consequently at $b \to 0, \infty$) to-leading order can be easily evaluated analytically by replacement $\ln x$ by $\ln x_0$ since logarithm is a very slowly varing function. We are especially interested in the case when the UV cut-off x_0 is very big, of course (it should go to infinity at the final stage, see below). Omitting all intermediate analytical calculations, in the $b \to 0$ limit to-leading order, one obtains

$$\Omega_g(x_0; b) \sim_{b \to 0} -\frac{9}{4\pi^2} x_0 \Big\{\frac{\ln x_0 - c}{\ln x_0 + 3c}\Big\} b, \tag{10}$$

which clearly shows first that there is no singularity in the effective potential (9) at unphysical point $x_0 = 1$ and second (which much more important) that in order for the effective potential as a function of the parameter b to approach zero from below, one has to parametrize the UV cut-off x_0 as follows:

$$\ln x_0 = \lambda c, \qquad \lambda > 1. \tag{11}$$

At the sime time, as a function of the parameter b it linearly diverges in the $b \to \infty$ limit. Thus, as a function of b the effective potential (9) has a minimum if the UV cut-off is big enough, (11). As a function of the UV cut-off x_0 itself, it approches zero from above and diverges as $\sim -x_0$ at infinity.

The minimization of the effective potential (9) as a function of the parameter b, $\partial \Omega_g(x_0; b)/\partial b = 0$, yields the following "stationary" condition,

$$x_0 = 4 \int_0^{x_0} dx\, x \frac{\ln x}{(3b + x)\ln x + 3cx}. \tag{12}$$

To-leading order of the UV cut-off λ, it can be analytically evaluated as follows

$$(\lambda - 1)(\lambda + 3)z_0 = 4\lambda^2 \ln\Big\{1 + \frac{(\lambda + 3)z_0}{3\lambda}\Big\}, \tag{13}$$

where for further convenience, we introduce the variable $z_0 = x_0/b$. Its solution in the $\lambda \to \infty$ limit is

$$z_0^{min}(\lambda \to \infty) = \Big\{\frac{x_0}{b}\Big\}_{min}(\lambda \to \infty) = 2.2, \tag{14}$$

i. e. the point $z_0^{min} = 2.2$ is the point of accumulation of minimums in the UV limit of the effective potential (9). This effect shows that the parameter b goes to infinity in the same way as the UV cut-off x_0 in order for their ratio to go to the above displayed finite limit. This means that initial mass scale parameter μ is a "bar" one. It depends on the UV cut-off λ, i. e. $\mu = \mu(\lambda)$. In what follows we will introduce the UV renormalized μ, denoted as $\tilde{\mu}$, and consequently the UV renormalized $\tilde{b}$. To-leading order in the $\lambda \to \infty$ limit the effective potential (9), on account of the "stationary" condition (13), becomes

$$\Omega_g(\lambda, z_0^{min}) = e^{2\lambda c} \frac{1}{2\pi^2} \Big\{\frac{3}{4}(z_0^{min})^{-1} - \ln(1 + \frac{3\lambda}{z_0^{min}(\lambda + 3)})\Big\}. \tag{15}$$

Thus in the $\lambda \to \infty$ limit the effective potential and consequently the vacuum energy density are badly divergent because of overall exponential λ-factor, $e^{2\lambda c}$.

3 Discussion

Though the vacuum energy density remains badly divergent, nevertheless our method makes it possible to establish a finite relation between two scale parameters in our model. Indeed, from (14) it follows that the parameter b diverges exactly as the UV cut-off $x_0 = e^{\lambda c}$ itself in order to get final result in the $\lambda \to \infty$ limit. So let us introduce now the finite b and μ as follows $\tilde{b} = b(\lambda)e^{-\lambda c}$ and $\tilde{\mu} = \mu(\lambda)e^{-\lambda c/2}$, respectively. Then

$$\tilde{b} = \tilde{\mu}^2/\Lambda_{YM}^2 = (z_0^{min})^{-1} = 0.4545, \tag{16}$$

so one obtains

$$\tilde{\mu} = 0.6742\Lambda_{YM}. \tag{17}$$

Let us underline that our numerical results (16) and (17) are obtained in a gauge invariant way, i.e., these numbers do not explicitly depend on a gauge choice.

The main problem now is the numerical value of the YM asymptotic scale parameter, $\Lambda_{YM} \equiv \Lambda$ (for brevity and further aims). Since it can not be experimentally determined in the YM theory alone, so its numerical value is to be taken from lattice approach. Unfortunately, lattice estimates of the QCD flavorless asymtotic scale parameter strongly depend on the renormalization scheme chosen for calculation. In Ref. [12] in the $\overline{MS}$ scheme it is expressed in terms of Sommer's scale r_0, namely $\Lambda_{\overline{MS}} = 0.636(54)/r_0$. Than (17) numerically becomes $\tilde{\mu} = 0.4288r_0^{-1}$. In physical units, using $r_0 = 0.5\ fm$, one gets, $\tilde{\mu} = 171.5\ MeV$. However, as was emphasized in the above-mentioned Ref. [12], the conversion to physical units is rather ambiguous in the pure gauge theory. In Ref. [13] the YM scale parameter was calculated first in the $M\tilde{O}M$ scheme and then it was matched to the $\overline{MS}$ scheme. The corresponding numerical results (here and everywhere below up to central values, for simplicity) are: $\Lambda_{M\tilde{O}M} = 0.88\ GeV$ and $\Lambda_{\overline{MS}} = 0.31\ GeV$. Substituting these values into (17), one arrives at the following estimates, $\tilde{\mu}_{M\tilde{O}M} = 593.3\ MeV$ and $\tilde{\mu}_{\overline{MS}} = 210\ MeV$. The most recent lattice calculation of the flavorless QCD asymptotic scale parameter is being presented in Ref. [14] are: $\Lambda_{MOM} = 0.361(6)\ GeV$ and $\Lambda_{M\tilde{O}M} = 0.349(6)\ GeV$ while $\Lambda_{\overline{MS}} = 0.412\ GeV$. Substituting this result into (17) one gets, $\tilde{\mu}_{\overline{MS}} = 277.7\ MeV$. In any case, our solid result is Eq. (17) since, as was underlined above, the numerical value of the flavorless QCD asymptotic scale parameter still remains rather ambiguous in the pure gauge theory.

Acknowledgements

One of the authors (V.G.) is grateful to Prof. H. Ejiri for kind hospitality during his stay at RCNP, Osaka University. He also would like to thank M. Polikarpov, M. Chernodub, A. Ivanov and especially V.I. Zakharov for useful and critical discussions. The present work was supported by Special COE grant of the Ministry of Education, Sience and Culture of Japan.

References

1. P. van Baal (Ed.), Confinement, Duality, and Nonperturbative Aspects of QCD, NATO ASI Series B: Physics, vol. 368;
   K-I. Aoki, O. Miymura, T. Suzuki (Eds.), Non-Perturbative QCD, Structure of the QCD Vacuum, Prog. Theor. Phys. Suppl. **131**, 1 (1998)
2. D.J. Gross, F. Wilczek, Phys. Rev. D **8**, 3633 (1973)3;
   S. Weinberg, Phys. Rev. Lett. **31**, 494 (1973);
   H. Georgi, S. Glashow, Phys. Rev. Lett. **32**, 438 (1974)
3. V. Gogohia, Gy. Kluge, H. Toki, T. Sakai, Phys. Lett. B **453**, 281 (1999)
4. W. Marciano, H. Pagels, Phys.Rep., **C36** (1978) 137
5. C.D. Roberts, A.G. Williams, Prog. Part. Nucl. Phys. **33**, 477 (1994);
   V. Gogohia, Gy. Kluge, M. Prisznyak, hep-ph/9509427
6. N. Brown, M.R. Pennington, Phys. Rev. D **39**, 2723 (1989);
   D. Atkinson et al., J. Math. Phys. **25**, 2095 (1984)
7. B.A. Arbuzov, Sov. Jour. Part. Nucl. **19**, 1 (1988);
   A.I. Alekseev, B.A. Arbuzov, Mod. Phys. Lett. A **13**, 1747 (1998)
8. V. Gogohia, H. Toki, T. Sakai, Gy. Kluge, Inter. Jour. Mod. Phys. A **15**, 45 (2000), see also hep-ph/9810516
9. V. Gogohia, H. Toki, Phys. Lett. B **466**, 305 (1999)
10. V. Gogohia, H. Toki, Phys. Rev. D **61**, 036006 (2000), see also hep-ph/9908301
11. J.M. Cornwall, R. Jackiw, E. Tomboulis, Phys. Rev., D **10**, 2428 (1974)
12. M. Luscher, hep-lat/9802029
13. C. Parrinello et al., Nucl. Phys. B (Proc. Suppl.), **63A-C**, 245 (1991)
14. Ph. Boucaud et al., hep-ph/9810322

GeV Photons at SPring8/RCNP and Quark Nuclear Physics

HIROSHI TOKI, MIHO TAKAYAMA and YOSHIAKI KOMA
Research Center for Nuclear Physics (RCNP), Osaka University
Mihogaoka 10-1, Ibaraki, Osaka 567-0047, Japan
E-mail: toki@rcnp.osaka-u.ac.jp

Abstract

We have started a research project of studying the dynamics of quarks and gluons using the GeV photon facility called LEPS at SPring8/RCNP. We would like to discuss the physics to be achieved with these facilities from a theoretical view point. We call the research field of describing hadrons and nuclei in terms of quarks and gluons as Quark Nuclear Physics.

1. Introduction

We have started the GeV photon project called LEPS at SPring8. We are recieving a strong enthusiasm from physicists all over the world. Hence, although we did not get the GeV photon beam yet, we start to talk about the next generation photon facility called PEARL. Let me therefore start out with the energetics of LEPS and PEARL. The photon energy of LEPS is $E_\gamma \sim 3GeV$. On the other hand, the energy of PEARL is $E_\gamma \sim 10GeV$. The usual excersize for me is to get the Mandelstam variable s.

$$s = (E_\gamma + M)^2 - E_\gamma^2 = 2E_\gamma M + M^2 \tag{1}$$

$E_\gamma = 3, 5$ and $10GeV$ correspond to $\sqrt{s} = 2.6, 3.3$ and $4.6GeV$. In order to see what do these energies mean, calculate the threshold energy of various systems;

s=0 hadrons system is $M(\pi) + M(p) = 1.1GeV$.
s=1 hadron system is $M(K) + M(\Lambda) = 1.6GeV$.
c=1 hadron system is $M(D) + M(\Lambda_c) = 4.2GeV$.

What then we can do with these GeV photons? Since the glueball mass is $E(g) \sim 1.5GeV$, we need $E_\gamma \sim 3GeV$. Hence PEARL is suited for the extensive study of these objects. The charm hadrons need $E_\gamma \sim 10GeV$. The charm hadrons are out of reach with these machines. Hence, energywise the GeV photons produced by LEPS and PEARL are suited for the quark nuclear physics caused by light SU(3) quarks and gluons.

The real photons could be considered as the virtual vector meson beams. This fact should be kept in mind in comparison with the virtual photons to be produced by electron beams. It is very important to use this feature in designing experiments with the photon beams.

2. Quark Nuclear Physics

Quark Nuclear Physics is the research field of describing hadrons and nuclei in terms of quarks and gluons. Here, confinement and chiral symmetry breaking are the most essential phenomena for quark nuclear physics. QCD is supposed to provide the dynamics of quarks and gluons. In fact, QCD is able to describe the above essential phenomena using the lattice QCD. On the other hand, the perturbative description of QCD provides the high momentum phenomena.

Although this is the case, we would like to develope the basic picture for quark nuclear physics. This is very important for the motivation of development of physics. For nuclear physics, it was the shell model, which changed the situation of physics associated with nucleus completely. Suppose we want to describe a nucleus with about 100 nucleons. Such a system was impossilbe to solve. However, once physicists noticed the magic number and the important degrees of freedom for the description of the excitation spectra as those nucleons in the valence orbits, they were able to describe nuclear excitations, which are caused by about 100 particles. The shell model played a key role for the developement of nuclear physics.

What is the basic picture (model) for quark nuclear physics then? There are various candidates; Dual Ginzburg-Landau (DGL) theory, Instanton model, QCD sum rule, gluonic $1/q^4$ model and so on. All these models have some truth in describing the low energy observables. Here we shall discuss the DGL theory, which has been studied in various directions in our group. The DGL theory is demonstrated to describe both confinement and chiral symmetry breaking and at the same time has a strong connection with QCD.

3. Dual Ginzburg-Landau Theory

The DGL theory is based on the assumption that our vacuum is in the superconductor like state; dual superconductor. In the superconductor, the magnetic field H is disliked by the superconducting matter due to the Meissner effect. Hence, if the magnetic field is stronger than the critical amount, and the superconductor is the second-kind, then the magnetic field does not spread over the matter, but rather is confined in a vortex like configuration. We assume that our vacuum does not like the (color) electric field. Hence, when a quark and an anti-quark are placed in the dual superconductor, the color electric field produced by the quark cannot

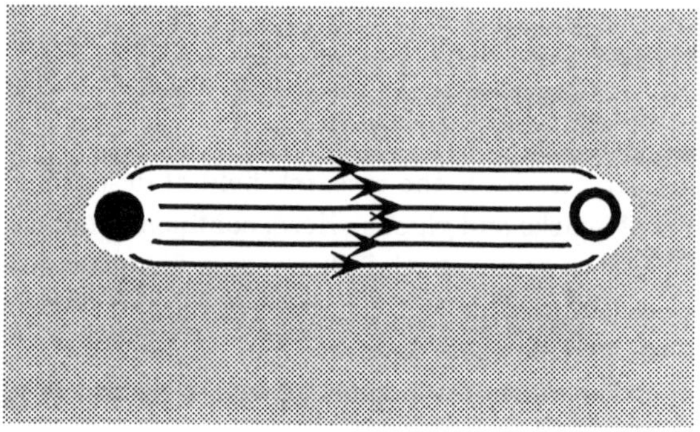

Fig.1: The schematic view of quark confinement. The circles at both ends denote quark and antiquark, while the lines with arrow are abelian color-electric field confined by the dual Meissner effect.

spread over the space but rather flows into the anti-quark in the vortex like configuration without disturbing much the dual superconductor. This property causes confinement of quarks, since quarks cannot exist alone but need a partner. The schematic view of quark confinement is shown in Fig.1.

This dual superconductor picture is very attractive. However, there should be two important ingredients in this picture. One is the abelian dominance and the other is the color monopole. On the other hand, the fundamental theory of strong interaction, QCD, is a non-abelian gauge theory. QCD has non-abelian gluons and does not have color monopoles.

A miracle happens in QCD by the choice of a special gauge. 't Hooft is the one who introduced the abelian gauge by fixing the gauge of the non-abelian gluons and showed that in this case there appear color monopoles.[1] The lattice QCD demonstrates the apearance of monopoles and the signature of their condensation as the appearance of long and complicated monopole trajectries in the abelian gauge. The abelian dominance needs the study of the correlations of charged gluons and those of the non-charged gluons (abelian gluons) in the abeilan gauge. The recent lattice QCD calculations showed that the correlatin length is very short for the charged gluons, while it is large for the abelian gluons in the maximally abelian (MA) gauge.[2] These facts indicate that the picture of quark confinement in terms of the dual-superconductor could have support from QCD.

This consideration then provides the effective lagrangian, where the color monopole couples with dual gluon fields and the quarks interact with abelian gluons. The Higgs term provides condensation of color monopoles. The resulting lagrangain has a similarity to the Ginzburg-Landau theory and hence the name dual Ginzburg-Landau theory is used for this model.[3,4]

The DGL theory provides clear predictions as

1. There should appear color monopole glueball around 1.5GeV. This is a Higgs particle of strong interaction.
2. The string tention for gluons is 3 times more than that for quark-antiquark string tention due to the abelian property of the long range physics.
3. We get the probabilities of hadronization with respect to the quark masses.
4. Various others.

4. Photon-nucleon collision

What happens when a GeV photon collides with a nucleon? My favorite figures show the total cross sections for vastly different systems being plotted as functions of the square root s; $\sqrt{s}$. All the cross sections could be described with just two energy dependent terms;

$$\sigma = As^{0.08} + Bs^{-0.45} \tag{2}$$

It is amazing to see that the photon proton cross section also follows the same s dependence. This convinces the importance of the virtual vector mesons in photon beam.

These two s dependence is described in terms of two Regge trajectories, one is made of vector mesons constituted with quarks and the other made of glueballs of gluons. The slope of the vector mesons is well known from the meson spectra. This information provides the string tention between quark and anti-quark. On the other hand, we do not know the glueball states at all. Hence, the determination of the glueball states is important to get the regge slope and hence the string tention between gluons.

GeV photons may provide this information by studying carefully the s dependence and the t dependence. In particular, if we observe ϕ meson, which consists mainly of s quarks, we can emphasize the gluon exchange process due to the OZI suppression of the mesons made of quarks. The precise measurement of the t dependence should provide the slope of the glueball trajectory. It would be nice then to extract the string tention from this information. In addition, we can predict the position of the 2^+ glueball state. The energy dependence of the phi production around $2GeV$ may have contributions of the 0^+ glueball.[5]

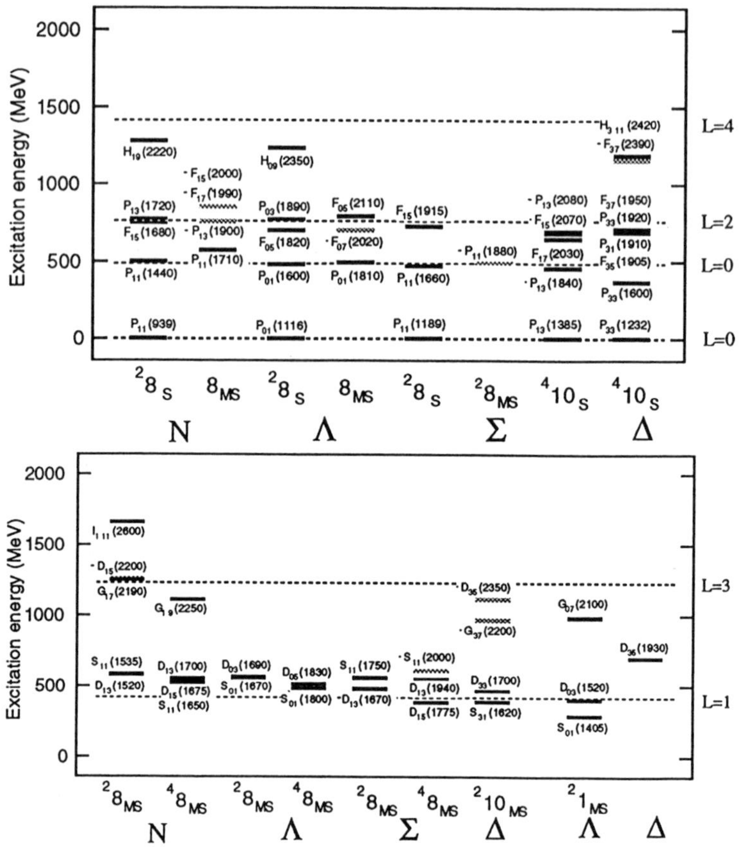

Fig.2: SU(3) flavor baryon spectra for even parity states (upper) and odd parity states (lower). The dashed lines with the angular momentum L are the predictions of the DOQ model[6].

5. baryon spectra

Let us look now at the baryon spectrum. The baryon spectrum, which includes all the known SU(3) baryons, seems to show no regularity in the spectrum. The ground state baryons have been studied by Gell-Mann and Okubo. The Gell-Mann-Okubo mass formula provided the existence of the SU(3) quarks, which leads to the quark model.

It would be then very interesting to measure the excited states with respect to these ground states. This is done in Fig.2. Those states without the corresponding ground state baryons are shifted downward by 200MeV to consider the spin effect, which is seen as the mass difference between

$\Lambda(1116)$ and $\Sigma(1385)$. Surprisingly the spectra show very good systematics as seen in Fig.2.

These simple spectra indicate rather a simple dynamics is acting for the excitation of quarks. In fact, when we take a deformed oscillator quark model, we find the ground state is sphenical and the first and the second excited state are deformed. Once a state is deformed, there should appear a rotational band on top of the intrinsic deformed state. This was demonstrated in terms of the deformed oscillator quark (DOQ) model by Hosaka et al.[6] and the results are indicated by dashed lines in Fig.2.

6. glueball

The mesons have the quark structure with the gluon vortex, which is shown in Fig.3. If we circle the gluon vortex and let the quark and antiquark annihilate, then we get the donuts-type configuration as shown in Fig.3. We can calculate this system using the DGL thoery. We find the energy of this configulation to be around $E = 1.6GeV$.[7] The rotational symmetry has to be recovered in order to get the actual masses of various spin states. In any case, we find glueballs in the mass range of about $1.6GeV$.

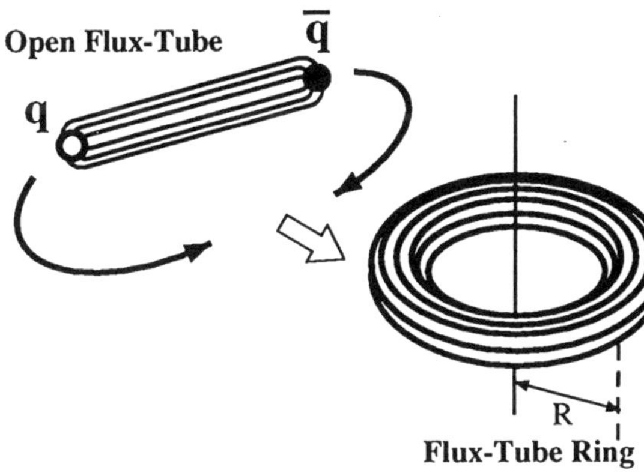

Fig.3: The ordinary meson and the donuts-type glueball.

7. conclusion

GeV photon facilities as LEPS and PEARL at SPring8/RCNP provide a good opotunity to study hadrons and nuclei in terms of quarks and gluons, the research field of which is called quark nuclear physics.

We would like to mention at this place that the QNP is a very special research field. Generally, experiments and/or observations provide motivations to develope theories/models to describe the phenomena. Theories then predict new experiments, which check the validity of the theories. In this way we can find the true picture of the phenomena. In QNP in addition to this excersize, we have QCD which tells which theory is collect. Hence, QNP provides an unique oppotunity for us to understand the microscopic world from the most fundamental theory from first principle.

1. G. 't Hooft, *Nucl. Phys.* **B190** (1981) 455.
2. K. Amemiya and H. Suganuma, Proc. Int. Symp. on *Innovative Computational Methods in Nuclear Many-Body Problems (INNO-COM 97)*, World Scientific (1998) 284.
3. H. Suganuma, S. Sasaki and H. Toki, *Nucl. Phys.* **B435** (1995) 207.
4. S. Sasaki, H. Suganuma and H. Toki, *Prog. Theor. Phys.* **94** (1995) 373.
5. T. Nakano and H. Toki, Proc. Int. Work. on *Exciting Physics with New Accelerator Facilities (EXPAF)*, World Scientific (1998) 48. and T. Nakano et al., *Nucl. Phys.* **A629** (1998) 559c-566c.
6. A. Hosaka, M. Takayama and H. Toki, *Mod. Phys. Lett.* **A13**(1998) 1699.
7. Koma, H. Suganuma and H. Toki, Proc. Int. Symp. on *INNOCOM 97*, World Scientific (1998) 315.; hep-ph/9902441.

INTERACTION OF NUCLEON RESONANCES IN A CONSTITUENT QUARK MODEL

M. ARIMA, I. MAEDA
Department of Physics, Osaka City University, Osaka 558-8585, Japan
E-mail: arima@ocunp1.hep.osaka-cu.ac.jp

K. MASUTANI
Faculty of Engineering, Yamanashi University, Kofu 400-8511, Japan
E-mail: masutani@grape.kkb.yamanashi.ac.jp

The influence of dynamical coupling with nucleon resonances on the nucleon-nucleon interaction is examined by using a resonating group method with a constituent quark model. When the one-pion-exchange-potential is used for the quark-quark interaction, the strong tensor interaction emerges large attractive force due to the coupling with the $^7D_1$-wave ($^5D_0$-wave) $\Delta\Delta$ state on the $^3S_1$ ($^1S_0$) channel of nucleon-nucleon scattering.

1 Introduction

A constituent quark model is one of the successful effective models to examine low-energy hadron physics. Although the relation with the fundamental theory (QCD) has not been clear, this model provides us valuable information on baryons. For example, we can reproduce the mass spectrum by using the non-relativistic kinematics for the constituent quarks in the effective confinement potential with the one-gluon-exchange potential (OGEP) for the quark-quark interaction.[1,2] Recent arguments on the chiral symmetry suggests the use of Goldstone boson exchange potential (OBEP) as the quark-quark interaction.[3,4]

The constituent quark model was applied to the dynamical calculation of nucleon-nucleon (NN) interaction by Oka, Yazaki [5,6], and this formalism has been developed by many researchers.[7] The six-quark wave function is described by using the resonating group method (RGM). The NN potential and the phase shifts for the NN scattering were calculated by the generating coordinate method (GCM), and the short-range repulsion was explained by taking account of the Pauli principle and the quark-quark interaction.

For the RGM wave functions, only the ground state configuration of quarks is usually taken into account. The influence of the orbital excitation or quarks has been considered to be small in the RGM calculation with the OGEP.[8] But the recent adiabatic calculation suggested that, when we use the OBEP as the quark-quark interaction, some important configurations are not incorporated in the RGM wave functions.[9,10]

We consider the dynamical effect of coupling with nucleon resonances on the NN interaction, and calculate the phase shifts. We also improve technical aspects of the RGM formalism and directly calculate potentials for the two-nucleon system including nucleon resonances. Our motivation is that we hope to approach the systematic understanding of baryons by applying the constituent quark model to single-baryon and two-baryon system.

2 Formalism

The standard definition of the RGM equation in momentum space is

$$\int \frac{d^3k}{(2\pi)^3} H(\vec{k}',\vec{k})\chi(\vec{k}) = E\int \frac{d^3k}{(2\pi)^3} N(\vec{k}',\vec{k})\chi(\vec{k}) \,, \tag{1}$$

where two baryons A and B are interacting with each other and the relative motion is described by the wave function χ. The hamiltonian- and the norm-kernels are defined as usual;

$$\begin{bmatrix} H(\vec{k}',\vec{k}) \\ N(\vec{k}',\vec{k}) \end{bmatrix} = \int d^3\tau \left([\phi_A\cdot\phi_B]^{s'} e^{i\vec{k}'\cdot\vec{R}}\right)^{\dagger} \begin{bmatrix} \hat{H} \\ \hat{1} \end{bmatrix} \mathcal{A}\left\{[\phi_A\cdot\phi_B]^{s} e^{i\vec{k}\cdot\vec{R}}\right\} \,, \tag{2}$$

where an antisymmetrizer $\mathcal{A}$ is given by

$$\mathcal{A} \equiv (1-\sum_{i\in A, j\in B} P_{ij})(1-P_{AB}) \,. \tag{3}$$

The typical form of the RGM wave function appears in Eq. (2), where the wave function of baryon A (B) is denoted by ϕ_A (ϕ_B) which is antisymmetrized with respect to three-quark permutation.

We use the non-relativistic quark model for the hamiltonian

$$\hat{H} = \sum_i^6 \frac{\hat{p}_i^2}{2m_q} - \sum_{i<j}^6 a\lambda_i^c\cdot\lambda_j^c r_{ij} + \sum_{i<j}^6 \hat{V}_{qq;ij} \,. \tag{4}$$

The linear form with the strength parameter a is assumed for the effective confinement potential. For the quark-quark interaction, we consider two possibilities separately: The first one is the OGEP which includes the color-coulomb, the spin-spin and the tensor interactions with the strength parameter α_s. And second one is the one-pion-exchange potential (OPEP) which includes the spin-spin and the tensor interactions with the strength parameter (the π-quark coupling constant) $g_{\pi qq}$. It is worth noting that the OGEP depends on the color-operator, while the OPEP on the isospin-operator. The δ-function piece in these interaction is smeared out by introducing a gaussian function $f(r)$ with the range parameter κ: $f(r) \propto e^{-\kappa r^2}$.

The harmonic oscillator eigen functions are used for the orbital part of baryons without the configuration mixing due to the quark-quark interaction. Note that the harmonic oscillator parameter, β, depends on the confinement strength.

Using the harmonic oscillator eigen function also for the relative wave functions, we can rewrite the RGM equation in the form of the Schrödinger equation. After doing a partial wave expansion, we obtain

$$(E_{\rm rel} - K)\chi^{Js'}_{l';i}(k') = \sum_{sl;j} \int \frac{dkk^2}{(2\pi)^3} \bar{U}^{Js's}_{l'l;ij}(k',k)\chi^{Js}_{l;j}(k) \,. \tag{5}$$

Because we consider the coupling with nucleon resonances, this is a coupled channel equation. As for the model space of two baryon channels included in this coupled-channel calculation, we set the truncation point at $2\hbar\omega$. Within this energy many observed resonances are included.

The non-local potential for two baryon system, $\bar{U}$ is given by

$$\bar{U}^{Js's}_{l'l}(k',k) = \sum_{n'n} C^*_{n'l'}(k')C_{nl}(k)\left\{(\bar{E}-\epsilon_{n'l'})\bar{N}^{Js's}_{l'l} - \bar{V}^{Js's}_{l'l} + \bar{W}^{Js's}_{l'l}\right\} \,. \tag{6}$$

The momentum dependence is only in the function $C_{nl}(k)$ defined as

$$C_{nl}(k) = (-1)^n\sqrt{\frac{\pi}{2}}R_{nl}(\sqrt{\tfrac{2}{3}}k) \,. \tag{7}$$

$\bar{N}$ is a part of the norm kernel, and $\bar{V}$ and $\bar{W}$ are the potential parts of the hamiltonian kernel. The exchange kernels $\bar{N}$ and $\bar{V}$ are given by

$$\begin{bmatrix} \bar{N}^{Js's}_{l'l} \\ \bar{V}^{Js's}_{l'l} \end{bmatrix} \propto i^{l'-l} \int d^3\tau \Big([\varphi_{n'l'}[\phi_A\cdot\phi_B]^{s'}]^J\Big)^\dagger \begin{bmatrix} \hat{1} \\ \hat{V}_{qq} \end{bmatrix} \mathcal{A}'\{[\varphi_{nl}[\phi_A\cdot\phi_B]^s]^J\} \,. \tag{8}$$

And the direct kernel $\bar{W}$ is also given by this same equation without the antisymmetrizer $\mathcal{A}' \equiv 1 - \mathcal{A}$. Because we use the harmonic oscillator eigen functions both for the three-quark clusters and the relative motion, we can use shell model technique to calculate these kernels directly without the GCM.

The direct kernel includes the direct potential between two baryons which is absent if the OGEP is used for $\hat{V}_{qq}$ due to the color operator and the color-singlet structure of each baryon. On the other hand, the OPEP has this direct potential which contributes significantly to the NN potential.

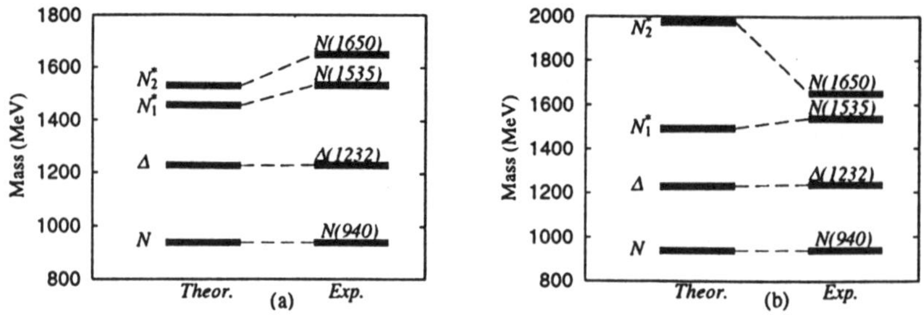

Figure 1: Mass spectrum of single baryons calculated by using (a) the OGEP and (b) the OPEP for the quark-quark interaction. Experimental data are taken from the Particle Data Table[11].

It takes very long time if we include all possible channels below the truncation energy. Thus we reduce Eq. (5) to the effective one for the NN channel by the projection operator method, and we consider only the first order correction due to the coupling with nucleon resonances in the following calculation.

3 Results

First we use the OGEP for the quark-quark interaction. All parameters are determined to reproduce the mass differences between N and Δ, and N and the first negative parity nucleon resonance. The value of parameters are summarized in table 1. The quark mass is fixed at 300 MeV and β is determined to minimize the nucleon mass. The mass spectrum of nucleon resonances obtained by using these parameters are also shown in figure 1a in comparison with values taken from the particle data table.

Figure 2 shows the phase shifts of ${}^3S_1$- and ${}^1S_0$-channels of NN scattering. For the ${}^3S_1$ channel, we do not include the SD mixing in the NN channels because we want to show clearly the effect of coupling on the phase shifts. We start with the no-coupling case, that is, all couplings with nucleon resonances are switched off. Then we gradually increase the number of coupled channels. The contribution of the $\Delta\Delta$ states is small but attractive. This is consistent with the GCM results. The contribution of the $1\hbar\omega$ excitations is very small

Table 1: Parameters in our model with the OGEP

m_q (MeV)	a (MeV/fm)	β (fm^{-2})	α_s	κ (fm^{-2})
300	83.6	4.00	1.05	10.0

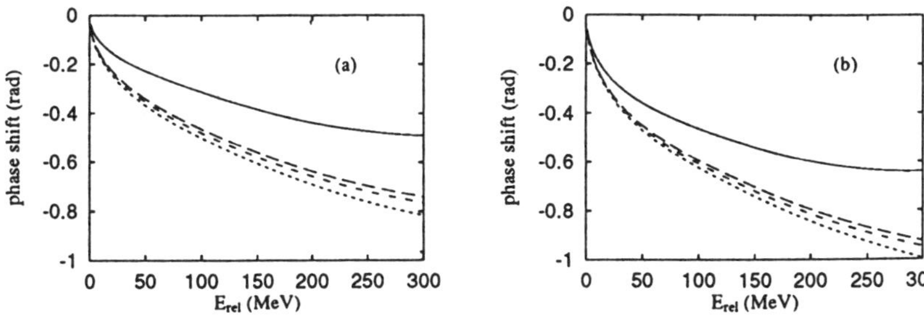

Figure 2: The phase shifts of NN scattering for (a) the $^3S_1$ and (b) the $^1S_0$ channels when the OGEP is used for the quark-quark interaction. The dotted curve stands for the no-coupling case. The short-dashed (long-dashed) curve is obtained by including the coupling with $\Delta\Delta$ ($\Delta\Delta + 1\hbar\omega$) states. The solid curve is the result of our full calculation.

and negligible both for $^3S_1$ and $^1S_0$ channels. The contribution of the $2\hbar\omega$ excitations is larger than that of the $\Delta\Delta$ states. Although the phase shifts are affected by the coupling with resonances, they are still repulsive and their energy-dependence does not change so much. This result is consistent with the suggestion by adiabatic calculations.[8]

Next we consider the OPEP for the quark-quark interaction. All parameters are determined as before, and their values are summarized in table 2. The mass spectrum calculated by these parameters are shown in figure 1b. Note the value of pion-quark coupling constant is larger than the standard value, about 2.9. This is because we try to reproduce the $N\Delta$ mass difference in the first order perturbation with the OPEP.

Figure 3 show the $^3S_1$ and the $^1S_0$ phase shifts for the NN scattering. Also in this case, we do not include the SD mixing. The effect of coupling is remarkable in this case. For $^3S_1$ channel, the coupling with $\Delta\Delta$ states (especially the $^7D_1$-wave $\Delta\Delta$ state) changes the repulsive phase shift into the attractive one in the low energy region. For the $^1S_0$ channel, the large attractive contribution is produced by the coupling with the $^5D_0$-wave $\Delta\Delta$ state. The attractive effect of the $2\hbar\omega$ excitation is also remarkable for these channels. We have checked that the origin of this drastic change is the strong tensor interaction in the direct part of the OPEP.

Table 2: Parameters in our model with the OPEP

m_q (MeV)	a (MeV/fm)	β (fm^{-2})	$g_{\pi qq}$	κ (fm^{-2})
300	95.0	3.89	7.00	3.40

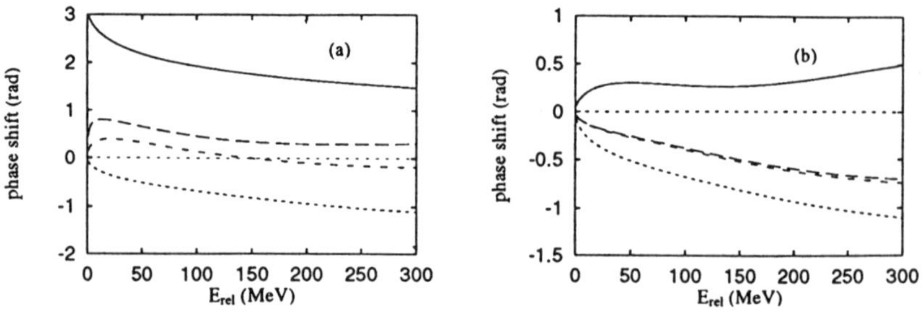

Figure 3: Same as Fig. 2 but for the OPEP as the quark-quark interaction.

4 Summary and Conclusion

We consider the NN interaction and the internal excitation of nucleon by using the RGM formalism. Calculating the NN potential directly without the GCM, we obtained the phase shifts of $^3S_1$ and $^1S_0$ channels. And we examine the effect of coupling with nucleon resonances in the dynamical calculation. The coupling effect on the OGEP case is small. On the other hand, for the OPEP, the large effect of coupling with nucleon resonances is observed. We find that this is not due to the exchange terms, but to the direct term of the OPEP when we consider the first order correction to the NN potential.

References

1. A. De Rùjula, H. Georgi and S. L. Glashow, *Phys. Rev.* **D12**, 147 (1975).
2. N. Isgur and G. Karl, *Phys. Rev.* **D18**, 4187 (1978).
3. L. Ya. Glozman, D. O. Riska, *Phys. Rep.* **268**, 263 (1996).
4. H. Kowata, M. Arima and K. Masutani, *Prog. Theor. Phys.* **100**, 339 (1998).
5. M. Oka and K. Yazaki, in *Quarks and Nuclei, International review of Nuclear Physics*, ed. W. Weise (World Scientific, Singapore, 1984).
6. M. Oka and K. Yazaki, Prog. Theor. Phys. **66**, 556 (1981); **66**, 572 (1981).
7. A. Faessler, F. Fernandez, G. Lübeck and K. Shimizu, *Phys. Lett.* **B112**, 201 (1982).
8. S. Ohta, M. Oka, A. Arima and K. Yazaki, *Phys. Lett.* **B119**, 35 (1982).
9. Fl. Stancu, S.Pepin and L. Ya. Glozman, *Phys. Rev.* **C56**, 2779 (1997).
10. D. Bartz and Fl. Stancu, *Phys. Rev.* **C59**, 1756 (1999).
11. Particle Data Group, *Phys. Rev.* **D54**, 1 (1996).

Physics with LEPS at SPring–8

T. Nakano
RCNP, Osaka University,
10-1 Mihogaoaka, Ibaraki
Osaka 567-0047, JAPAN
E-mail: nakano@rcnp.osaka-u.ac.jp

The GeV photon beam at SPring–8 is produced by backward-Compton scattering of laser photons from 8 GeV electrons. The maximum energy of the photon will be above 3 GeV, and the beam intensity will be 10^7 photons/sec. Polarization of the photon beam will be 100 % at the maximum energy with fully polarized laser photons. We report the outline of the quark nuclear physics project with this high-quality high-intensity beam.

1 LEPS Facility

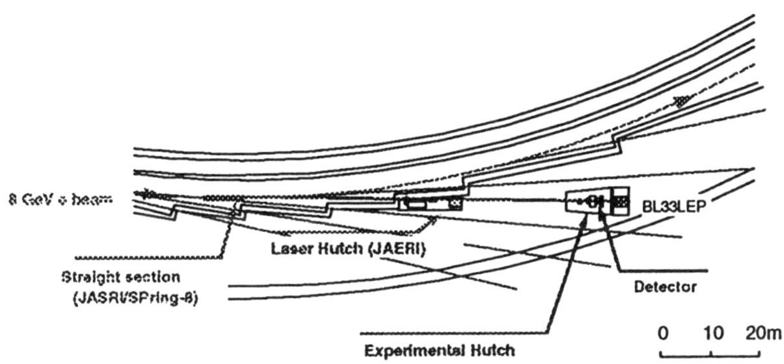

Figure 1: Plan view of the Laser-Electron Photon facility at SPring–8 (LEPS).

The Spring-8 facility is the most powerful third-generation synchrotron radiation facility with 61 beamlines. We use a beamline, BL33LEP (Fig. 1), for the quark nuclear physics studies exclusively. The beamline has a 7.8-m long straight section between two bending magnets. Polarized laser photons are injected from a laser hutch toward the straight section where Backward-Compton scattering (BCS)[1] of the laser photons from the 8 GeV electron beam takes place. The BCS photon beam is transfered to the experimental hutch, 60 m downstream of the straight section. Figure 2 shows the maximum energy of the BCS photon as a function of a laser-photon energy for the cases of 8 GeV and 6 GeV electron beams. The maximum energy depends more strongly on the electron-beam energy than on the laser-photon energy. SPring–8 is the only facility where a high-intensity BCS photon beam above 2 GeV is obtainable.

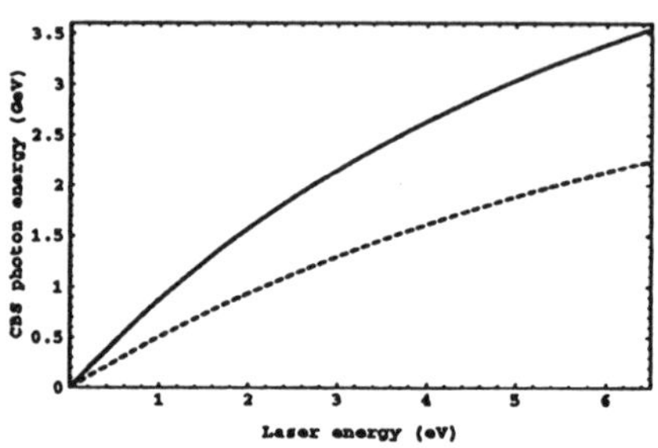

Figure 2: Energies of a BCS photon with a 8 GeV electron beam (solid line) and a 6 GeV electron beam (dashed line)

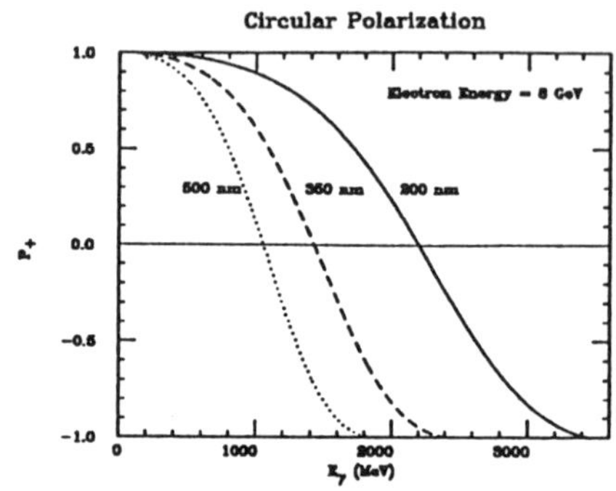

Figure 3: Relation between polarization and photon energy. Laser photons at various wavelengths are assumed to have 100 % circular polarization.

If laser lights are 100 % circular polarized, a backward-Compton-scattered photon is also polarized. The polarization drops as the photon energy decreases as shown in Figure 3. However, an energy of laser photons is easily changed so that the polarization remains reasonably high in the energy region of interest. The intensity, position, and polarization of the laser lights which do not interact with the electron beam are monitored at the end of the beamline.

The beam energy is determined by measuring the energy of a recoil electron with a tagging counter. The tagging counter is located at the exit of the bending magnet after the straight section. It consists of multi-layers of a 0.1 mm pitch silicon strip detector (SSD) and plastic scintillator hodoscopes. Electrons in the energy region of 4.5 – 6.5 GeV are detected by the counter. The corresponding photon energy is 1.5 – 3.5 GeV. The energy resolution (RMS) of 15 MeV for the photon beam is limited by the energy spread of the electron beam and an uncertainty of a photon-electron interaction point.

In the first stage, we use a conventional Ar laser. A 25 W Ar laser has a 2 W output at 351 nm (3.5 eV). The maximum energy of the BCS photon with this laser light is 2.4 GeV. If we use a multi-line mode around 351 nm, the laser output power is about 5 W. The corresponding intensity of the BCS photon beam is expected to be 10^7 photons/sec.

2 Physics Motivation

In this section, we briefly review a physics motivation of a ϕ photoproduction experiment, for which the detector system is optimized.

It is well known that all the hadron-hadron total cross sections (including a γp total cross section) in a wide energy range are reproduced very well in terms of two s–dependent terms with $s^{-0.5}$ and $s^{0.08}$ dependences, where s is the Mandelstam s variable [2]. The Reggeon exchange model clearly suggests that the $s^{-0.5}$ term originates from the ρ meson ($T = 1$, $J^{\pi} = 1^{-}$.) trajectory. On the other hand, the $s^{0.08}$ term requires the introduction of an unobserved Regge trajectory (Pomeron trajectory) [3], whose $\alpha(t = 0)$ has to be 1.08. One can identify the Pomeron trajectory with a gluon trajectory since the first particle state on the trajectory appears at $m^2 \sim 4$ GeV^2 with $J = 2$ (2^{++} glueball).

At high energies, diffractive photo-production of a vector meson from a proton target is well described as a pomeron-exchange process in the framework of the Regge theory and of the Vector Dominance Model (VDM) [4]; a high energy photon converts into a vector meson and then it is scattered from a proton by an exchange of the pomeron [5]. Within QCD, interactions between hadrons can be due to quark-exchange and gluon-exchange. Phenomenologically, we know that the quark-exchange processes can be simulated by meson-exchange.

For the ϕ production, the meson-exchange contribution is small because the couplings of the ϕ meson to other non-strange mesons are weak due to the OZI suppression. Precise measurements of $d\sigma/dt$ and spin observables of the ϕ production near the production threshold may reveal the existence of another gluon-exchange trajectory (a 0^{++} glueball trajectory) whose contribution falls off rapidly at high energies [6]. One can also address the question concerning the $s\bar{s}$ components in nucleon. An interference between a pomeron-exchange amplitude and a small amplitude due to a direct knockout of $s\bar{s}$ in the nucleon may cause a large asymmetry of the production cross sections between the spin parallel and anti-parallel configurations of the polarized photon and polarized nucleon [7].

3 Detector

The LEPS detector (Fig. 4) consists of charged-particle tracking counters, a dipole magnet, and a time-of-flight wall. The design of the detector is optimized for a ϕ photoproduction in forward angles. It has a large forward acceptance and good coverage of momentum transfer t for the photoproduction as shown in Figure 5.

The opening of the dipole magnet is 135-cm wide and 55-cm height. The length of the pole is 60 cm, and the field strength at the center is 1 T. The vertex detector consists of 2 planes (x- and y-) of single-sided SSDs (SVTX) and 5 planes (x,x',y,y',u) multi-wire drift chamber (DC1), which are located upstream of the magnet. The stereo ambiguity (pairing ambiguity) for two-track events are solved with DC1. The thickness of each SSD is 300μm and it results in angle spread (σ) of $\sim$ 1 mrad. The corresponding deterioration in 2-kaon invariant mass and momentum transfer t measurements are 300 KeV and 2×10^{-3}GeV^2 respectively. Two sets of MWDCs (DC2 and DC3) are located downstream of the magnet. The active area size of DC2 and DC3 is 200cm(W) $\times$ 80cm(H). Each set has 5 planes: x,x',y,y', and u(v) in order to solve both left-right ambiguity and stereo ambiguity locally. Figure 6 shows the momentum resolution as a function of a momentum.

The identification of momentum analyzed particles is performed by mea-

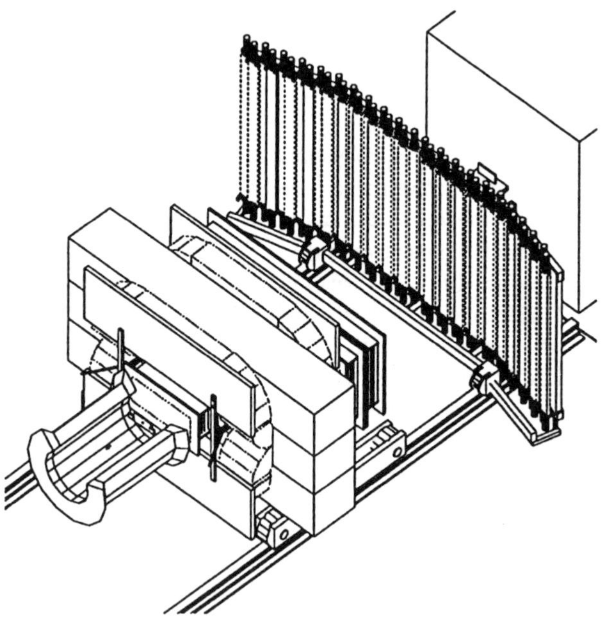

Figure 4: The LEPS detector setup.

suring a time of flight from a target to the TOF wall. The start signal for the TOF measurement is provided by a RF signal from the 8-GeV ring, where electrons are bunched at every 2 nsec (508MHz) with a width (σ) of 12 psec. Since the speeds of the both electron beam and a laser-electron photon are same, the arrival time of the laser-electron photon at the target is synchronized with the RF signal. A stop signal is provided by the TOF wall consisting of 40 2m-long plastic scintillation bar with a cross section of 4cm (t) $\times$ 12 cm (w). The resolution of the TOF counter is less than 100 psec. When the TOF wall is located at 3 m from the magnet, a flight-length of a charged particle is about 4 m. In this case, the detector has a capability of 4-σ K/π separation up to 2 GeV/c.

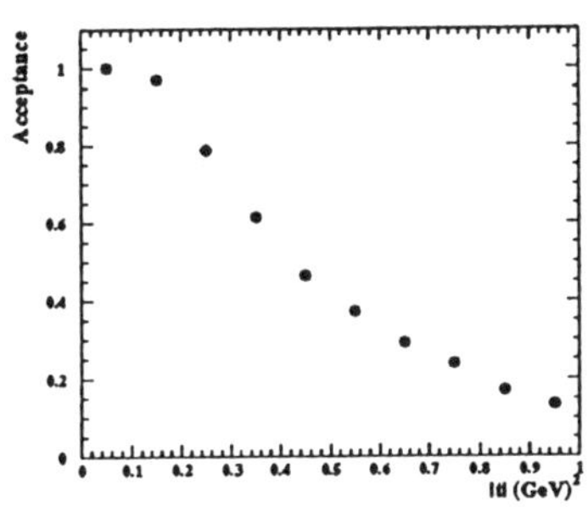

Figure 5: Geometrical acceptance of the dipole magnet for the reaction $\gamma + p \rightarrow \phi + p$, $\phi \rightarrow K^+ + K^-$ at $E_\gamma = 2.4$ GeV as a function of a momentum transfer $|t|$. The both kaons are required to go through the magnet. The distance between a target and the magnet center is 90 cm.

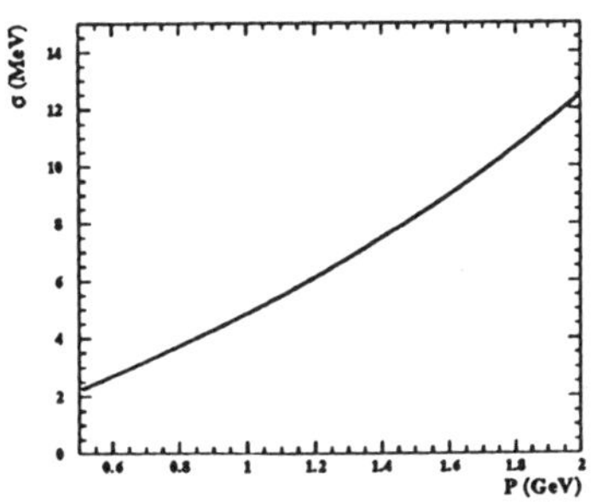

Figure 6: The overall momentum resolution (σ MeV) as a function of a momentum in GeV.

A ϕ meson is identified through the reconstruction of the KK invariant mass. The typical mass resolution is 600 keV, which is much smaller than the ϕ width. The measurement error of a momentum transfer is about 0.01 GeV^2, and it mainly comes from the measurement error of the incident photon energy.

The elastic ϕ photoproduction from a proton and a quasi-free ϕ photoproduction from a nucleon are separated by reconstructing a recoiled target mass from the incident photon energy and the energy-momentum of a ϕ. Figure 7 shows the reconstructed mass distribution by a simulation. The narrow peak at the proton mass corresponds to the elastic production, and the broad tail

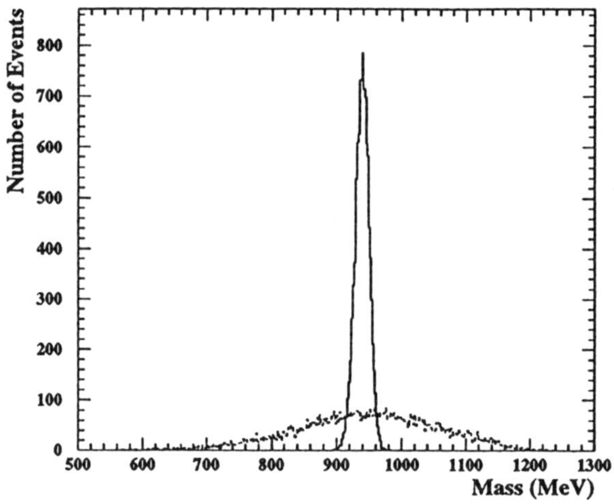

Figure 7: The target mass distribution reconstructed from the incident gamma energy and the ϕ momentum for ϕ photoproduction from a proton (solid) and a nucleus (dashed).

corresponds to the quasi-free production. The broadening is due to a Fermi momentum of a nucleon in a nucleus.

The beam intensity and profile during experiments are monitored by a beam monitor consisting of a pre-radiator to convert a photon into an electron-positron pair, a thin plastic scintillation counter, and SSD counters.

For future upgrade, developments of an aerogel cherenkov counter, a gamma counter, and a polarized target are in progress.

4 Schedule

The construction of the beamline and the laser hutch were completed in September of 1998. The laser system was installed in the winter shutdown period of 1998-1999. The construction of the experimental hutch and the detector system will be completed by the end of March, 1999. The commissioning of the Laser-Electron Photon beam will be carried out from July in 1999. The first test experiment will start in 2000.

References

1. R.H. Milburn, Phys. Rev. Lett. 10 (1963) 75.
2. A. Donnachie and P. V. Landshoff, Phys. Lett. B 296 (1992) 227.
3. I.Y. Pomeranchuk, Sov. Phys. 7, (1958) 499.
4. J.J. Sakurai, Ann. Phys. 11 (1960) 1; J.J. Sakurai, Phys. Rev. Lett. 22 (1969) 981.
5. T.H. Bauer *et al.*, Rev. Mod. Phys. 50 (1978) 261.
6. A. I. Titov, T.-S. H. Lee, and H. Toki, nucl-th/9812074.
7. A. I. Titov, Y. Oh, and S. N. Yang, Phys. Rev. Lett. 79 (1997) 1634; A. I. Titov, Y. Oh, and S. N. Yang, Phys. Rev. C58 (1998) 2429.

NON-PERTURBATIVE CHIRAL CORRECTIONS FOR LATTICE QCD

A. W. THOMAS, D. B. LEINWEBER
Department of Physics and Mathematical Physics and
Special Research Centre for the Subatomic Structure of Matter
University of Adelaide
Adelaide SA 5005, AUSTRALIA
E-mail: athomas,dleinweb@physics.adelaide.edu.au

D. H. LU
Department of Physics, National Taiwan University,
Taipei, TAIWAN
E-mail: dhlu@phys5.phys.ntu.edu.tw

We explore the chiral aspects of extrapolation of observables calculated within lattice QCD, using the nucleon magnetic moments as an example. Our analysis shows that the biggest effects of chiral dynamics occur for quark masses corresponding to a pion mass below 600 MeV. In this limited range chiral perturbation theory is not rapidly convergent, but we can develop some understanding of the behaviour through chiral quark models. This model dependent analysis leads us to a simple Padé approximant which builds in both the limits $m_\pi \to 0$ and $m_\pi \to \infty$ correctly and permits a consistent, model independent extrapolation to the physical pion mass which should be extremely reliable.

1 Introduction

At present the only known way to calculate the properties of QCD directly is through the formulation of lattice gauge theory on a discrete space-time lattice. While the lattice formulation of QCD is well established [1], there have recently been a number of exciting advances in lattice action improvement which are greatly facilitating the reduction of systematic uncertainties associated with the finite lattice volume and the finite lattice spacing. However, direct simulation of QCD for light current quark masses, near the chiral limit, remains computationally intensive. In particular, present lattice calculations of masses in full QCD are limited to equivalent pion masses of order 500 MeV or more – although there have been recent preliminary results from CP-PACS using improved actions at a mass as low as 300 MeV [2]. For nucleon magnetic moments the situation is much worse, with the best calculations still being made within quenched QCD for pion masses greater than 600 MeV. This is clearly beyond the range of validity of chiral perturbation theory (χPT).

In view of this situation, the present approach of calculating the properties of QCD using quark masses away from the chiral regime and extrapolating to the physical world is likely to persist for the foreseeable future. It is therefore vital to understand the quark mass dependence of hadronic observables calculated on the lattice and how to connect these calculations with the physical world. A major difficulty in this endeavour is the rapid rise of the pseudoscalar mass for small increases in the quark mass away from the chiral limit – with $m_\pi^2 \propto \bar{m}$, where $\bar{m}$ is the light quark mass.

Historically, lattice results were often linearly extrapolated with respect to m_π^2, particularly in exploratory calculations. More recently the focus has turned to chiral perturbation theory (χPT), which provides predictions for the leading nonanalytic quark-mass dependence of observables in terms of phenomenological parameters[4,5]. Indeed, it is now relatively standard to use χPT in the extrapolation of lattice simulation data for hadron masses and decay constants[1]. On the other hand, earlier attempts to apply χPT predictions for the quark-mass dependence of baryon magnetic moments failed, as the higher order terms of the chiral expansion quickly dominate the truncated expansion as one moves away from the chiral limit. To one meson loop, χPT expresses the nucleon magnetic moments as[6]

$$\mu_N = \mu_0 + c_1\, m_\pi + c_2\, m_\pi^2 \log m_\pi + c_3\, m_\pi^2 + \cdots \,, \tag{1}$$

where μ_0 and c_3 are fitted phenomenologically and c_1 and c_2 are predicted by χPT. The $m_\pi^2 \log m_\pi$ term quickly dominates as m_π moves away from the chiral limit, making contact with the lattice results impossible.

As a result of these early difficulties, lattice QCD results for baryon magnetic moments[7,8,9] remain predominantly based on linear quark mass (or m_π^2) extrapolations of the moments expressed in natural magnetons. This approach systematically underestimates the measured moments by 10 to 20%. Finite lattice volume and spacing errors are expected to be some source of systematic error. However, χPT clearly indicates the linear extrapolation of the simulation results is also suspect. It is therefore imperative to find a method which can bridge the void between the realm of χPT and lattice simulations.

We report such a method, which provides predictions for the quark mass (or m_π^2) dependence of nucleon magnetic moments well beyond the chiral limit – for more details we refer to Ref. [10]. The method is motivated by studies based on the cloudy bag model (CBM), a chiral quark model which preserves the correct leading non-analytic behaviour of chiral perturbation theory while providing what should be a fairly reliable transition to the regime of large pion mass. These studies suggest a Padé approximant which incorporates both the leading nonanalytic structure of χPT and Dirac-moment mass dependence in the heavy quark-mass regime. We apply this method to the existing lattice data as an illustration of how important such an approach will be to the analysis of future lattice QCD calculations.

2 The Cloudy Bag Model

The linearized CBM Lagrangian, with pseudoscalar pion-quark coupling (to order $1/f_\pi$), is given by [11,12]

$$\mathcal{L} = [\overline{q}(i\gamma^\mu \partial_\mu - m_q)q - B]\,\theta_V - \frac{1}{2}\overline{q}q\delta_S + \frac{1}{2}(\partial_\mu \pi)^2 - \frac{1}{2}m_\pi^2\pi^2 - \frac{i}{2f_\pi}\overline{q}\gamma_5\tau\cdot\pi q\delta_S, \quad (2)$$

where B is the bag constant, f_π is the π decay constant, θ_V is a step function (unity inside the bag volume and vanishing outside) and δ_S is a surface delta function. In a lowest order perturbative treatment of the pion field, the quark wave function is not affected by the pion field and is simply given by the MIT bag model. Our calculation is carried out in the Breit frame with the center-of-mass correction for the bag performed via Peierls-Thouless projection. The detailed formulas for calculating nucleon electromagnetic form factors in the CBM are given in Ref. [13].

In the CBM, a baryon is viewed as a superposition of a bare quark core and bag plus meson states. Both the quark core and the meson cloud contribute to the baryon magnetic moments. These two sources are balanced around a bag radius, $R = 0.7 - 1.1$ fm [14]. A large bag radius suppresses the contributions from the pion cloud, and enhances the contribution from the quark core. The CBM reproduces the leading non-analytic behavior of χPT, which has as its origin the process shown in Fig. 1(c), along with the other contributions in the model.

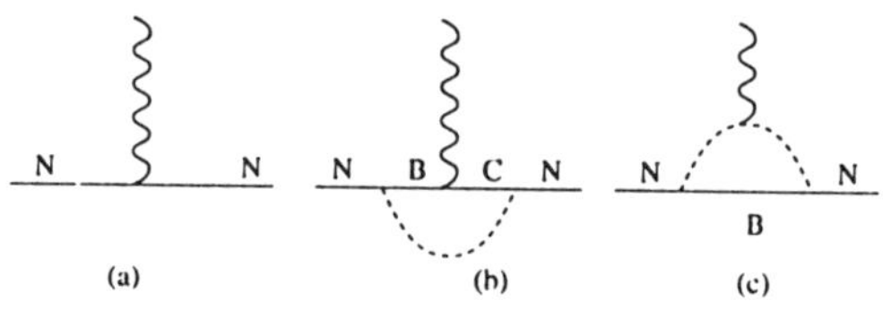

Figure 1: Schematic illustration of the processes included in the CBM calculation. B and C denote intermediate state baryons and include N and Δ.

For the πNN vertex, we take a phenomenological, monopole form, $u(k) = (\Lambda^2 - \mu^2)/(\Lambda^2 + k^2)$, where k is the loop momentum and Λ is a cut-off parameter.

As current lattice simulations indicate that m_π^2 is approximately proportional to m_q over a wide range of quark masses [15], we scale the mass of the quark confined in the bag as $m_q = (m_\pi/\mu)^2\, m_q^{(0)}$, with $m_q^{(0)}$ being the current quark mass corresponding to the physical pion mass (μ). In order to obtain a first idea of the behaviour within the model between the lowest mass lattice point and the experimental data, the parameters R_0, Λ and $m_q^{(0)}$ were tuned to reproduce the experimental moment while accommodating the lattice data. It turns out that the best fits are obtained with $m_q^{(0)}$ in the range 6 to 7 MeV for a bag radius of 0.8fm and Λ of order 600-700 MeV, which are all quite satisfactory.

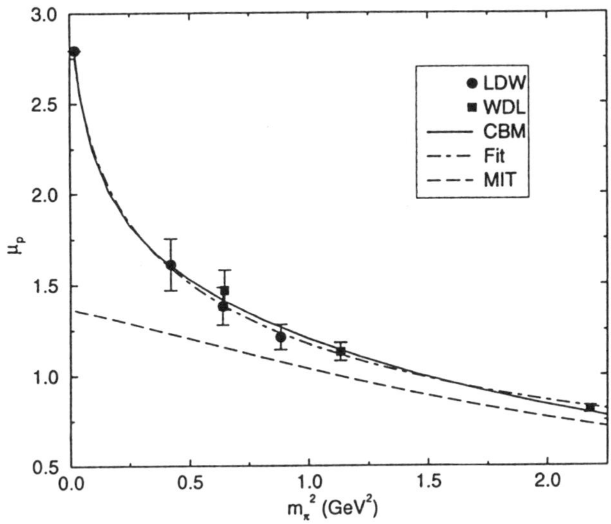

Figure 2: The proton magnetic moment as calculated in lattice QCD (• LDW Ref.[7], ■ WDL Ref.[8]), the cloudy bag model (CBM) and the MIT bag model (MIT). Also illustrated is a fit of the simple analytic form given in Eq. (3) to the CBM results. The point at the physical value of m_π^2 is the experimental measurement and is used to constrain the parameters of the CBM.

Inspection of the results of this calculation, in Fig. 2, show clearly that the pion cloud contribution to the nucleon magnetic moments decreases very quickly, becoming quite small for large quark masses – especially in the range corresponding to the current lattice calculations. In particular, the total pionic correction at the first lattice data point (for m_π around 600 MeV) is of the order of only 10-15% of the total. Since the present lattice data is based on a quenched calculation, which gives incorrect chiral contributions, this is quite good news. It suggests that in using this data as the basis for a correct chiral extrapolation to the physical mass quenching should not induce a major error.

It is also clear from Fig. 2 that below $m_\pi = 600$ MeV the behaviour of the magnetic moment is highly non-linear. One clearly needs to account for such behaviour in a reliable manner if we are to make believable extrapolations of the lattice data. Rather than rely on a model dependent extrapolation method, we chose to investigate whether it was possible to find a simple phenomenological form, *with a sound physical basis*, which could do the job. The successful conclusion to that search is described next.

3 Encapsulating Formula

After considerable effort we found that the following simple Padé approximant, which builds in both the linear behaviour in m_π as $m_\pi \to 0$ (i.e., a square root branch point in $\bar{m}$) and the Dirac moment for large m_π^2 (i.e., $\mu_N \sim 1/\bar{m}$), was able to reproduce the behaviour found in the CBM calculations:

$$\mu_N(m_\pi) = \frac{\mu_N^{(0)}}{1 + \frac{\alpha}{\mu_N^{(0)}} m_\pi + \beta m_\pi^2} . \qquad (3)$$

The fit for the proton case is also shown in Fig. 2. Even more remarkable is that the fit parameter α turned out to be quite close to the value required by χPT (e.g., $\alpha = 4.54$ for the proton, compared with $\alpha = 4.41$ in χPT).

This result encourages us to propose that Eq.(3), with α taken from χPT, should be used as the method of extrapolation in future analyses of lattice data for nucleon magnetic moments – and suitably generalized, for all members of the baryon octet. As an example, we show in Fig.3 the result of two parameter fits to the proton and neutron data. The nucleon magnetic moments at the physical pion mass, obtained from this extrapolation, are

$$\mu_p = 2.85(22)\ \mu_N \quad \text{and} \quad \mu_n = -1.90(15)\ \mu_N, \qquad (4)$$

which agree surprisingly well with the experimental measurements, 2.793 and $-1.913\ \mu_N$ respectively. We note that the data required to do a fit of the lattice results in which covariances are taken into account is no longer available. As such, the uncertainties quoted here should be regarded as indicative only.

4 Conclusion

We have explored the quark mass dependence of nucleon magnetic moments. Quark masses beyond the region appropriate to chiral perturbation theory

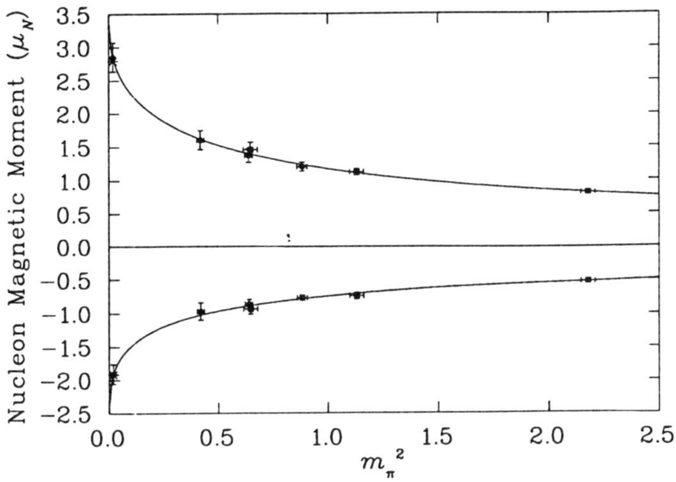

Figure 3: Extrapolation of lattice QCD magnetic moments for the proton (upper) and neutron (lower) to the chiral limit. The curves illustrate a two parameter fit of Eq. (3) to the simulation data in which the one-loop corrected chiral coefficient of m_π is taken from χPT. The experimentally measured moments are indicated by asterisks.

have been explored using the cloudy bag model (CBM) which reproduces the leading nonanalytic behaviour of χPT while modeling the internal structure of the hadron under investigation. We find that the predictions of the CBM are succinctly described by a simple formula which reproduces the leading nonanalytic behavior of χPT in the limit $m_\pi \to 0$ *and* provides the anticipated Dirac moment behavior in the limit $m_\pi \to \infty$. As an example we applied this encapsulating formula to the existing lattice data for nucleon magnetic moments, leading to surprisingly accurate predictions in comparison with the observed values. It will be interesting to see how the fit parameters change as finite volume and lattice spacing artifacts are eliminated in future simulations and whether the level of agreement seen in this investigation is maintained or improved. We strongly advocate the use of the Padé approximant given in Eq. (3) in future lattice QCD investigations of octet baryon magnetic moments.

Acknowledgements

This work was supported by the Australian Research Council.

References

1. *Lattice '98,* Proc. of the International Symposium, edited by T. DeGrand, C. DeTar, R. Sugar, and D. Toussaint, Boulder, CO, 1998, Nucl. Phys. B (Proc. Suppl.), (1998).
2. S. Aoki et al. (CP-PACS), hep-lat/9902018.
3. D. B. Leinweber and T. D. Cohen, Phys. Rev. D **47**, 2147 (1993), hep-lat/9211058; A. W. Thomas, Australian J. Phys. **44**, 173 (1991).
4. M. F. L. Golterman, Acta Phys. Polon. B **25**, 1731 (1994).
5. J. N. Labrenz and S. R. Sharpe, Nucl. Phys. B (Proc. Suppl.) **34**, 335 (1994).
6. E. Jenkins, M. Luke, A. V. Manohar, and M. J. Savage, Phys. Lett. B **302**, 482 (1993), *ibid.* B **388**, E866 (1996).
7. D. B. Leinweber, R. M. Woloshyn, and T. Draper, Phys. Rev. D **43**, 1659 (1991).
8. W. Wilcox, T. Draper, and K. F. Liu, Phys. Rev. D **46**, 1109 (1992).
9. S. J. Dong, K. F. Liu, and A. G. Williams, Phys. Rev. D **58** 074504 (1998).
10. D. B. Leinweber et al., hep-lat/9810005.
11. S. Théberge, A. W. Thomas and G. A. Miller, *Phys. Rev.* D **22** (1980) 2838; D **23** (1981) 2106(E).
12. A. W. Thomas, in *Advances in Nuclear Physics, Vol. 13 (1984) 1*, edited by J. Negele and E. Vogt, New York, 1984, Plenum Press; G. A. Miller, in *Int. Rev. Nucl. Phys.* **2** (1984) 90.
13. D. H. Lu, A. W. Thomas, and A. G. Williams, Phys. Rev. C **57**, 2628 (1998).
14. S. Théberge and A. W. Thomas, Nucl. Phys. A **393** (1983) 252.
15. S. Aoki et al., CP-PACS results for quenched QCD spectrum with the Wilson action, in *Lattice '97,* Proc. of XVth International Symposium on Lattice Field Theory, edited by C. T. H. Davies et al., Edinburgh, UK 1997, Nucl. Phys. B (Proc. Suppl.), (1998), hep-lat/9710056.

Partial Chiral Restoration at Finite Baryon Density

T. HATSUDA

Physics Department, Kyoto University, Kyoto 606-8502, Japan
E-mail: hatsuda@ruby.scphys.kyoto-u.ac.jp

Spectral enhancement near $2m_\pi$ threshold in the $I=J=0$ channel in nuclei is shown to be a distinct signal of the partial restoration of chiral symmetry. The relevance of this phenomenon with possible detection of $2\pi^0$ and 2γ in hadron-nucleus and photo-nucleus reactions is discussed.

1 Introduction

Low density theorem in QCD and various model calculations suggest that partial restoration of chiral symmetry takes place in nuclear medium.[1] If the quark condensate decreases substantially in nuclear matter, fluctuation of the condensate becomes large in accordance with the general wisdom of statistical physics.[2] This implies that there is a softening of a collective excitation in the scalar-isoscalar channel with the following physical consequences[3]: (i) partial degeneracy of the scalar-isoscalar particle (traditionally called the σ-meson) with the pion, and (ii) decrease of the decay width of σ due to the phase space suppression caused by (i) in the reaction $\sigma \rightarrow 2\pi$. Although σ appears only as a broad resonance in the vacuum and is hard to be distinguished from the background[4], it was suggested that it may appear as a sharp resonance at finite temperature (T) because of (i) and (ii) discussed above.[3,5,6]

This talk is based on our recent work[7] in which we demonstrate that the spectral enhancement associated with the partial chiral restoration takes place also at finite baryon density close to $\rho_0 = 0.17\text{fm}^{-3}$. As possible experiments to detect this softening, we will discuss the production of the neutral-dipion ($2\pi^0$) and diphoton (2γ) in reactions with heavy nuclear targets. We also mention the relevance of the softening and the recent measurement of the near-threshold $\pi^+\pi^-$ production in π^+-nucleus reactions.[8]

2 Basic idea

Let us first describe the general aspects of the spectral enhancement near the two-pion threshold. Consider the propagator of the σ-meson at rest in the medium : $D_\sigma^{-1}(\omega) = \omega^2 - m_\sigma^2 - \Sigma_\sigma(\omega;\rho)$, where m_σ is the mass of σ in the tree-level, and $\Sigma_\sigma(\omega;\rho)$ is the loop corrections in the vacuum as well as in the

Table 1: Parameters for m_σ^{peak} = 550, 750 MeV taken from ref.6.

	m_σ^{peak} (MeV)	$\sqrt{-\mu^2}$ (MeV)	$\lambda/4\pi$	$h^{1/3}$ (MeV)
(I)	550	284	5.81	123
(II)	750	375	9.71	124

medium. The corresponding spectral function reads

$$\rho_\sigma(\omega) = -\frac{1}{\pi}\frac{\mathrm{Im}\Sigma_\sigma}{(\omega^2 - m_\sigma^2 - \mathrm{Re}\Sigma_\sigma)^2 + (\mathrm{Im}\Sigma_\sigma)^2}. \tag{1}$$

Near the two-pion threshold, the phase space factor gives $\mathrm{Im}\Sigma_\sigma \propto \theta(\omega - 2m_\pi)\ (1 - 4m_\pi^2/\omega^2)^{1/2}$ in the one-loop order. On the other hand, partial restoration of chiral symmetry indicates that m_σ^* ("effective mass" of σ defined by $\mathrm{Re}D_\sigma^{-1}(\omega = m_\sigma^*) = 0$) approaches to m_π. Therefore, there exists a density ρ_c at which $\mathrm{Re}D_\sigma^{-1}(\omega = 2m_\pi)$ vanishes even before the complete σ-π degeneracy takes place. At $\rho = \rho_c$, the spectral function is solely dictated by the imaginary part of the self-energy;

$$\rho_\sigma(\omega \simeq 2m_\pi) = -[\pi\ \mathrm{Im}\Sigma_\sigma]^{-1} \propto \theta(\omega - 2m_\pi)\ (1 - \frac{4m_\pi^2}{\omega^2})^{-1/2}. \tag{2}$$

This implies that, even if there is no sharp resonance in the scalar channel in the vacuum, there arises a mild (integrable) singularity just above the threshold in the medium. This is a general phenomenon associated with the partial restoration of chiral symmetry.

3 Linear sigma model

Let us now evaluate $\rho_\sigma(\omega)$ in a *toy* model, namely the SU(2) linear σ-model:

$$\mathcal{L} = \frac{1}{4}\mathrm{tr}[\partial M \partial M^\dagger - \mu^2 M M^\dagger - \frac{2\lambda}{4!}(MM^\dagger)^2 - h(M + M^\dagger)], \tag{3}$$

where tr is for the flavor index and $M = \sigma + i\vec{\tau}\cdot\vec{\pi}$. Although the model is not a precise low energy representation of QCD,[9] it is known to describe the pion dynamics qualitatively well up to 1GeV.[10]

The coupling constants μ^2, λ and h, which are determined in the vacuum to reproduce f_π = 93 MeV, m_π = 140 MeV as well as the s-wave π-π scattering phase shift in the one-loop order, are recapitulated for two characteristic cases in Table 1 in which m_σ^{peak} is defined as a peak position of $\rho_\sigma(\omega)$.

Let us now consider the σ-meson embedded in nuclear matter. The interaction of M with the nucleon N in SU(2) chiral symmetry is modeled as

$$\mathcal{L}_I(N, M) = -g\chi\bar{N}U_5N - m_0\bar{N}U_5N, \tag{4}$$

where we have used a polar representation $\sigma + i\vec{\tau}\cdot\vec{\pi}\gamma_5 \equiv \chi U_5$ for convenience. The first term in (4) with g is a standard chiral invariant coupling in the linear σ model. The second term with a new parameter m_0 is also chiral invariant and non-singular, but is not usually taken into account in the literatures.

After the dynamical breaking of chiral symmetry in the vacuum ($\langle\sigma\rangle_{\rm vac} \equiv \sigma_0 \neq 0$), the coupling of $\tilde{\sigma} = \sigma - \sigma_0$ and π with N is dictated by $g_s \equiv g$ and $g_p \equiv g_s + m_0/\sigma_0$), respectively. Because of m_0, the standard constraint $g_s = g_p$ can be relaxed without conflicting with chiral symmetry. g_p is constrained as $g_p = m_N/\sigma_0 \simeq m_N/f_\pi = g_{\pi N} = 13.5$, while g_s is independent from g_p and can be treated as a free parameter. This freedom partially avoid the well-known problem that $g_s = g_p$ combined with eq.(3) does not reproduce the known nuclear matter properties in the mean-field level. [11]

4 One-loop estimate

Let us parametrize the chiral condensate in the medium as

$$\langle\sigma\rangle \equiv \sigma_0 \ \Phi(\rho). \tag{5}$$

In the linear density approximation, $\Phi(\rho) = 1 - C\rho/\rho_0$ with $C = (g_s/\sigma_0 m_\sigma^2)\rho_0$. The plausible value of $\Phi(\rho = \rho_0)$ is $0.7 \sim 0.9$. [1]

The one-loop corrections to the self-energy for σ can be read off from the diagrams in Fig.1: $\Sigma_\sigma(\omega;\rho) = \Sigma^A_{\rm vac} + \Sigma^B_{\rm vac} + \Sigma_{MF}(\rho) + \Sigma_{ph}(\rho)$. $\Sigma^A_{\rm vac}$ ($\Sigma^B_{\rm vac}$) corresponds to Fig.1A and Fig.1B , respectively. The explicit form of $\Sigma^{A+B}_{\rm vac}$ is given in Appendix A of ref.6. $\Sigma_{MF}(\rho)$ corresponds to the mean field correction in the nuclear matter (Fig.1C). $\Sigma_{ph}(\rho)$ is a correction from the nuclear particle-hole excitation. We take only the density dependent part in these diagrams and neglect the problematic vacuum-loops of the nucleon. [11]

The leading term in the mean-field part is easily estimated as

$$\Sigma_{MF}(\rho) = \lambda\sigma_0 \left(\langle\sigma\rangle - \sigma_0\right) = -\lambda\sigma_0^2 \left(1 - \Phi(\rho)\right), \tag{6}$$

Leading term in the particle-hole part (Fig.1D) starts from $O(\rho^{5/3})$ and is not more than a few % of $\Sigma_{MF}(\rho)$ at $\rho = \rho_0$. This is in contrast to the case of the pion, where both $\Sigma_{MF}(\rho)$ and $\Sigma_{ph}(\rho)$ are proportional to ρ and cancel with each other due to chiral symmetry. Because of this cancellation, we can

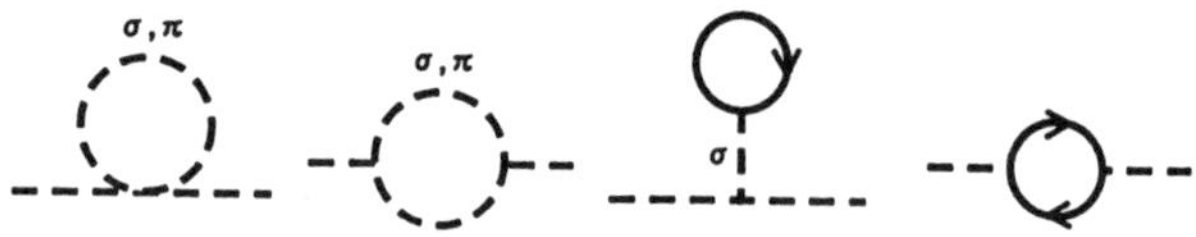

Figure 1: One-loop self energy. The dashed (solid) line denotes σ or π (N).

neglect the two-loop contribution related to the medium modification of the low-momentum pion near the two-pion threshold.

Up to the order we are considering, $\mathrm{Im}\Sigma_\sigma$ solely comes from $\mathrm{Im}\Sigma^B_{\mathrm{vac}}$, since there is no Landau damping and scalar-vector mixing for the σ-meson at rest in nuclear matter. For $2m_\pi \leq \omega \leq 2m_\sigma$,

$$\mathrm{Im}\Sigma_\sigma(\omega;\rho) = \mathrm{Im}\Sigma^B_{\mathrm{vac}} = -\frac{\lambda^2}{96\pi}\sigma_0^2\sqrt{1-\frac{4m_{0\pi}^2}{\omega^2}}. \tag{7}$$

As we have already discussed, the threshold peak is expected to be prominent when $\mathrm{Re}D_\sigma^{-1} = \omega^2 - m_\sigma^2 - \mathrm{Re}\Sigma_\sigma = 0$. In terms of Φ, this condition is rewritten as $\Phi(\rho_c) = 0.74$ (for case (I) in table 1), and 0.76 (for case (II) in Table 1). The numbers in the right hand side are rather insensitive to the parameters in Table 1 as far as the physical quantities in the vacuum are fixed. In the linear density formula $\Phi(\rho) \simeq 1 - 0.2\rho/\rho_0$, we obtain $\rho_c \simeq 1.25\rho_0$. The spectral functions together with $\mathrm{Re}D_\sigma^{-1}(\omega)$ for two cases (I) and (II) are shown in Fig.2. The characteristic enhancements just above the $2m_\pi$ threshold are seen for $\rho \simeq \rho_c$.

We emphasize here that there are two reason behind the enhancement: partial restoration of chiral symmetry ($m_\sigma^* \to m_\pi$), and the cusp structure of $\mathrm{Re}D^{-1}(\omega = 2m_\pi$. Both features can be seen in the lower panels of Fig.2: Although the cusp is not prominent at zero density, it eventually hits the real axis at $\rho = \rho_c$ because $\mathrm{Re}D^{-1}(\omega)$ increases associated with $m_\sigma^* \to 2m_\pi$. This is a general phenomena for systems where the internal symmetry is partially restored in the medium. Another important observation is that, even at densities well below the point where m_σ^* and m_π are degenerate, one can expect a large enhancement of $\rho_\sigma(\omega \simeq 2m_\pi)$.

5 Experiments

To confirm the threshold enhancement associated with the partial chiral restoration, measuring $2\pi^0$ and 2γ with hadron/photon beams off the heavy nuclear targets should be most appropriate. By measuring $\sigma \to 2\pi^0 \to 4\gamma$, [12] one

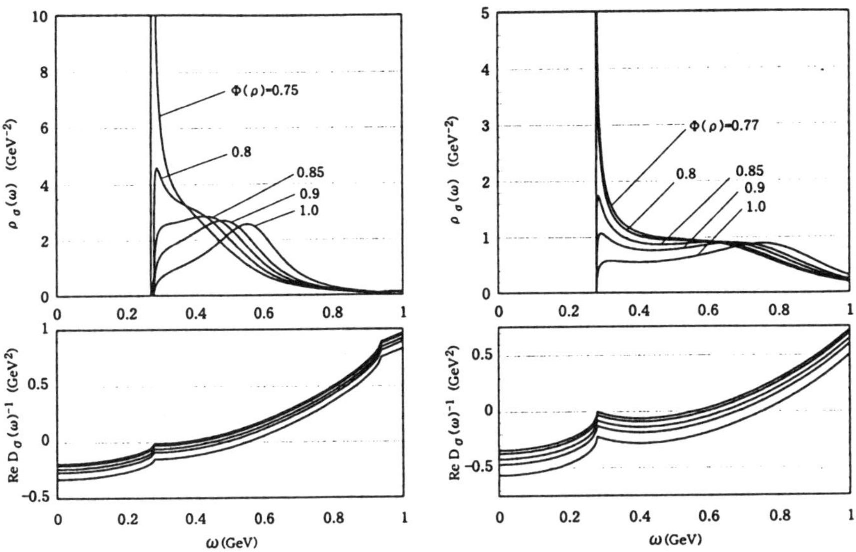

Figure 2: Left: Spectral function for σ and the real part of D_σ^{-1} for several values of $\Phi = \langle\sigma\rangle/\sigma_0$ with $m_\sigma^{peak} = 550$ MeV (case (I) in Table 1). In the lower panel, Φ decreases from bottom to top. Right: Same for $m_\sigma^{peak} = 750$ MeV (case (II) in Table 1.)

can avoid the possible $I = J = 1$ background from the ρ meson inherent in the $\pi^+\pi^-$ measurement. Measuring the electromagnetic decay $\sigma \to 2\gamma$ is also important because of the small final state interactions. [3] There is also a possibility that one can detect dilepton through the scalar-vector mixing in matter: $\sigma \to \gamma^* \to e^+e^-$. [13] To enhance the production cross section of the critical fluctuation in the σ-channel, (d, ^{3}He) reactions may be useful as the case of η and ω mesic nuclei.[14] To cover the spectral function in the range $2m_\pi < \omega < 750$ MeV, the incident kinetic energies of d in the laboratory system is estimated as 1.1GeV $< E <$ 10 GeV.

Recently CHAOS collaboration[8] measured the $\pi^+\pi^\pm$ invariant mass distribution $M^A_{\pi^+\pi^\pm}$ in the reaction $A(\pi^+, \pi^+\pi^\pm)X$ with the mass number A ranging from 2 to 208: They observed that the yield for $M^A_{\pi^+\pi^-}$ near the $2m_\pi$ threshold is close to zero for $A = 2$, but increases dramatically with increasing A. They identified that the $\pi^+\pi^-$ pairs in this range of $M^A_{\pi^+\pi^-}$ is in the $I = J = 0$ state. Attempts so far in hadronic models without considering the partial chiral restoration failed to reproduce this enhancement. [15,16] On the other hand, A dependence of the the invariant mass distribution of the CHAOS data near $2m_\pi$ threshold has close resemblance to our model calculation in Fig.2.

which suggests that this experiment may already provide a hint about how the (partial) restoration of chiral symmetry manifests itself at finite density.

6 Summary

In summary, we have shown that the spectral function in the $I = J = 0$ channel has a large enhancement near the $2m_\pi$ threshold even at nuclear matter density due to the partial chiral restoration. Detection of the dipion and diphoton spectral distribution in the reactions of hadron/photon with heavy nucleus is suitable to confirm the idea of partial chiral restoration in nuclei.

References

1. T. Hatsuda and T. Kunihiro,.Phys. Rep. **247**, 221 (1994). G. E. Brown and M. Rho, Phys. Rep. **269**, 333 (1996).
2. P. C. Hohenberg and B. I. Halperin, Rev. Mod. Phys. **49**, 435 (1977). S. Sachdov, `cond-mat/9705266`.
3. T. Hatsuda and T. Kunihiro, Phys. Lett. **145**(1984) 7; Phys. Rev. Lett. **55**, 158 (1985); Phys. Lett. **B185**, 304 (1987).
4. Particle Data Group, Eur. Phys. J. **C** 363, 390 (1998).
5. H. A. Weldon, Phys. Lett. **B274**, 133 (1992). C. Song and V. Koch, Phys. Lett. **B404**, 1 (1997).
6. S. Chiku and T. Hatsuda, Phys. Rev. **D58**, 076001 (1998).
7. T. Hatsuda, T. Kunihiro and H. Shimizu, Phys. Rev. Lett. **82**, 2840 (1999).
8. CHAOS Collaboration, Phys. Rev. Lett. **77**, 603 (1996).
9. J. Gasser and H. Leutwyler, Ann. Phys. **158**, 142 (1984).
10. L. H. Chan and R. W. Haymaker, Phys. Rev. **D7**, 402 (1973); ibid. **D10**, 4170 (1974).
11. B. D. Serot and J. D. Walecka, Int. J. Mod. Phys. **E6**, 515 (1997).
12. H. Shimizu, in Proceedings of *XV RCNP Osaka International Symposium, Nuclear Physics Frontiers with Electro-weak Probes (FRONTIER 96)*, (Osaka, March 7-8, 1996), p.161.
13. T. Kunihiro, Prog. Theor. Phys. Supplement **120**, 75 (1995).
14. R.S. Hayano, S. Hirenzaki and A. Gillitzer, `nucl-th/9806012`.
15. R. Rapp, J. W. Durso and J. Wambach, Nucl. Phys. **A596**, 436 (1996).
16. H. C. Chiang, E. Oset, and M. J. Vicente-Vacas, Nucl. Phys. **A644**, 77 (1998).

CONSTRUCTION OF THE DUAL GINZBURG-LANDAU THEORY FROM THE LATTICE QCD

H. Suganuma, K. Amemiya, H. Ichie and Y. Koma

Research Center for Nuclear Physics (RCNP), Osaka University, Mihogaoka 10-1, Ibaraki, Osaka 567-0047, Japan
E-mail: suganuma@rcnp.osaka-u.ac.jp

We roughly review the QCD physics and then introduce recent topics on the confinement physics. In the maximally abelian (MA) gauge, the low-energy QCD is abelianized owing to the effective off-diagonal gluon mass $M_{\rm off} \simeq 1.2\rm GeV$ induced by the MA gauge fixing. We demonstrate the construction of the dual Ginzburg-Landau (DGL) theory from the low-energy QCD in the MA gauge in terms of the lattice QCD evidences on infrared abelian dominance and infrared monopole condensation.

1 What is the Quark Confinement Physics ?

About two centuries ago, Dalton and Avogadro introduced the idea of "atoms" and "molecules" into the modern science. Since then, to find out the elementary object becomes one of the most important subject in the modern science.

According to the experimental progress, the "elementary object" has been changed. The atom is divided into a nucleus and electrons, and the nucleus is divided into nucleons... Then, to separate out the more fundamental object becomes one of the most central philosophy in the particle physics.

Nowadays, the nucleon is recognized to consist of quarks and gluons. Nevertheless, the individual quark cannot be separated from the nucleon, but is confined inside the nucleon. We can never observe the raw quark directly, and quarks are strongly bound each other. Then, the Quark Physic is a sort of many-body physics. This is the most different point from the other subjects in the elementary particle physics.

Furthermore, such a curious quark feature is deeply related to the property of the QCD vacuum, which is not a large "empty box" but is regarded as the "condensed matter" with a finite values of the gluon condensate and the quark condensate. Therefore, the Quark Physics requires the total sense of the particle physics, many-body physics and condensed matter physics. (So, the Quark Physics looks like a jigsaw puzzle. We need individual precise knowledge keeping the total balance.)

2 What is the Strategy of the Quark Physics ?

To seek the elementary law is one of the most important aims in the particle physics. However, the aim of the Quark Physics is quite different, because we already know the elementary law of QCD. Surely, QCD tells us the "rule" of the elementary interaction between quarks and gluons, but to solve QCD is another difficult problem.

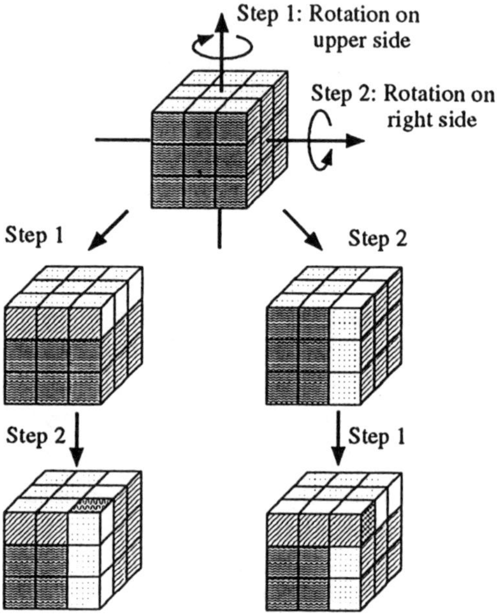

Figure 1: The difficulty and the interest of the Rubik cube originates from non-commutable operations based on the nonabelian nature of the rotational group. The configuration after Step 1 and Step 2 depends on the order of these operations. Can you find a nonabelian nature of QCD in the Rubik cube ?

Then, the QCD physic is similar to the Rubik Cube ! The rule is simple, but to solve is rather difficult. This difficulty is also based on the non-commutable procedures of the "nonabelian" rotational process as shown in Fig.1. Here, a kind of "locality" of the rotational process leads to variety and extremely large number of configurations, which makes this puzzle difficult and interesting. Thus, one may be able to find interesting analogy between the Rubik cube and QCD, which is characterized by the nonabelian local symmetry.

The present strategy for the Quark Physic seems to be the both-side attack from QCD and hadron phenomena. This is similar to the Nuclear Physics strategy. In fact, we already know the experimental data of the nuclear force between nucleons, which is the elementary law of nuclear dynamics. Nevertheless, we need several phenomenological models such as the liquid drop model, the shell model, the cluster model, the interacting boson model to understand the many aspects of the nuclear properties. (No nuclear physicist would say that nuclear physics is to solve the Schrödinger equation of nucleons.) Similarly in the nuclear physics, the linkage of the "both sides", QCD and hadron phenomena, is one of the most important aims in the Quark Physics.

3 Why is the QCD Physics difficult ?

The main difficulty of QCD originates from the nonabelian nature and the strong-coupling nature in the infrared region below 1 GeV. In the ultraviolet region, the QCD coupling becomes weak and then we can use the perturbation technique, where the nonabelian part can be treated as the perturbative interaction. But, in the infrared region, the nonabelian and the strong-coupling nature becomes significant.

To see the difficulty on the nonabelian property of QCD, let us consider the simple electro-magnetic system. In the ordinary electro-magnetism, we can individually consider the partial electro-magnetic field formed by each charge, and the total electro-magnetic field can be obtained by adding these individual solutions. Here, additivity of the solution plays the key role, and this additivity originates from the linearity of the field equation, $\partial_\mu F^{\mu\nu} = j^\nu$, in the electro-magnetism.

On the other hand, owing to the nonabelian nature, the QCD field equation becomes nonlinear as

$$\partial_\mu G^{\mu\nu} + ig[A_\mu, G^{\mu\nu}] = j^\nu, \tag{1}$$

and then it is difficult to solve it even at the classical level, because the solution does not hold the additivity. So, we cannot divide the color-electromagnetic field into each part formed by each quark. Instead, the QCD system is to be analyzed as a whole system. (Furthermore, the QCD vacuum is the nontrivial medium with the gluon condensate.) For the analysis of QCD, the nonabelian nature provides a serious difficulty.

4 Abelianization of QCD - Infrared Abelian Dominance in the Maximally Abelian Gauge

The nonabelian nature is one of the characteristic features of QCD. However, by taking the maximally abelian (MA) gauge in QCD, one can make the nonabelian (off-diagonal) ingredients of QCD inactive for the infrared QCD properties such as quark confinement and chiral-symmetry breaking. We call these phenomena as the infrared abelian dominance in the MA gauge.[1-15]

In the Euclidean QCD, the MA gauge is defined so as to minimize the total amount of off-diagonal gluons, $R_{\rm off} \equiv \int d^4x \sum_{\mu,\alpha} |A_\mu^\alpha(x)|^2$, by the SU($N_c$) gauge transformation. Here, $A_\mu^\alpha(x)$ denotes the off-diagonal gluon in the Cartan decomposition, $A_\mu = \vec{A}_\mu(x) \cdot \vec{H} + A_\mu^\alpha E^\alpha$. In the MA gauge, by removing the off-diagonal gluons, QCD can be well approximated as an abelian gauge theory like the electro-magnetism keeping the essence of the infrared QCD properties. This approximation is called as abelian projection.

Owing to this remarkable feature of MA gauge fixing, the gluon field can be approximated to be abelian as $A_\mu^a(x)T^a \simeq \vec{A}_\mu(x) \cdot \vec{H}$ for the argument on long-distance physics. Accordingly, the field equation of the abelian-projected QCD becomes linear like the Maxwall equation,

$$\partial_\mu F^{\mu\nu} = j^\nu, \quad \partial_\mu \tilde{F}^{\mu\nu} = k^\nu, \qquad (2)$$

with the color-electric current j^μ and the color-magnetic current k^μ. Thus, the additivity on color-electromagnetic fields $F^{\mu\nu}$ works in the abelian-projected QCD in the MA gauge. This is the most attractive point of the MA gauge.

5 Origin of Infrared Abelian Dominance - Mass Generation of Off-diagonal Gluons in the MA Gauge

QCD in the MA gauge exhibits the following interesting properties.

- In the MA gauge, the nonabelian SU(N_c) gauge symmetry is reduced into the abelian U(1)$^{N_c-1}$ gauge symmetry with the global Weyl symmetry.[11]
- In the MA gauge, the off-diagonal gluon amplitude is strongly suppressed. As the result, there appears a strong randomness of the off-diagonal gluon phase in the MA gauge, which is found to be mathematical origin of abelian dominance for confinement.[10,12]
- As a remarkable fact, the color-magnetic monopole appears as the topological object in the MA gauge.[2,3,7,10,12]
- Infrared abelian dominance holds for quark confinement and dynamical chiral-symmetry breaking in the MA gauge.

As the physical origin of infrared abelian dominance, we recently find out effective mass generation of off-diagonal gluons in the MA gauge using the lattice QCD.[10,13] The effective off-diagonal gluons mass is measured as $M_{\rm off} \simeq 1.2$

GeV from the behavior of the off-diagonal gluon propagator $G_{\mu\nu}^{\rm off}$ in the SU(2) lattice QCD as shown in Fig.2.

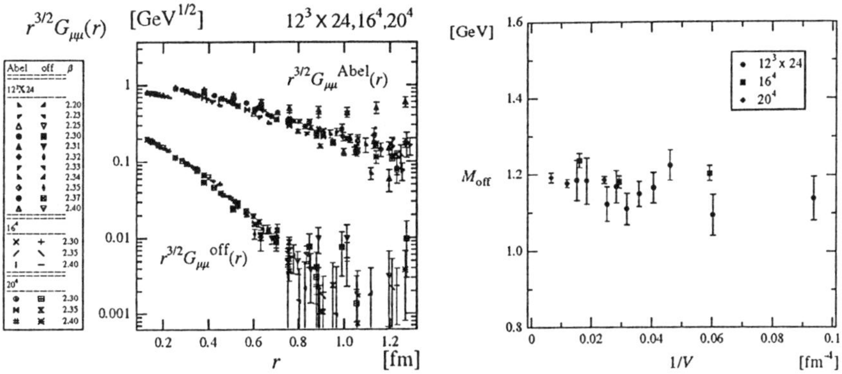

Figure 2: (a) The logarithmic plot of $r^{3/2}G_{\mu\mu}^{\rm off}(r)$ and $r^{3/2}G_{\mu\mu}^{\rm Abel}(r)$. The slope corresponds to the effective mass. (b) The off-diagonal gluon mass $M_{\rm off}$ v.s. inverse lattice volume.

In the MA gauge, the off-diagonal gluon behaves as a massive field with a large mass about 1.2 GeV and then becomes inactive for the long-distance physics, similar to the negligible contribution of the massive weak boson to the long-distance force in the Weinberg-Salam model. This is the essence of the infrared abelian dominance in the MA gauge.

6 Appearance of Color-Magnetic Monopoles in MA gauge in QCD

In the MA gauge, the nonabelian SU(N_c) gauge symmetry is reduced into the abelian $\mathrm{U}(1)^{N_c-1}$ gauge symmetry. Owing to this partial gauge fixing, the color-magnetic monopole appears as the topological object corresponding to the nontrivial homotopy group, $\Pi_2(\mathrm{SU}(N_c)/\mathrm{U}(1)^{N_c-1}) = Z^{N_c-1}$, as was first pointed out by 't Hooft in 1980.[2]

You may suspect the reality of the monopole in QCD. Then, do you know the following fact ? If the Weinberg-Salam model had simpler gauge symmetry $\mathrm{SU}(2)_L$ instead of $\mathrm{SU}(2)_L \times \mathrm{U}(1)$, this theory predicted the existence of the magnetic monopole with the mass about 100 GeV ! So, the experimentalists might make much effort to create and to observe the magnetic monopole in order to get the Nobel prize !! Unfortunately (?), the symmetry of the Weinberg-Salam model is $\mathrm{SU}(2)_L\times$ U(1), and therefore this theory predicted the neutral current instead of magnetic monopoles.

Grand Unified Theory (GUT) actually predicts the quite heavy magnetic monopole (GUT monopole) with the mass about 10^{16} GeV. The appearance of color-magnetic monopoles in the MA gauge in QCD has similar topological origin to the GUT monopole. Actually in the lattice QCD simulation in the MA gauge, there appears a global network of the monopole world-line covering the whole system as shown in Fig.3(a).[10,15]

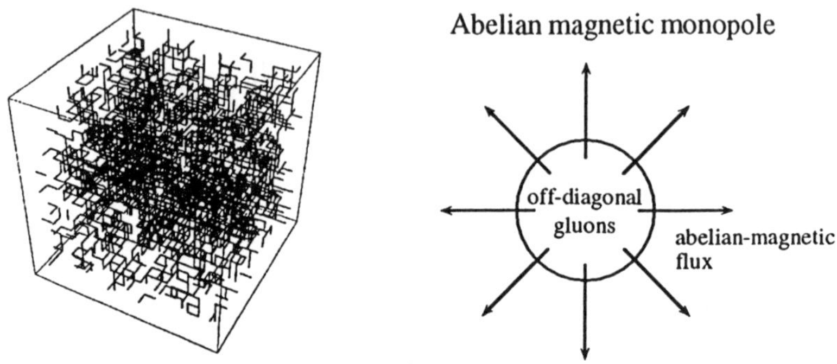

Figure 3: (a) The $\mathbf{R}^3$ projected world-line of the color-magnetic monopole in the MA gauge in the SU(2) lattice QCD with $16^3 \times 4$ and β=2.2 (confinement phase). (b) The structure of the color-magnetic monopole. There remains large off-diagonal gluon component near the monopole center.

Here, even in the MA gauge, where the off-diagonal gluon element is strongly suppressed, around monopoles, there remains large off-diagonal gluon component. This off-diagonal-gluon-rich region around the monopole provides an "intrinsic size" and the structure of the monopole as shown in Fig.3(b), like the 't Hooft-Polyakov monopole.[10,12,15] (The instanton is also an important topological object appearing in the nonabelian gauge manifold, and obviously the instanton needs full SU(2) components for existence. Therefore, instantons tend to appear in the off-diagonal-gluon-rich region around the monopole world-line in the MA gauge, which predicts the local correlation among monopoles, instantons and off-diagonal gluons.[10,15])

Using the lattice QCD, we find that the dual gluon B_μ is massive in the MA gauge, which is an numerical evidence of monopole condensation and the dual Higgs mechanism.[10,15] Then, quark confinement seems to be interpreted with monopole condensation, which was first proposed by Nambu in 1974.[1] Here, a color-singlet monopole appears as the dual Higgs particle,[11] and its fluctuation can be observed as a scalar glueball with $J^{CP} = 0^{++}$.

In fact, the dual superconductor scenario for quark confinement predicts a massive scalar glueball, like the Higgs particle in the Weinberg-Salam model.

From Lattice QCD to Dual Ginzburg-Landau Theory

— Infrared Effective Theory directly based on QCD —

QCD : $SU(3)_c$ Nonabelian Gauge Theory

↓ Maximally Abelian (MA) Gauge Fixing

(partial gauge fixing) [G.'t Hooft . NPB190('81)455]

$U(1)_3 \times U(1)_8$ Abelian Gauge Theory+ QCD-monopole

↓ Lattice QCD studies

⇓

· Abelian Dominance

Only diagonal gluon is relevant for NP-QCD

$\pi_2\,[SU(3)/\{U(1)_3 \times U(1)_8\}] = Z^2_\infty$

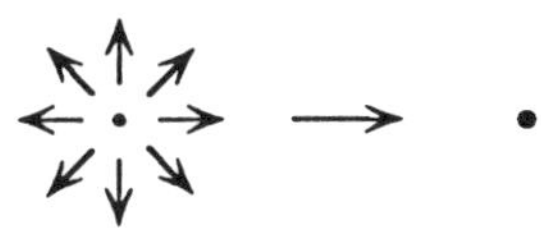

· Monopole Condensation

[cf. GUT - monopole]

Dual Ginzburg-Landau Theory

$$\mathcal{L}_{DGL} = -\frac{1}{2}\mathrm{tr}(\partial_\mu B_\nu - \partial_\nu B_\mu)^2 + \mathrm{tr}[i\hat{\partial}_\mu + gB_\mu, \chi]^\dagger [i\hat{\partial}_\mu + gB_\mu, \chi]$$
$$-\lambda\,\mathrm{tr}(\chi^\dagger\chi - v^2)^2$$

$B_\mu = B^3_\mu T^3_{\rm dual} + B^8_\mu T^8_{\rm dual}$: dual gluon field

$\chi = \chi_\alpha E^{\rm dual}_\alpha$: QCD-monopole field

$g = \frac{4\pi}{e}$: dual gauge coupling constant

λ : coupling of monopole self-interaction

v : imaginary mass of monopole → monopole condensate

Dual Gauge Symmetry is spontaneously broken instead of Gauge Symmetry

Figure 4: Construction of the dual Ginzburg-Landau (DGL) theory from the lattice QCD in the maximally abelian (MA) gauge.

7 Summary and Outlook

To conclude, the lattice QCD in the MA gauge exhibits *infrared abelian dominance* and *infrared monopole condensation*, and therefore the dual Ginzburg-Landau (DGL) theory[3-15] can be constructed as the infrared effective theory directly based on QCD in the MA gauge.

Historically, an interesting interpretation of phenomena has eventually lead to new finding on the particle as follows.

- To interpret the origin of strong nuclear force, Yukawa predicted existence of mesons.
- Nambu and Goldstone proved that spontaneous chiral-symmetry breaking leads to inevitable existence of light pseudoscalar particles (pions).
- From the classification of hadrons, Gell-Mann and Zweig found quarks.
- In near future, from further studies on the quark-confinement mechanism, new important finding may be done on the existence or the feature of the monopole (a scalar glueball).

Acknowledgments

We would like to thank Professor Yoichiro Nambu for his useful suggestions. We acknowledge also Professors Hiroyasu Ejiri and Hiroshi Toki for their continuous warm encouragements at RCNP, Osaka University.

1. Y. Nambu, *Phys. Rev.* **D10**, 4262 (1974).
2. G. 't Hooft, *Nucl. Phys.* **B190**, 455 (1981).
3. H. Suganuma, S. Sasaki and H. Toki, *Nucl. Phys.* **B435**, 207 (1995).
4. H. Suganuma, S. Sasaki, H. Toki, H. Ichie, *PTP Suppl.* **120**, 57 (1995).
5. S. Sasaki, H. Suganuma and H. Toki, *Prog. Theor. Phys.* **94**, 373 (1995); *Phys. Lett.* **B387**, 145 (1996); *Prog. Theor. Phys. Suppl.* **129**, 227 (1997).
6. H.Ichie,H.Suganuma,H.Toki, *Phys.Rev.* **D52**, 2994 (95); **D54**, 3382 (96).
7. M. Fukushima, H. Suganuma, H. Toki, *Phys. Rev.* **D60**, 94504 (1999).
8. S. Umisedo, H. Suganuma and H. Toki, *Phys. Rev.* **D57**, 1605 (1998).
9. H.Monden, H.Ichie, H.Suganuma, H.Toki, *Phys. Rev.* **C57**, 2564 (1998).
10. H.Suganuma, H.Ichie, A.Tanaka, K.Amemiya, *PTP Suppl.***131**, 559 (98).
11. H. Ichie and H. Suganuma, *Phys. Rev.* **D60**, 77501 (1999).
12. H. Ichie and H. Suganuma, *Nucl. Phys.* **B548**, 365 (1999); *Nucl. Phys.* **B**.
13. K. Amemiya and H. Suganuma, *Phys. Rev.* **D60**, 114509 (1999).
14. Y. Koma, H. Suganuma and H. Toki, *Phys. Rev.* **D60**, 74024 (1999).
15. H. Suganuma, K. Amemiya, H. Ichie, A. Tanaka, *Nucl. Phys.* **A** in press.

Primakov Production of the σ Meson

H. Shimizu[a] and T. Matsumura
Department of Physics
Yamagata University
Yamagata 990-8560
Nippon
E-mail: hshimizu@rcnp.osaka-u.ac.jp

The Primakov process provides an interesting method of testing the existence of the σ meson. The feasibility of an experiment with the new method of utilizing the inverse process of $\sigma \to 2\gamma$ is discussed. The experiment is planning to be conducted by using a 2.4 GeV laser-electron photon beam at SPring-8.

1 Introduction

The low-mass scalar meson, called the σ meson, was first predicted in order to account for the intermediate range attractive force in the nucleon-nucleon interaction. There has been, however, no clear experimental evidence of the existence of the σ meson up to now. The strong coupling of the σ meson to the 2π channel makes its width large. As for the S wave $\pi - \pi$ phase shifts, also no clear resonance behavior has been observed irrespective of strong attractive force. On the other hand, from a point of view of chiral symmetry in QCD, the σ meson is thought to be the chiral partner of the π mesons. And the σ and the π mesons get degenerate in mass when chiral symmetry is restored in the Wigner phase. In our real world where chiral symmetry is spontaneously broken, the σ meson acquires its dynamical mass, which is predicted to be about 600 MeV according to a QCD-motivated effective theory[1]. The σ meson is, anyway, such a meson as has a mass of $\simeq$600 MeV with a broad width like $\simeq$500 MeV[2], if it exists. This broadness of the width makes things difficult in search for this meson experimentally[3].

We propose to search for the σ meson via the $\sigma \to 2\gamma$ process. Observation of this process would give a solution for the long-standing puzzle about the existence of the σ meson. However, it is difficult to observe the $\sigma \to 2\gamma$ process since the branching ratio of this process is estimated to be about 10^{-5}. In such a case as a meson decays into a channel with a small branching ratio, the inverse process is of great advantage in the study of the decay channel of the meson. In the present case we can make use of the Primakov effect[4] to produce the σ meson in the inverse process, $2\gamma \to \sigma$. Namely, 2 γs in

[a] present address: Research Center for Nuclear Physics, Osaka University, Osaka 567-0047.

the initial channel are an incident photon and a virtual photon of the electric field of the target nucleus. In this experiment a 2.4 GeV laser-electron photon (LEP) beam is employed and $\pi^+\pi^-$ and $\pi^0\pi^0$ events will be detected in the final channel.

2 Primakov production

The differential cross section for Primakov production of a meson is given by

$$\frac{d\sigma_P}{d\Omega} = \Gamma_{\gamma\gamma}\frac{8\alpha Z^2}{m^3}\frac{\beta^3 E_\gamma^4}{Q^4}|F_{em}(Q)|^2 \sin^2\theta. \quad (1)$$

Here $\alpha = 1/137$ is the fine structure constant, Z the charge number of the target nucleus; E_γ is the incident photon energy; m, β, and θ are the mass, velocity, and the direction of the meson; Q is the momentum transfer to the

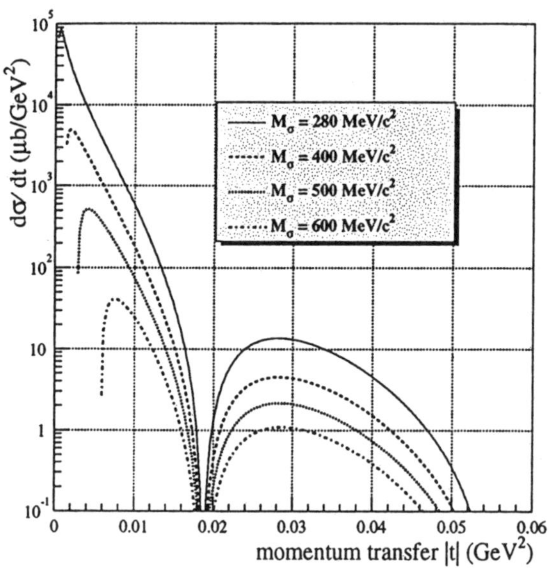

Figure 1: The differential cross section of Primakov production with a Pb target at 2.4 GeV.

nucleus; $F_{em}(Q)$ is the electromagnetic form factor. $\Gamma_{\gamma\gamma}$ denotes the partial decay width to 2γ channel. Eq. (1) shows a strong dependence of E_γ, Q and m on the cross section. The lower limit of Q decreases as E_γ goes up when the meson mass produced in this process is set to be a certain value. Fig. 1 shows the differential cross section of Primakov production of the σ meson for several masses with a Pb target at the incident energy of 2.4 GeV. At this energy, the production cross section goes down by 1 order as the mass increases every 100 MeV. Therefore it is difficult to see the "peak" of σ. But we can investigate the "tail" of σ in detail.

We apply Eq. (1) for σ production with a large mass width. The Breit-Wigner formula is assumed to represent the mass with a non-negligible width. Fig. 2 shows the differential cross section $d\sigma_P/dW$ of σ Primakov production as a function of W for the final state of $2\pi^0$, where W denotes $\sqrt{s}$ of the outgoing 2π system. The similar result has been obtained for the $\pi^+\pi^-$ channel. Here

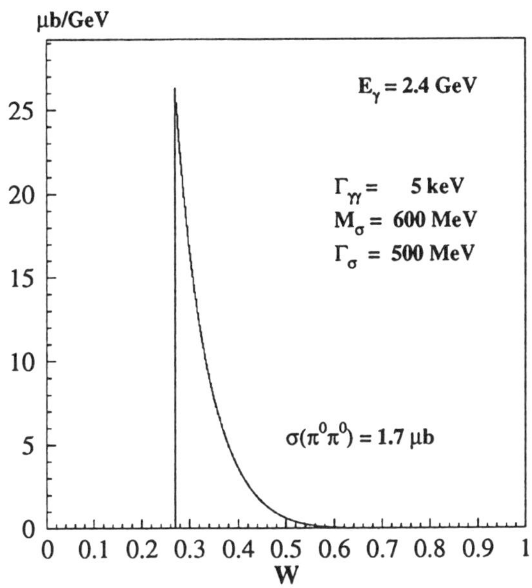

Figure 2: The W dependence of the cross section of Primakov production with a Pb target at 2.4 GeV.

the mass and the width are set to be 600 MeV and 500 MeV, respectively, and the partial decay width of the $\sigma \rightarrow 2\gamma$ process is assumed to be 5 keV for the whole σ mass region.

The behavior of the production cross section is similar to that for the two photon process in e^+e^- collider experiments[5] since Primakov production is also another two-photon process. In the case of the $\gamma\gamma \to \pi\pi$ process in a collider experiment, the two pion invariant mass distribution, $d\sigma_{ee}/dW$, is written by

$$\frac{d\sigma_{ee}}{dW} = L^{ee}_{\gamma\gamma}(W)\sigma_{\gamma\gamma}, \tag{2}$$

where $L^{ee}_{\gamma\gamma}(W)$ denotes the photon-photon luminosity function and $\sigma_{\gamma\gamma}$ is the total cross section for the resonance state R in the reaction $\gamma\gamma \to R \to \pi\pi$. By analogy with the collider case, the differential cross section of Primakov production may be represented as

$$\frac{d\sigma_P}{dW} = L^P_{\gamma}(W)\sigma_{\gamma\gamma}, \tag{3}$$

where $L^P_\gamma(W)$ is a single photon luminosity function which reflects the electric field of the target. Thus we can make a connection between the differential cross section of Primakov production and $\sigma_{\gamma\gamma}$. The photon field of the target nucleus appears as a superposition of almost real transverse photons in the two photon center of momentum system. Namely, the collision of two real photons is realized through the Primakov process. This situation is much clear than in the case of the two photon interaction in e^+e^- collider experiments.

According to a selection rule governing the decay of a meson into two photons, the $J = 1$ state cannot be produced as a resonance from the initial state of two photons. This is an advantage in search for the σ meson through the $\pi^+\pi^-$ channel, which is not appropriate in the case of the hadron-hadron interaction since there is a large background coming from $\rho^0 \to \pi^+\pi^-$ decay.

3 Experiment

In the $\pi^+\pi^-$ measurement, the Laser-Electron Photon Spectrometer (LEPS) system will be utilized. The LEPS comprises a dipole magnet placed just after the target, drift chambers, SSD, and TOF counters. This detector system was designed so as to cover the forward angular region. And the geometrical acceptance for $\pi^+\pi^-$ events in Primakov production is estimated to be 0.98. However, a huge number of e^+e^- events have to be rejected. This will be done with a horizontal bar of plastic scintillator, first of all. Consequently the acceptance goes down to 0.62, which is still good enough. Second, a small gas Cherenkov counter will be placed just after the target and another larger gas Cherenkov counter will also be installed after the magnet to reject e^+e^- events.

For the $\pi^0\pi^0$ channel, two separated forward γ counters will be employed. The first forward counter covers the polar angle from 5° to 30° with full azimuthal angles. The second forward counter will be placed downstream in order to detect γs at the angle down to $1°$. In this experimental setup, the energy and position information of γs emitted into the region from $4°$ to $5°$ cannot be obtained, but the total acceptance is estimated to be 0.47, which is acceptable. If we employ PWO crystals for the γ counters we expect the mass resolution of $2\pi^0$ to be about 5% for the mass region, $300MeV < M_{4\gamma} < 500MeV$.

4 Summary

We are proposing an experiment of Primakov production of the σ meson using a LEP beam at SPring-8. In this experiment, the $\pi\pi$ invariant mass distributions will be measured at $|t| < 0.01GeV^2$. The LEPS system will be employed to detect $\pi^+\pi^-$ events. And for the $\pi^0\pi^0$ channel, two forward γ detectors will be used. This is the first experiment trying to find the signal of the σ meson in the $\sigma \to 2\gamma$ channel with a fixed target. At the same time this is the first experiment to investigate the $\gamma\gamma \to \pi\pi$ process through Primakov production.

Acknowledgment

We acknowledge valuable discussion on the Primakov effect with K. Imai and H. Yamazaki. We would like to thank T. Kunihiro and T. Hatsuda for their useful comments from a theoretical point of view.

References

1. T. Hatsuda and T. Kunihiro, *Phys. Rev. Lett.* **55**, 158 (1985); *Phys. Rep.* **247**, 221 (1994).
2. R. Kaminski et al., *Phys. Rev.* D **50**, 3145 (1994).
3. D. Alde etal., *Phys. Lett.* B **397**, 350 (1997).
4. H. Primakov, *Phys. Rev.* **81**, 899 (1951).
5. H. Marsiske et al., *Phys. Rev.* D **41**, 3324 (1990); J. Boyer et al., *Phys. Rev.* D **42**, 1350 (1990).

Some aspects of the ϕ photoproduction at a SPring-8 energy region

A.I. Titov [1,2a], T.-S.H. Lee [3b], H. Toki [1c], and O. Streltsova [2]
[1] *Research Center for Nuclear Physics, Osaka University, Osaka 567-0047, Japan*
[2] *Bogoliubov Laboratory of Theoretical Physics, JINR, Dubna 141980, Russia*
[3] *Physics Division, Argonne National Laboratory, Argonne,IL 60439, USA*

Abstract

The structure of the ϕ photoproduction amplitude in the $\sqrt{s} \sim 2-5$ GeV region is analyzed based on Pomeron-exchange and meson-exchange mechanisms. The differences between these two competing mechanisms are shown to have profound effects on various density matrices which can be used to calculate the cross sections as well as various single and double polarization observables.

Traditionally, the $\gamma p \to \phi p$ reaction is considered as a tool to study the Pomeron exchange dynamics, implying that (i) the vector meson dominance model [1,2] is valid, and (ii) other hadronic mechanisms are suppressed by OZI rule. Together with an examination of the conventional Pomeron trajectory with an intercept $\alpha(0) \sim 1.08$ [1,2,3], it has been suggested that this reaction can be used to probe more exotic processes associated with additional trajectories including the glueball trajectory [4]. In order to learn about the Pomeron structure and possible manifestation of more exotic dynamics, such as the hidden strangeness in proton (cf. [5,6] for references), one should first fix the "non-strange" background determined by the conventional mechanisms. Namely, we need to evaluate as accurately as possible the leading amplitudes due to the pseudo-scalar (π, η)-exchange and the direct ϕ-radiation from the nucleon legs. The η-exchange and direct ϕ-radiation mechanisms are of current interest. In particular, the coupling constant $g_{\eta NN}$ is not well established and it is subject to further investigations. Similarly, our understanding of ϕNN coupling is still limited.

In this talk we present some results of our recent analysis of the ϕ photoproduction amplitudes in a few GeV region with different models for diffractive (Pomeron exchange) amplitudes and predict all spin density matrix elements that are needed to evaluate various observables associated with the K^+K^- angular distributions from the decay of the produced ϕ meson [7,8]. We consider the $(\pi, \eta)-$exchange, and the direct $\phi-$radiation amplitudes using the parameters *predetermined* in previous works constrained by SU(3) symmetry.

[a]E-mail address : atitov@thsun1.jinr.ru
[b]E-mail address : lee@anph09.phy.anl.gov
[c]E-mail address : toki@rcnp.osaka-u.ac.jp

This will allow us to pin down the parameters associated with an additional Pomeron trajectory that will be introduced later.

A microscopic model for vector-meson photo- and electro-production at high energies based on the Pomeron-photon analogy has been proposed by Donnachie and Landshoff [1] (DL-model). It can be shown that the data at high energies can be described by a Pomeron trajectory with an intercept $\alpha(0) \sim 1.08$. This trajectory will be called the P_1 trajectory in this report. It was suggested [4] that a different Pomeron trajectory, as inspired by the glueball ($J^\pi = 0^+$, $M_{\rm gb}^2 \simeq 3$ GeV) predicted by Lattice QCD calculations and a QCD-based Dual Landau-Ginsburg model[9], could dominate the cross section at low energies by having a negative intercept $\alpha(0)$. This Pomeron trajectory, called P_2, will be also considered below. Follow Donnachie and Landshoff, the invariant amplitude can be written as

$$T_{fi}^{\rm P_n} = M_0^{\rm P_n} \varepsilon_{\phi\mu}^* \mathcal{F}_\alpha^{P_n} \Gamma^{P_n \alpha,\mu\nu} \varepsilon_{\gamma\nu}, \tag{1}$$

where $\varepsilon_{\gamma\nu}$ and $\varepsilon_{\phi\mu}$ are respectively the polarization vectors of the photon and ϕ meson, $\mathcal{F}_\alpha^{P_n}$ describes the Pomeron-nucleon vertex, and $\Gamma^{P_n\alpha,\mu\nu}$ is associated with the Pomeron–vector-meson coupling which is related to the $\gamma \to q\bar{q}$ vertex Γ_ν and the $q\bar{q} \to \phi$ vertex V_μ. We follow the DL-model and assume that the P_1-nucleon coupling is photon-like. For the P_2-nucleon vertex, we assume a scalar coupling. Accordingly, we have $\mathcal{F}_\alpha^{P_n} = \bar{u}(p')\, g_\alpha^{P_n} u(p)$, $g_\alpha^{\rm P_1} = \gamma_\alpha$, $g_\alpha^{\rm P_2} = 1$. The Pomeron-vector meson couplings in Eq.(1) read[6]: $\Gamma^{\rm P_1 \alpha,\mu\nu} = g^{\alpha\nu} k^\mu - k^\alpha g^{\mu\nu}$, and $\Gamma^{\rm P_2 \mu\nu} = (k^\mu q^\nu - k \cdot q\, g^{\mu\nu})/M_\phi$, where the transversality conditions are fulfilled. The strength factors $M_0^{\rm P_{1,2}}$ in Eq.(1) take the conventional form of Regge parameterization

$$M_0^{\rm P_i} = C_i\, F(t)\, F_i(s,t)\, e^{-i\frac{\pi}{2}\alpha_i(t)} \left((s - s_i)/s_0\right)^{\alpha_i(t)}, \tag{2}$$

where the trajectories are taken to be

$$\begin{aligned} \alpha_1(t) &= \alpha_1(0) + \alpha_1' t = 1.08 + 0.25\, t, \\ \alpha_2(t) &= -0.75 + 0.25\, t, \quad s_0 = \alpha'^{-1}. \end{aligned} \tag{3}$$

As discussed above, $\alpha_1(t)$ is chosen to fit the data at large s and $\alpha_2(t)$ is inspired by the ($J^\pi = 0^+$,$M_{gb}^2 \sim 3$ GeV^2) glueball, predicted by Lattice QCD calculations and a QCD-based Dual Landau-Ginsburg model[9]. The overall form factor in Eq.(2) is $F(t) = F_\phi(t) \cdot F_N(t)$ with $F_N(t) = (M_N^2 - 2.8t)\,/\,((4M_N^2 - t)(1 - (t/0.7))^2)$ and $F_\phi(t) = \exp[B(t - t_{\rm max})/2]$. The correcting functions $F_{1,2}(s,t)$ for Eq.(2) are $F_i^{-2} = {\rm Tr}\{[\mathcal{F}^i\Gamma^i] \cdot [\mathcal{F}^i\Gamma^i]^\dagger\}/4$. The strength factor C_1 in Eq.(2) is chosen to reproduce $d\sigma/dt|_{\theta=0}$ at large s, where the cross section is entirely due to P_1

trajectory. The strength factor C_2 is chosen to reproduce $d\sigma/dt|_{\theta=0}$ at low energies where all mechanisms are important. The slope B is chosen to reproduce the available data of angular distributions, $d\sigma/dt$. The phase of the Pomeron exchange amplitudes is fixed and controlled by the exponential factors in (2). At $t = 0$, P_1 amplitude is almost purely imaginary and negative, while the P_2 amplitude is complex.

We calculate the (π, η)-exchange by using the standard $\pi(\eta)NN$ and $\phi\gamma\pi(\eta)$ effective Lagrangians [10]. For the πNN coupling we use the standard value $g_{\pi NN} = (4\pi \cdot 14)^{1/2}$. The status of the ηNN coupling is not so clear. The value of $g^2_{\eta NN}/4\pi$ reported in the literatures varies from 0 to about 7. The largest value comes from fits to NN scattering data, using one boson exchange potential models [11]. But many other studies [12] favor a smaller value, $g^2_{\eta NN}/4\pi \sim 0.3 - 1.0$. In our calculation we accept the SU(3) symmetry prediction based on the most recent value of $F/D = 0.575$ [13]. This leads to $g_{\eta NN} = (0.99 \cdot 4\pi)^{1/2}$ which is close to its "overall" averaged value. Later, we will discuss how our prediction of polarization observables can be used to pin down the value of $g_{\eta NN}$. The magnitude of the coupling constant $g_{\phi\gamma\pi(\eta)}$ can be estimated through the decay widths of $\phi \to \gamma\pi$ and $\phi \to \gamma\eta$: $|g_{\phi\gamma\pi}| = 0.14$ and $|g_{\phi\gamma\eta}| = 0.707$. We also employ the SU(3) symmetry to determine their signs. We find that their signs are the same but are opposite to the sign of $g_{\omega\gamma\pi}$, which is found to be positive in the pion photoproduction Since $g_{\eta pp}$ and $g_{\pi^0 pp}$ (here we include the sign due to the matrix element of the isospin factor) have the same sign, we expect a *constructive* interference between π- and η-exchange amplitudes in the $\gamma p \to \phi p$ reaction. The situation changes in $\gamma n \to \phi n$ reaction. In this case the isospin property determines that $g_{\pi^0 nn} (= -g_{\pi^0 pp})$ is of opposite sign of $g_{\eta nn} (= g_{\eta pp})$ and hence the corresponding interference becomes *destructive*.

We evaluate the ϕ direct radiation amplitude by using the following interaction Lagrangians

$$\begin{aligned}
\mathcal{L}_{\gamma NN} &= -e\left(\bar{N}\gamma_\mu \frac{1+\tau_3}{2} N A^\mu - \frac{\kappa_N}{2M_N}\bar{N}\sigma^{\mu\nu} N \partial_\nu A_\mu\right), \\
\mathcal{L}_{\phi NN} &= -g_{\phi NN}\left(\bar{N}\gamma_\mu N\phi^\mu - \frac{\kappa_\phi}{2M_N}\bar{N}\sigma^{\mu\nu} N\partial_\nu\phi_\mu\right),
\end{aligned}$$

with $\kappa_{p(n)} = 1.79(-1.91)$. The resulting direct and crossed amplitudes are real and their signs are controlled by the constants $g_{\phi NN}$ and κ_ϕ. Here we choose $g_{\phi NN} = -0.24$ and $\kappa_\phi = 0.2$, which are obtained from an estimate [14] of the OZI-violation contributions due to $\phi\pi\rho$ couplings and the interaction of ϕ with the kaon cloud of the nucleon. The ϕNN vertices in the direct and crossed terms may be dressed by form factors $F^\phi_s(s)$ and $F^\phi_u(u)$, respectively. But this results in strong violation of gauge invariance with respect to the photon and ϕ meson fields. One can use the gauge invariant prescription of[15] but our result would then have an additional depending on the cutoff parameter. To avoid this problem we use the prescription of[16] and simply set $F_s^\phi = F_u^\phi = 1$. Our prediction of the contribution from the direct ϕ−radiation may therefore be considered as an upper limit.

With the (π, η)-exchange and direct ϕ-radiation amplitudes defined above, our first task is to determine the parameters of Pomeron-exchange amplitudes. We first observe that the meson-exchange and ϕ−radiation amplitudes only have appreciable contributions in the low energy region $\sqrt{s} <$ about 5 GeV. Because of a negative intercept $\alpha(0) = -0.75$, the contribution from the second Pomeron P_2 also drops very fast as energy increases. Consequently, the parameters of the first Pomeron P_1 can be fixed by fitting the cross sections in the high energy region. We find that the data of $d\sigma/dt(\theta = 0)$ up to $\sqrt{s} \sim 100$ GeV can be reproduced by choosing $C_1 \sim 2. - 2.3$ and $s_1 \sim 1.5 - 2$ GeV^2. The uncertainties are due to the quality of the data. In considering the data in the low energy region, we therefore still have some freedom in adjusting the parameters of P_1.

<u>Model A</u>. In this model we consider only P_1, pseudoscalar meson(π, η) exchange amplitudes and the direct ϕ radiation. The model has three parameters: C_1, s_1 and the slope B. The first two are fixed by fitting the energy dependence[17,18,19] of $d\sigma/dt$ at $t = t_{\text{max}}$ $(\theta = 0)$: $C_1 = 2.09$, $s_1 = 1.6$ GeV^2. The slope $B_1 = 1.7$ GeV^{-2} is found by fitting the data of angular distribution of $d\sigma/dt$ at $E_\gamma = 2$ GeV[17].

<u>Model B</u>. This model is obtained by extending Model A to include the P_2 amplitude. The slop parameter B_2 is chosen to be the same as B_1 of P_1 in order to maintain the same fit to the differential cross section $d\sigma/dt$ at $E_\gamma = 2$ GeV. The model then has two additional parameters: s_2 and C_2 for specifying the strength of the P_2 amplitude. In fact, they are not independent: decreasing s_2 leads to some increase of $|C_2|$, therefore for simplicity, we choose $s_2 = 0$. Because P_1 and P_2 amplitudes have different energy dependence, the energy dependence of $d\sigma/dt$ in the low energy region is sensitive to the ratio $\xi = C_2/C_1$. The parameter s_1 obviously also controls the energy dependence in the low energy region. For simplicity, we assume that the threshold behavior of P_1-pomeron amplitude is defined by the natural scale $\sqrt{s_1} = M_N + M_\phi$. The model then only has two parameters C_1 and ξ. We find that the data can be fitted by setting $\xi = -0.55$, $C_1 = 2.34$. We find that this value of the ratio ξ is perhaps the maximum amount of the P_2 amplitude the data can accommodate. With a larger magnitude of ξ, the predicted energy-dependence of the cross section has a shoulder-type behavior at $W \sim 2.5 - 3$ GeV, which might be in

contradiction with the available data.

Model C. This model is obtained by extending Model A to include scalar meson exchange amplitude[8]. We find that this additional amplitude has very small effect on the unpolarized cross sections. The data can be fitted by slightly modifying the P_1 parameters to $C_1 = 2.12$, $s_1 = 1.5$ GeV^2.

Our result for unpolarized cross sections for the model B is shown in Fig.1. we see that both the energy dependence (left panel) and angular distribution (right panel) of the unpolarized differential cross sections for the proton target. can be described (thick solid curves) well. The same conclusion comes for the models A,C. So, all models describe the unpolarized data with approximately equal quality and at the present level of the data accuracy it is difficult to give preference to one of them.

Our predictions for spin-density matrix elements ρ^0_{00} and Reρ^0_{10} which are responsible for K^+K^- angular distribution[20] with unpolarized photon beam at $E_\gamma = 2$ GeV are shown in Fig.2. In Fig.3 we show ρ^1_{00} and Imρ^3_{10} which are responsible for the angular distributions with linear and circular polarized beam, respectively. From the Figs.2,3 one can see that some of spin-density matrix elements are sensitive to the models and the corresponding K^+K^- angular distributions may be used for understanding the role of P_2 amplitudes at low energy.

We now point out that the isotopic effect in ϕ photoproduction can be verified by examining polarization observables. As an example, we show in Fig.4 our prediction of the ϕ decay asymmetry[20]

$$\Sigma_\phi = (\sigma_\parallel - \sigma_\perp)/(\sigma_\parallel + \sigma_\perp), \tag{4}$$

where $\sigma_\parallel(\sigma_\perp)$ are the cross sections induced by photons polarized in the direction parallel (perpendicular) to the $\gamma - \phi$ plane. These two cross sections refer to measurements of the symmetric K^+K^- pairs produced in the ϕ decay. Our prediction of Σ_ϕ at $E_\gamma = 2$ GeV are shown in Fig.4. We find that the predicted isotopic effect at forward angles is largely due to the $\pi - \eta$ interference. Consequently, the predicted energy-dependence at $\theta = 0$ shown in the left panel of Fig.5 can be used to explore the relatively undetermined ηNN coupling constant. To facilitate such a study, we show the dependence of $\Sigma_\phi(\theta = 0)$ on $g_{\eta NN}$ in the right panel of Fig.5. In this calculation, the strength C_2 of the P_2 Pomeron is adjusted to maintain the same fit to the $\gamma p \longrightarrow \phi p$ data as $g_{\eta NN}$ is varied in the calculation. At large angles, the cross sections are dominated by the direct $\phi-$radiation amplitude. Hence a measurement of Σ_ϕ at large angles can be used to study the ϕNN coupling.

In summary, we have developed a model to describe the ϕ photoproduction at low energies, $\sqrt{s} \sim 2 - 5$ GeV. We have found a strong dependence

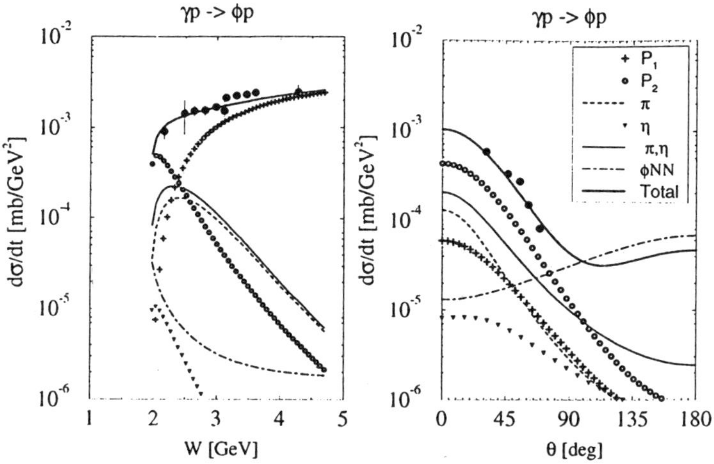

Figure 1: Left panel: differential cross section for the $\gamma p \to p\phi$ reaction at $t = t_{\max}$ ($\theta = 0$) as a function of the total energy distribution $W = \sqrt{s}$. Data are taken from [17,18,19] Right panel: differential cross section at $E_\gamma = 2$ GeV (W=2.15 GeV). Data are taken from [17]

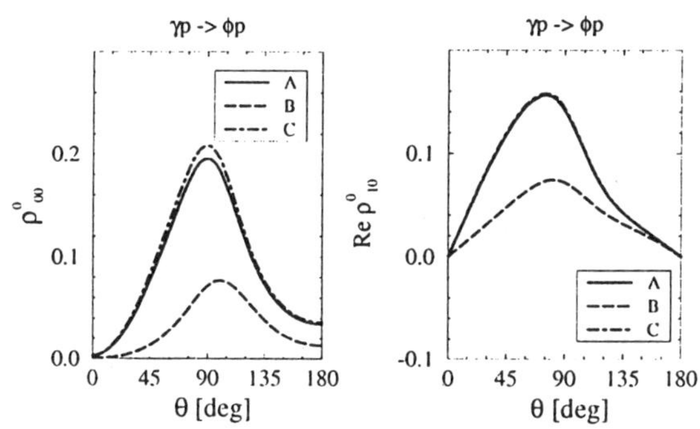

Figure 2: Spin-density matrix element ρ^0_{00} (left panel) and Reρ^0_{10} (right panel) for the different models.

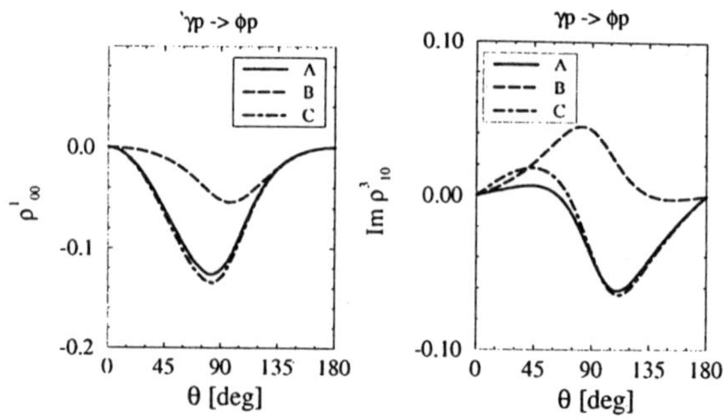

Figure 3: Spin-density matrix element ρ^1_{00} (left panel) and Imρ^3_{10} (right panel) for the different models.

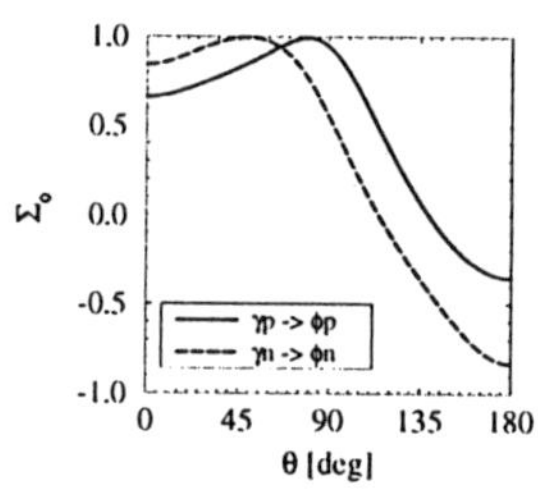

Figure 4: The ϕ decay asymmetry Σ_ϕ, defined in Eq.(4), as function of θ at E_γ=2 GeV.

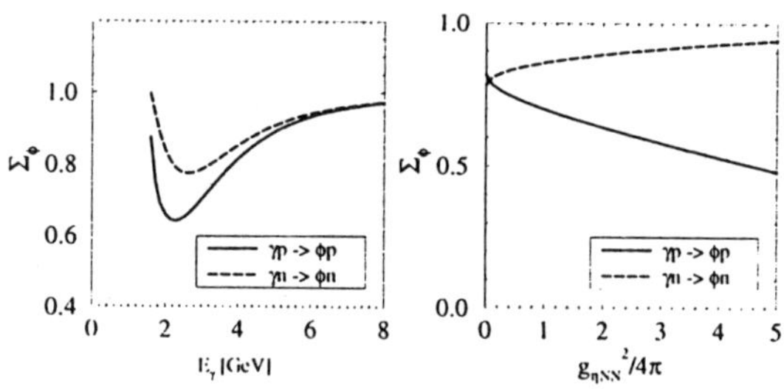

Figure 5: The ϕ decay asymmetry Σ_ϕ, defined in Eq.(4), at $\theta = 0$ as a function of the photon energy (left panel) and as a function of the ηNN coupling constant (right panel).

of the polarization observables on the underground dynamics, indicating the possibility of using the ϕ photoproduction polarization observables to explore P_2 exchange mechanism and the ηNN and ϕNN couplings. Experiments for measuring such observables at LEPS of SPring-8 (Japan) and TJNAF (USA) are highly desirable.

Acknowledgments.

We gratefully acknowledge fruitful discussions with H. Ejiri, M. Fujiwara, T. Hotta, T. Nakano, Y. Oh, E. Oset. A.I.T. appreciates support of the COE professorship program and the warm hospitality of Research Center of Nuclear Physics of Osaka University. This work is also partially supported by the U.S. Department of Energy, Nuclear Physics Division, under Contract NO. W-31-109-ENG-38.

1. A. Donnachie and P. V. Landshoff, Nucl. Phys. **B267**, 690 (1986); Phys. Lett. B **185**, 403 (1987); Phys. Lett. B **296**, 227 (1992);
2. M. A. Pichowsky and T.-S. H. Lee, Phys. Rev. D **56**, 1644 (1997).
3. P. V. Landshoff and O. Nachtmann, Z. Phys. C **35**, 405 (1987).
4. T. Nakano and H. Toki, in Proc. of Intern. Workshop on Exciting Physics with New Accelerator Facilities, SPring-8, Hyogo, 1997, World Scientific Publishing Co. Pte. Ltd.,1998, p.48.
5. J. Ellis, M. Karliner, D.E. Kharzeev, M.G. Sapozhnikov, Phys. Lett. B **353**, 319 (1995).
6. A.I. Titov, Y. Oh, S.N. Yang, Morii, Phys. Rev. C **58**, 2429 (1998).
7. A.I. Titov, T.-S. H. Lee, H. Toki, nucl-th/9812074
8. A.I. Titov, T.-S. H. Lee, H. Toki and O. Streltsova, e-print aps1999mar05-001, Phys. Rev. C (*in print*)
9. H. Suganuma, S. Sasaki and H. Toki, Nucl. Phys. **B435**, 207 (1995).
10. R.A. Williams, Phys. Rev. C **57**, 223 (1998).
11. R. Brockmann and R. Machleidt, Phys. Rev. C **42**, 189 189 (1989).
12. B. Krusche et al., Phys. Rev. Lett. **74**,3736 (1995).
13. F.E. Close and R.D. Roberts, Phys. Lett. B **316**, 165 (1993).
14. U.-G. Meissner, V. Mull, J. Speth, J.W. Van Orden, Phys. Lett. B **408** 381 (1997).
15. H. Haberzettl, Phys. Rev. C **56**, 2041 (1997); nucl-th/9804051.
16. K. Ohta, Phys. Rev. C **40**, 1335 (1989).
17. H. Ballam et al., Phys. Rev. C **7**, 3150 (1974).
18. H. Besh et al., Nucl. Phys. **B70**, 257 (1973).
19. J. Behrend et al., Nucl. Phys. **B144**, 22 (1978).
20. K. Schilling, P. Seybothg, G. Wolf Nucl. Phys. **B15**, 397 (1970).

PHYSICS OF SU(3) BARYONS

A. HOSAKA

Numazu College of Technology, Numazu, Shizuoka 410-8501, Japan
E-mail: hosaka@la.numazu-ct.ac.jp

M. TAKAYAMA, H. TOKI

Research Center for Nuclear Physics (RCNP), Osaka University,
Ibaraki, Osaka 567-0047, Japan

We study properties of baryons of the SU(3) light flavor sector. After a quick review on the ground states we discuss excited states in detail. We emphasize an interesting possibility of deformed baryons for excited states, which provides a remarkably simple and good description for observed mass spectrum. Electromagnetic couplings are then investigated for further study of deformed baryons.

1 Introduction

Baryons contain three valence quarks as their constituents. However, the dynamics how current quarks generate physical baryons is not fully understood due to non-perturbative nature of QCD. In this contribution, we attempt to draw an intuitive picture for baryon structure using effective models.

Up to the present date, various effective models have been invented, e.g., non-relativistic quark models [1], bag models [2], soliton models [3], their hybrid models [4] and so on. Nevertheless, our present understanding might not be fully satisfactory, as remarked by Iachello recently; the baryon system looks very unusual since the $1/2^+$ states appear as the first excited states rather than the $1/2^-$ state [5].

Despite such a situation, we would still like to propose a simple picture especially for excited states. It is based on a phenomenological observation that many SU(3) baryon levels look like rotational bands. As a consequence we imagine that excited baryons are deformed in space. The idea of deformed baryons is not new [6,7]. However, we have realized that it works remarkably well essentially for all spin-flavor multiplets [8]. At this point, we should make a comment on another successful description in the chiral quark model [9]. Our point of view is to consider that the deformation is primarily important as is common to all spin-flavor channels. Effects of chiral mesons can be taken into account after this major dynamics is included appropriately.

2 Ground state baryons

We know empirically that nucleons interact strongly with mesons. The most important meson is the pion which appears as the Nambu-Goldstone boson of spontaneously broken chiral symmetry. Chiral symmetry also governs how they interact with nucleons. Therefore, we can not ignore the role of chiral symmetry and the pion. On the other hand, we also know that quarks exist inside baryons. Indeed, phenomenological quark models are quite successful for the description of baryons [1]. From these observations, it seems that there are two important ingredients for low energy baryon physics: quarks and mesons.

As an example of models which incorporate the two degrees of freedom, we consider the chiral bag model [4]. This is a hybrid model of the quark bag and a mesonic (dominantly pionic) soliton. The bag radius R is the parameter which determines the ratio of the mixing of these two models, interpolating the two limits: the bag model when $R \to \infty$ and the Skyrme model when $R \to 0$. A detailed study was performed for static properties of nucleons [4]. It turned out that an optimal description was obtained at medium size of $R \sim 0.6$ fm. At this point the nucleon consists of a core of quarks and of meson clouds around it. Indeed, both degrees of freedom are important for ground state properties. Through the interactions with the pion, the quarks here look like constituent quarks rather than the current quarks. Hence we should emphasize once again the important role of chiral symmetry. A similar observation was made in the chiral quark soliton model, which is a partially bosonized version of the NJL model [10].

3 Masses of excited baryons and the DOQ model

Let us consider masses of excited baryons. The data points include 49 out of 50 states of three and four stars and several one and two star states [11]. We arrange experimental data as follows: (1) Masses are measured from the ground states of the corresponding spin-flavor states. By doing so we expect that spin-flavor dependence of excited states is dictated by the Gell-Mann-Okubo mass formula. (2) Masses of $^2 8_{MS}$, $^4 8_{MS}$ for positive parity, and of $^4 8_{MS}$ for negative parity are reduced by 200 MeV. Such an energy shift is expected from the spin-spin interactions for the $^4 8_{MS}$ states and from some spatial correlations for $^2 8_{MS}$ positive parity states [7].

Now the resulting mass spectra show up with a very simple systematics for all spin-flavor states as shown in Figs. 1 [8]. They have the following characteristic features: (1) There is degeneracy in different spin j states for a given orbital angular momentum L, implying that spin-orbit interactions are small. (2) $1/2^+$ states appear as the first excited state for all spin-flavor states. These

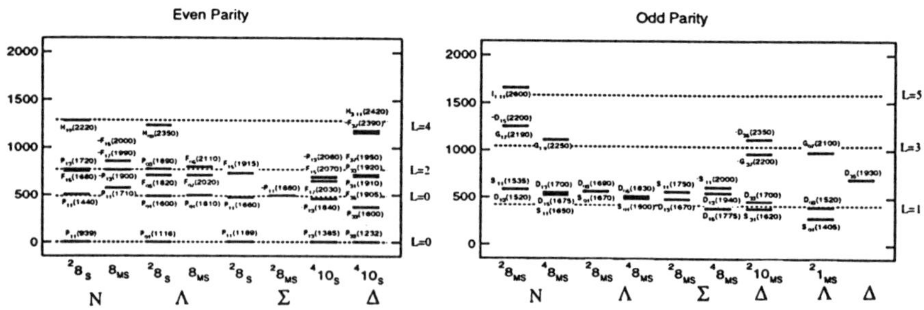

Figure 1: Observed baryon masses as compared with the prediction of the DOQ model with intruder states.

states correspond to the Roper resonance. (3) The level spacings of excited states become larger for higher L. Also the level spacings of the negative parity states are larger than those of the positive parity states.

The second and the third points are typical for rotational bands of deformed nuclei, and motivate us to consider deformed baryons.

In order to explain the above features we consider the deformed oscillator quark (DOQ) model. The model hamiltonian is

$$H_{DOQ} = \sum_{i=1}^{3} \left[\frac{p_i^2}{2m} + \frac{1}{2} m (\omega_x^2 x_i^2 + \omega_y^2 y_i^2 + \omega_z^2 z_i^2) \right] . \tag{1}$$

We ignore interactions of such as gluon and meson exchanges. Thus the dynamical content of (1) is simple such that only the deformation of the oscillator potential, $\omega_x \neq \omega_y \neq \omega_z$, is taken into account.

After removing the center of mass motion, we find an intrinsic energy

$$E_{int}(N_x, N_y, N_z) = (N_x + 1)\omega_x + (N_y + 1)\omega_y + (N_z + 1)\omega_z , \tag{2}$$

where N_x, N_y, N_z are the sum of principal quantum numbers for the internal degrees of freedom for the ρ and λ coordinates. Then we consider a variation with respect to ω's for a given set of (N_x, N_y, N_z). In doing so, we impose the condition of volume conservation in order to prevent the system collapsing[8].

We consider the variation for the prolate deformation along the z-direction when $N_x = N_y = 0, N_z \equiv N$, since the energy of the prolate shape is the lowest among various shapes. We define the deformation parameter $d = \omega_x/\omega_z$ which is the ratio of long to short axes of deformed states. Then the minimum energy and the deformation parameter are given by

$$E_{int}(N) = 3(N+1)^{1/3}\omega\,, \quad d(N) = N+1\,. \tag{3}$$

They give $E_{int}(0) = 3, E_{int}(1) = 3.78, E_{int}(2) = 4.33$ (in units of ω) when $d(0) = 1, d(1) = 2, d(2) = 3$, respectively. These energy values should be compared with $E(1) = 4, E(2) = 5$ of spherical states. Generally, the deformed states have lower energies for excited states $N \geq 1$.

From the deformed states one can construct physical states of good angular momentum by the standard cranking method [12]. For $N = 1(2)$, odd (even) parity states appear as a rotational band, whose energy is given by

$$E_{rot} = E_{int} - \frac{\langle L^2 \rangle}{2I} + \frac{L(L+1)}{2I}\,. \tag{4}$$

Here the momentum of inertia I and the expectation value of the angular momentum fluctuation $\langle L^2 \rangle$ of the intrinsic state are given in literatures [7,8]. The third term of (4) is the rotational kinetic energy of the rigid rotor. The second term is for the energy subtraction due to the angular momentum fluctuation and is crucially important to bring the mass of the $1/2^+$ excited states down close to the observed values.

The rigid rotor formula of (4) may not be good for baryonic system and so we should consider an improved formula. This can be done by considering intruder states from higher N states. Intuitively, this incorporates the softness of the system which stretches as it rotates faster. Let us consider several rotational bands with different N's such that energies are given as $E(N, L)$. Then, picking up the lowest one for a given L, one can construct improved energy spectra. This formula of intruders differs from the rigid rotor one in their high energy behavior as the latter increases as proportional to L.

Now we compare the theoretical prediction with experimental data. The only one parameter in the DOQ model ω is determined by the average of the excitation energies of the first $1/2^+$ excited states in the $^2 8$ multiplet. The resulting value is $\omega = 644$ MeV. The predicted energy levels are shown by the dashed lines in Figs. 1. It is remarkable to see that the predicted energy levels agree with observed levels. On the other hand, in Figs. 1 obviously there are many missing observed states in the region of $L = 4$ and $L = 3$. It would be interesting if those states are confirmed in experiments. Further detailed discussions for individual states are found in Refs. [8].

4 Electromagnetic transitions

For further study of the spatial deformation, we investigate electromagnetic transitions. The discussion below is restricted only to nucleons and their excited states since experimental data are available only for them. Experimentally, electromagnetic couplings to excited states are extracted from the pion photoproduction. In a resonance dominant model, amplitudes are decomposed into the Born terms and resonance contributions. Then we can learn not only the magnitude but also the signs of the resonance contributions relative to the Born terms [13]. In other words, we can determine the sign ϵ of the combination

$$\langle N|H_\pi|N^*\rangle \cdot \langle N^*|H_\gamma|N\rangle \equiv \epsilon|\langle N|H_\pi|N^*\rangle \cdot \langle N^*|H_\gamma|N\rangle| , \tag{5}$$

where H_π and H_γ are the interaction hamiltonian for the pion and photon couplings. Therefore, in a model calculation, both the electromagnetic and strong (pion) couplings have to be calculated simultaneously. Ignorance of the strong coupling part sometimes leads to incorrect results for the sign [14].

We adopt the non-relativistic form for the γNN and πNN couplings:

$$\begin{aligned} H_\gamma &= -e\vec{J}\cdot\vec{A}\,, \quad \vec{J} = \frac{1}{2m}\left(u_f^\dagger(i\overleftarrow{\nabla} - i\overrightarrow{\nabla})u_i + \vec{\nabla}\times(u_f^\dagger\vec{\sigma}u_i)\right)\tau_\mu\,, \\ H_\pi &= \frac{g_{\pi qq}}{2M_N}\vec{\sigma}\cdot\vec{\nabla} \end{aligned} \tag{6}$$

Matrix elements are then computed between the ground state and excited states of $N = 1, 2$, where the effect of the intruders are not considered here. Results for proton helicity amplitudes $A^p_{1/2}$ and $A^p_{3/2}$ are summarized in Table 1, where the DOQ results are compared with the conventional quark model results by Koniuk-Isgur (KI) [14] and with experimental data [11]. We note that the identification of the model states with physical states is only tentative, where possible mixing effects are ignored.

We have found that in many channels, theoretical results of the DOQ and KI are similar to each other within experimental uncertainties; there is only weak dependence on the deformation d. One exception is, however, seen in the coupling of the Roper; even the sign is not correctly reproduced. In the DOQ model, the absolute value becomes better but still it has the wrong sign. This problem was first discussed by Kubota and Ohta who considered relativistic corrections [15].

Here instead of discussing the effect of the relativistic corrections, we consider a limitation of the non-relativistic method. Since the orbital wave functions of both the Roper and the nucleon are the S-state, matrix elements of H_γ and H_π reduce to the same spin matrix element:

$$\langle N^*|H_\gamma|N\gamma\rangle \sim \langle \pi N|H_\pi|N^*\rangle \to \langle L=0|\sigma|L=0\rangle\,.$$

Therefore, we can not change the sign of (5), for it is the square of the same matrix elements. Hence we can not reproduce the sign of the photon-Roper

Table 1: Proton helicity amplitudes in units of $GeV^{-1/2} \times 10^{-3}$. The column indicated as DOQ is for the present results of the DOQ model, while the column indicated as KI is for the conventional results of Koniuk-Isgur.[14]

Proton		$A_p^{1/2}$ DOQ	KI	Exp	$A_p^{3/2}$ DOQ	KI	Exp	
1/2+	$^2S'_S$	109	22.6	-68± 5	–	–	–	$P_{11}(1440)$
	$^2S_{MS}$	-14.6	-15.9	+5±16	–	–	–	$P_{11}(1710)$
3/2+	^2D_S	70.9	111	52±39	-23.5	-36.7	-35±24	$P_{13}(1720)$
5/2+	^2D_S	-3.8	-5.9	-17±10	47.0	73.5	127±12	$F_{15}(1680)$
1/2−	$^2P_{MS}$	151	156	74±11	–	–	–	$S_{11}(1535)$
	$^4P_{MS}$	0	0	48±16	–	–	–	$S_{11}(1650)$
3/2−	$^2P_{MS}$	24.8	25.6	-23±9	138	143	163±8	$D_{13}(1520)$
	$^4P_{MS}$	0	0	-22±13	0	0	0±19	$D_{13}(1700)$
5/2−	$^4P_{MS}$	0	0	19 ± 12	0	–	19 ± 12	$D_{15}(1675)$

coupling as long as we use the non-relativistic hamiltonian. We need more delicate treatment to explain transition amplitudes for more information on the structure of baryons.

5 Summary and discussions

We have studied properties of SU(3) baryons based on effective models. For the ground states, constituent quarks and pions due to spontaneous breakdown of chiral symmetry appear to provide an intuitive picture for baryon structure. On the other hand, for excited states, we have seen that deformed baryons work remarkably well in explaining almost 80 % of the observed excited baryons. It would be interesting if they are indicating the change in the relevant dynamics from low to high energy regions.

We have investigated electromagnetic transitions from the deformed excited states to the ground state. So far, we have not been able to see clear indication of the deformation in the amplitudes we have studied. Moreover, the long-standing discrepancy in the coupling to the Roper can not be resolved within the framework of the DOQ model, or in general in the non-relativistic framework. It would be interesting to further study various transitions including not only electromagnetic but also strong interactions.

One interesting question is the microscopic origin of the deformed structure. We have shown that simple two-body interactions can not generate the strongly deformed states as we have seen [8]. Many-body dynamics including gluons and sea quarks would be responsible for it. For instance, the DGL theory for confinement and chiral symmetry breaking would be an interesting tool for such investigations[16].

References

1. N. Isgur and G. Karl, Phys. Rev. D18 (1978) 4187; *ibid.* D20 (1979) 1191.
2. T. DeGrand, R.L. Jaffe, K. Johnson and J. Kiskis, Phys. Rev. D12 (1975) 2060 and references therein.
3. G.S. Adkins, C.R. Nappi and E. Witten, Nucl. Phys. B228 (1983) 552; I. Zahed and G. E. Brown, Phys. Reports 142 (1986) 1.
4. A. Hosaka and H. Toki, Phys. Reports, 277 (1996) 65.
5. F. Iachello, Talk given at *The US-Japan joint workshop on probing hadron structure with polarized photons*, Hawaii, March (1999).
6. H. Toki, J. Dey and M. Dey, Phys. Lett. B133 (1983) 20.
7. M.V.N. Murthy, M. Brack, R.K. Bhaduri and B.K. Jennings, Z. Phys. C29 (1985) 385 and references therein.
8. A. Hosaka, M. Takayama and H. Toki, Mod. Phys.Lett. A13 (1998) 1699; M. Takayama, H. Toki and A. Hosaka, to appear in Prog. Theor. Phys. (1999).
9. L.Ya. Glozman and D.O. Riska, Phys. Reports 268 (1996) 263; Prog. Part. Nucl. Phys. 36 (1996) 275.
10. M. Wakamatsu and H. Yoshiki, Nucl. Phys. A524 (1991) 561; C.V. Christov et al, Prog. Part. Nucl. Phys. 37 (1996).
11. C. Caso et al. (Particle Data Group), Euro. Phys. J. C3 (1998) 1; R.M. Barnett et al. (Particle Data Group), Phys. Rev. D54 (1996) 1.
12. A. Bohr and B. Mottelson, "Nuclear Structure", Benjamin Inc. (1975).
13. *"Electromagnetic excitation and decay of hadron resonances"*, in *Electromagnetic interactions of hadrons, Ch.2*, edited by A. Donnachie and G. Shaw, Plenum (1978).
14. R. Koniuk and N. Isgur, Phys. Rev. D21 (1980) 1868.
15. T. Kubota and K. Ohta, Phys. Lett. B65, (1976) 374.
16. H. Suganuma, S. Sasaki and H. Toki, Nucl. Phys. B435 (1995) 207; H. Toki, H. Suganuma and S. Sasaki, Nucl. Phys. A577 (1994) 353c.

PROBING THE FEW-BODY SYSTEMS WITH BREMSSTRAHLUNG

N. Kalantar-Nayestanaki
Kernfysisch Versneller Instituut (KVI),
Zernikelaan 25, 9747 AA Groningen, The Netherlands
E-mail: nasser@kvi.nl

A series of bremsstrahlung measurements have been performed with the superconducting cyclotron, AGOR, at KVI. These measurements, on the proton-proton and proton-deuteron systems, aim to investigate the nucleon-nucleon interaction as the nucleons go off their mass shell. Cross sections and analyzing powers have been measured for all possible exit channels of both systems with an incident proton energy of 190 MeV, which is below the particle-production threshold. Some results of the real-photon production are presented for both systems and compared to theoretical predictions.

1 Introduction

Nucleon-nucleon bremsstrahlung is the most fundamental reaction used in studying the off-shell behavior of the nucleon-nucleon interaction as it involves the strong interaction between two nucleons, and the well-known electromagnetic interaction. In addition, the cross section for nucleon-nucleon bremsstrahlung is the basis of all calculations which deal with photon production in heavy-ion collisions. These calculations rely heavily on theoretical predictions of the nucleon-nucleon bremsstrahlung process, which resort to approximations such as the soft-photon approximation that contain no information on the underlying dynamics.[1] Experimentally, the huge elastic background has always hampered the efforts to perform a high-luminosity measurement because of the problems with the singles count rate. This, in turn, will make it difficult to get a clean bremsstrahlung signal out of measurements. Most of the first generation experiments performed in the 60's measured the outgoing protons at larger angles to avoid interference with beam and beam-related background at smaller angles.

Recently, the study of nucleon-nucleon bremsstrahlung below the particle-production threshold has been receiving increasing attention both theoretically and experimentally. The reason for this is the success of modern potential-model calculations on the one hand and the advent of high-precision detection systems on the other. For an extended review of this topic see [2] and the references therein. Polarized-proton beams and high-flux neutron beams have also delivered a boost to the experimental efforts. Cross-section and analyzing-power data from TRIUMF [3] clearly show the need for the inclusion of off-shell effects in the calculations. It should be noted that the TRIUMF data were normalized by a factor 2/3 by the authors of the paper.[3] Without this normalization, the cross-section data overshoots the calculations by about 50%. This normalization problem should, however, be resolved by experiments with higher accuracies. With even higher accuracies achievable nowadays, one can hope to study, in detail, the properties of the off-shell nucleon as it propagates in an intermediate state as well as the non-nucleonic degrees of freedom (such as the Δ–excitation) which can even admix at energies below the pion-production threshold. In this contribution, after a brief review of nucleon-nucleon bremsstrahlung in section 2, the experimental setup used in the KVI measurements is sketched in section 3. The analysis method along with the first results are discussed in section 4. A summary with some outlook are presented in the last section.

2 Nucleon-nucleon bremsstrahlung and the KVI experiment

The dominant ingredients in nucleon-nucleon bremsstrahlung are radiation from external and internal nucleon lines. The higher-order effects which have been investigated in the past few years include meson-exchange currents (MEC), the delta contributions and relativistic contributions. The Δ-isobar contribution in the intermediate state and the MEC effects have been studied in detail by several people.[4,5,6,7] Even though all these contributions are relatively small, precise measurements and selected kinematics should distinguish between models with or without them. For proton-proton bremsstrahlung, MEC is absent in the leading order. For proton-neutron bremsstrahlung, on the contrary, MEC contributions are the largest for higher photon energies. As a result of the absence of the electric MEC's in the proton-proton system, one is more sensitive to higher-order effects in this system.

The experimental situation on the proton-neutron bremsstrahlung is quite different than that of the proton-proton system. Generally speaking, the neutron beams lack the high quality and intensity required for precise measurements. Because of this, the cross sections presented in the literature have been inclusive. In the past, the proton-deuteron system was chosen for the investigation of the proton-neutron bremsstrahlung. The complication with this choice is, of course, the Fermi motion of the nucleons in the deuteron. No good calculations exist for this process. Any calculation attempting to describe the inclusive results which have been published in the literature[8,9] should treat all the possible outgoing channels of this reaction properly. These are: radiative capture, three- and four-body final states.

The KVI bremsstrahlung program with AGOR on few-body systems, set out to study and measure the cross sections and analyzing powers of all the above-mentioned reactions in a series of measurements specifically tuned for nucleon-nucleon bremsstrahlung. The measurements started with the proton-proton system and extended to the proton-deuteron system in which all the different final states were studied. In addition to coplanar geometry, non-coplanar geometries have been measured for a large number of angle combinations. Angular coverage of protons is from 6° to 26° and photons were measured from 60° to 175°. All the measurements are exclusive in the sense that all the outgoing particles were measured. Besides, all the measurements were carried out using polarized protons. In this paper, experimental results concerning the proton-proton system will be presented along with some recent theoretical calculations. In addition, first results on proton-deuteron system will be presented.

3 Experimental apparatus

To detect the outgoing hadrons at forward angles, the Small-Angle Large-Acceptance Detector, SALAD, was designed and built. This detector which was built for a precise measurement of the absolute cross sections and analyzing powers possesses an almost cylindrical symmetry and detects particles between 6° and 19° for the whole azimuthal angular range and proton energies of up to 130 MeV with a resolution of about 10%. The upper limit of the range of the polar angles increases to about 26° for a smaller region of the azimuthal angles. The large solid angle (about 400 msr) is mandated by the very small cross sections of the nucleon-nucleon bremsstrahlung process. To allow the beam free passage, our detector was designed with a central hole. The detector and the performance of various parts of it have been discussed in detail in several publications.[2,10,11,12] We used a liquid hydrogen/deuterium target with very thin synthetic foils for these measurements eliminating the need for a good target-window study.[13]

For the measurement of the position and the energy of the photons, we used the Two-Arm Photon Spectrometer, TAPS,[14] which was placed primarily at backward angles.

4 Results

For the analysis of the experiments, the fact that most measured channels were kinematically overdetermined, was used for background rejection. For instance, in the proton-proton case, five measured angles, namely the polar and azimuthal angles, θ and ϕ, of the two outgoing protons and the polar angle of the photon, θ_γ, and the incident proton energy were used to reconstruct the other parameters. By comparing these parameters to their measured counterparts most of the background left was rejected. In this section, some results are presented for the p+p and p+d systems.

For all the measurements performed, a fraction of the elastically scattered particles were recorded for normalization and calibration purposes. The elastic-scattering cross sections and analyzing powers for the p+p system are well known from literature. The measured elastic-scattering cross sections and analyzing powers were then compared to the calculations to obtain the normalization factor and the polarization of the beam.

As was mentioned earlier, our detection system is capable of measuring a very large number of angle combinations of the outgoing particles. Here, we only show a few percent of the available real-bremsstrahlung phase space. In figure 1, the typical results of the coplanar cross sections and analyzing powers are presented as a function of the polar angle of one of the protons, while the polar angle of the other proton and the polar angle of the photon are fixed. For comparison, a data point from an earlier TRIUMF experiment is also shown in the figure.[15]

Three different types of calculations were performed for our kinematics. The first one uses the Soft-Photon Approximation (SPA) of Liou, Timmermans and Gibson.[16] This calculation is consistent with the low-energy theorem and is relativistic and gauge invariant.. The second set of calculations is done by Martinus, Scholten and Tjon.[7] These relativistic, microscopic calculations are covariant, thereby including the negative-energy states. To see the higher order effects, a calculation called the "nucleonic calculation" is shown in which only the internal and external bremsstrahlung (rescattering) contributions are considered, see the dotted curve, and one which includes MEC, $N\bar{N}$ and Δ contributions as well (solid curve). As can be seen in the figure, there are sizable differences between the data and the calculations for the cross sections and analyzing powers (especially the ones shown in the upper-left panel). A third semi-relativistic, microscopic calculation is done by Eden and Gari which is based on the non-relativistic Ruhrpot potential (dash-dotted curve in Figure 1).[4] The big differences between the two microscopic calculations are not yet understood. In should be noted that Coulomb corrections are expected to be negligible for the kinematics considered here.

There are still many more regions of phase space which are being looked at. In particular, non-coplanar geometries (which have a larger phase-space factor thereby higher statistical accuracies) are interesting. These *first* precision measurements are of special interest for a couple of reasons: 1) one can probe completely different regions of phase space compared to the coplanar geometry; and 2) one obtains more observables (analyzing powers) for these

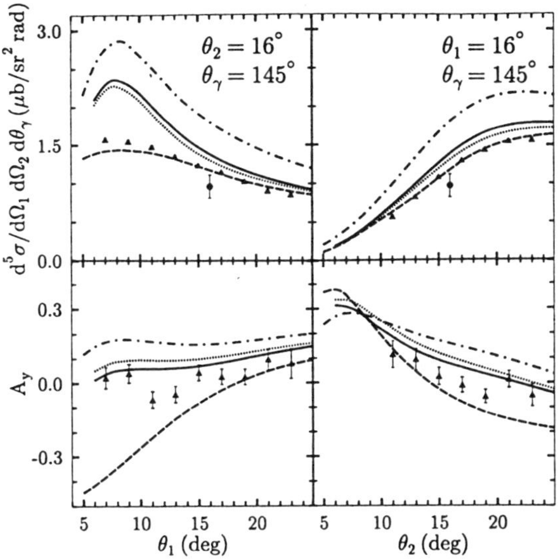

Figure 1: Cross sections and analyzing powers as a function of first (left) and second (right) proton polar angles, obtained in the "supercluster" geometry for $\theta_\gamma = 145°$. The second (first) proton angle is fixed at 16°, respectively. The calculations are SPA (dash), the "nucleonic calculation" (dotted), the full calculation including MEC, $N\bar{N}$ and the Δ as well (solid), and another full calculation by Eden and Gari (dash-dotted). The circle data point is from the TRIUMF experiment at 200 MeV.[15]

geometries.

Finally, the preliminary results for one of the proton-deuteron channels, namely the "coherent" channel in which the outgoing channel consists of a proton, a deuteron and a photon, is presented. Figure 2 shows the cross sections as a function of photon angle for a number of proton-deuteron angle combinations, indicated in the figure. Note that to this date, there are no calculations available in the literature.

5 Summary and outlook

In this contribution, the most complete bremsstrahlung measurements on the p+p and p+d systems have been outlined. All the measurements were done with the new superconducting cyclotron, AGOR, at KVI with a polarized proton beam with an energy of 190 MeV. The detection systems SALAD (specially built for these measurements) and TAPS were utilized for these investigations. A hole in the center of all detection systems allowed us to measure the outgo-

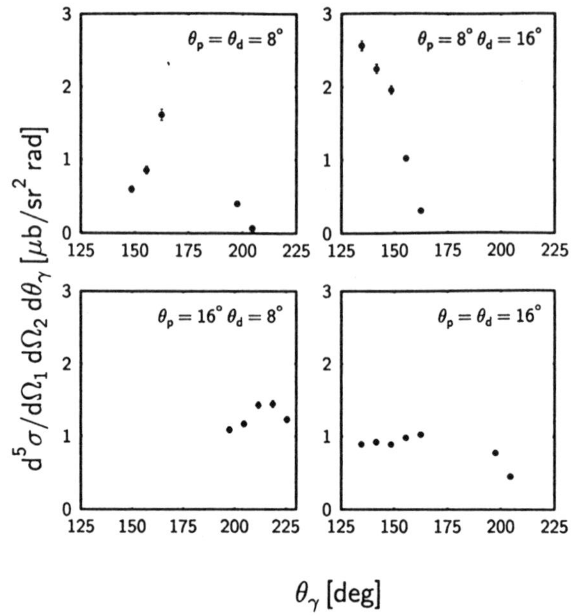

Figure 2: Preliminary *cross sections for the "coherent" proton-deuteron bremsstrahlung as a function of θ_γ for four different angle combinations (given in the figures).*

ing particles at very small angles enlarging the energy of the bremsstrahlung photons detected in coincidence. For the first time, liquid targets contained within very thin foils were used for these systems eliminating the need for a good target-window study. Since all the outgoing channels of these reactions (real and virtual photon production) are kinematically overdetermined, the background levels were reduced to a minimum allowing the extraction of cross sections and analyzing powers which are extremely clean. The elastic cross sections serve as a monitor of the integrity of the system and as a luminosity monitor.

Very high-accuracy cross sections are obtained for the real-photon production which do not agree with the most recent calculations which attempt to include all the possible effects. More calculations including the off-shell electromagnetic form factor of the nucleon are underway.[17] The results for the virtual-photon production which, for the first time, was measured in parallel have already been published.[18] Aside from our measurements, results from other laboratories are emerging on this system.[19,20]

For the proton-deuteron system, data with high accuracy have been obtained. These data should guide the theorists in their calculations for this simple system.

There is still a large part of the phase space which has been measured and is under analysis. Data with high statistical accuracies from non-coplanar geometries are being looked at which will certainly produce interesting physics and new observables, never measured before.

I would like to acknowledge the assistance of all those who helped build and commission SALAD, took part in all the preparations and data taking and are involved in the analysis. Here, I should thank my graduate students H. Huisman and M. Volkerts who are responsible for the analysis of the data presented here.

References

1. F.E. Low, *Phys. Rev.* **110**, 974 (1958).
2. N. Kalantar-Nayestanaki, *Nucl. Phys.* A **631**, 242c (1998).
3. K. Michaelian et al., *Phys. Rev.* D **41**, 2689 (1990).
4. J.A. Eden and M.F. Gari, *Phys. Lett.* B **347**, 187 (1995); *Phys. Rev.* C **53**, 1102 (1996).
5. F. de Jong et al., *Phys. Lett.* B **333**, 1 (1994); F. de Jong et al., *Phys. Rev.* C **51**, 2334 (1995).
6. M. Jetter and H.W. Fearing, *Phys. Rev.* C **51**, 1666 (1995).
7. G. Martinus et al., *Phys. Lett.* B **402**, 7 (1997); *Phys. Rev.* C **56**, 2945 (1997); *Phys. Rev.* C **58**, 686 (1998).
8. J.A. Pinston et al., *Phys. Lett.* B **249**, 402 (1990).
9. J. Clayton et al., *Phys. Rev.* C **45**, 1810 (1992).
10. N. Kalantar-Nayestanaki et al., submitted to *Nucl. Inst. and Meth.*
11. M. Volkerts et al., *Nucl. Inst. and Meth.* A **428**, 432 (1999).
12. S. Schadmand et al., *Nucl. Inst. and Meth.* A **423**, 174 (1999).
13. N. Kalantar-Nayestanaki et al., *Nucl. Inst. and Meth.* A**417**, 215 (1998).
14. O. Schwalb et al., *Nucl. Inst. and Meth.* A **295**, 191 (1990); H. Ströher, *Nucl. Phys. News Int.* Vol. **6**, No. 1, 7 (1996).
15. J.G. Rogers et al., *Phys. Rev.* C **22**, 2512 (1980).
16. M.K. Liou et al., *Phys. Lett.* B **345**, 372 (1995); *Phys. Rev.* C **54**, 1574 (1996); R. Timmermans, private communication.
17. S. Kondratyuk et al., *Phys. Lett.* B **418**, 20 (1998).
18. J.G. Messchendorp et al., *Nucl. Phys.* A **631**, 618c (1998); Phys. Rev. Lett. **82**, 2649 (1999).
19. R. Bilger et al., *Phys. Lett.* B **429**, 195 (1998).
20. K. Yasuda et al., *Phys. Rev. Lett.* in press.

Proton-proton bremsstrahlung at RCNP

K. Yasuda, T. Hotta, M. Kato, Y. Maeda, N. Matsuoka, T. Matsuzuka, Y, Mizuno, M. Nomachi, T. Noro, Y. Sugaya, K. Takahisa, K. Tamura, Y. Yokota, H. P. Yoshida, M. Yoshimura, and Y. Yuasa
Research Center for Nuclear Physics, Osaka University, Ibaraki 567-0047, Japan.

K. Imai, T. Murakami, J. Murata
Department of Physics, Kyoto University, Kyoto 606-8502, Japan.

M. Kawabata, I. Nakagawa, T. Tamae
Laboratory of Nuclear Science, Tohoku University, Sendai 982-0826, Japan.

H. Tsubota
Department of Physics, Tohoku University, Sendai 980-8578, Japan.

H. Akiyoshi
RIKEN, Institute of Physics and Chemical Research, Wako 351-0106, Japan.

Differential cross sections and analyzing powers for the proton-proton bremsstrahlung have been measured at 389 MeV incident energy. A two-arm spectrometer and a liquid hydrogen target system have been used to measure the proton-proton bremsstrahlung events with small background. At the present kinematical conditions, an enhancement of the cross sections due to the Δ current contribution is predicted. The measured cross sections are larger than the theoretical predictions including the Δ current contribution.

1 Introduction

Proton-proton bremsstrahlung ($pp\gamma$) is one of the most fundamental nucleon-nucleon (*NN*) inelastic scattering reactions. In this process only two particles in the final state are strongly interacting and a photon comes from the well understood electromagnetic process. Therefore we can study the hadronic process with less ambiguity than other inelastic *NN* scatterings, for example, the pion production, which includes three strongly interacting particles in the final state.

Since 1970's several measurements for the $pp\gamma$ reaction were performed mainly below the pion production threshold energy in order to investigate the off-shell behavior of the *NN* interaction [1,2]. In recent theoretical studies on the $pp\gamma$ reaction at the intermediate energy, it is predicted that the influence of more elaborate mechanisms on the $pp\gamma$ reaction should be included, such as meson-exchange currents, negative energy states and the Δ current [4,5,6,7]. Some of these calculations predict that the Δ current contribution may increase the cross section about 100% even at about 400 MeV incident proton energy, which is far below the Δ resonance region [5,6,7]. There were, however, no experimental data in this energy region. Precise experimental data were desired to investigate the Δ effect on the $pp\gamma$ process.

In this contribution we report the measurement of the $pp\gamma$ reaction at a proton incident energy of 389 MeV. We measured the two outgoing protons with two magnetic spectrometers. The spectrometers were placed at the most forward angles, which were 26°, on each side of the beam so as to be close to the kinematical condition that gives maximum Δ contribution. In this kinematical condition, the Δ current contribution is predicted to be large, especially at forward photon emission angles [5,7].

2 Experiment

The experiment was performed at the Research Center for Nuclear Physics (RCNP), Osaka University. A 389 MeV proton beam from the RCNP ring cyclotron was delivered to a liquid hydrogen target, which was developed by the Kyushu University group [12]. The target thickness was about 9 mm and the container windows were made of 12.5 μm thick aramid foil. Two outgoing protons were detected with the two-arm spectrometer, Grand Raiden (GR) [13] and Large Acceptance Spectrometer (LAS) [14], both at 26.0°. The four-momentum and invariant mass of the third outgoing particle were calculated with the measured proton momenta. In order to reduce the background caused by the beam halo, a beam halo monitor system which consisted of four plastic scintillators was set at about 100 cm upstream from the target. The beam was tuned to minimize the counting rate of these plastic scintillators. The beam intensity was monitored with a beam line polarimeter placed at upstream of the target, which counted the pp elastic scattering events from an aramid foil target. The beam intensity was about 15 nA. The luminosity was measured by observing the pp elastic scattering on the liquid hydrogen target. These protons were detected by the luminosity monitor, which consisted of plastic scintillators mounted vertically at the angle of 42.2°, at a distance of 33.8 cm from the target.

The momentum acceptances of the GR and LAS were limited to 4% and 20% respectively by a software cut. The $pp\gamma$ events can not be covered by one magnetic field setting, so that we have chosen 14 settings of the two-arm spectrometer in order to measure the angular distribution of the photon emission between 0° and 180°. Figure 1 shows the $pp\gamma$ phase space corresponding to the momenta of the two protons, with the geometrical acceptances of the two-arm spectrometer taken into account. The boxes in this figure correspond

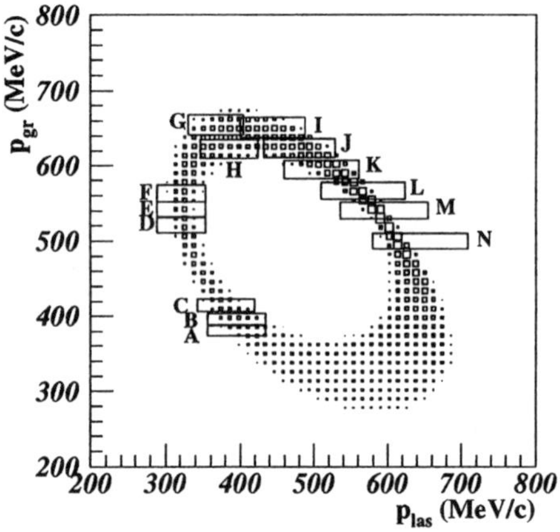

Figure 1: The $pp\gamma$ phase space as a function of the momenta of the two protons at our experimental setting. The boxes in this figure are the momentum acceptances for each magnetic field setting. This is obtained with a phase space calculation.

to the 14 magnetic field settings. The photon is emitted to the same side of the LAS. The limited momentum acceptance of the spectrometer reduces the phase space of the $pp\gamma$ events. This reduction was corrected with a phase space calculation.

At an incident energy of 389 MeV, there are large possible sources of background. One of them is the $pp \to pp\pi^0$ process, which at this energy has a rather large cross section of 48μb [16]. Protons from the pp elastic scattering from the target hydrogen and inelastic scattering from nuclei of the target container foils also cause large backgrounds. Using the two-arm spectrometer and the liquid hydrogen target, we suppressed these background and obtained the $pp\gamma$ events with a good signal-to-noise ratio [17] . The spectrometer has been set to detect protons from the $pp\gamma$ reactions, therefore the protons from the $pp\pi^0$ process do not hit the trigger scintillators because they have very different momenta from those of the $pp\gamma$ process. The pp elastic scattering events are not detected either, unless one of the protons is scattered from the wall and loses energy in the spectrometer. The protons from the A(p,p')

reactions also cause triggers. Two independently scattered protons may enter the spectrometer in an accidental coincidence. We estimated the number of these events from coincidences coming from two different beam bunches and it is subtracted. The A(p,$2p$) quasi elastic events from nuclei of the target container foils also contributed to the background. The magnitude of this contribution was estimated from the "empty target" runs, which were hydrogen gas target runs.

The squared missing mass spectra reconstructed from the detected two protons are shown in Fig. 2. These are obtained from the G and I magnetic field settings which correspond to about 70° and 120° photon emission angle respectively. A clear peak around 0 MeV^2 which corresponds to the $pp\gamma$ process is obtained. The background events due to the accidental coincidence and from the target container foils are also shown in the top figures. The difference spectra, given by open circles with statistical errors, are shown in the bottom figures. The figures show that the background has been estimated correctly and has been subtracted from the data. The experimental data are compared to the result of a Monte Carlo simulation. In this simulation, the events were generated with the phase space distribution, and the energy straggling and multiple scattering in the target and the momentum and angular resolutions of the spectrometer were taken into account. It is clear that the missing mass spectra were well accounted by the simulation.

3 Results and discussion

Figure 3 shows the coplanar $pp\gamma$ differential cross sections and analyzing powers as a function of the reconstructed photon emission angle θ_γ in the laboratory system. The error bars include statistical errors and systematic errors due to the error in the magnetic field setting of the spectrometer and the uncertainty of the missing mass gate. In addition to these errors, there are systematic errors for the absolute normalization of 5.9% in the cross section data. In order to verify the accuracy of the absolute normalization factor, we measured the pp elastic scattering cross section with a CH_2 target of 2.37 mg/cm^2 thickness. The measurements were done in three ways: the GR single-arm measurement, the LAS single-arm measurement, and the GR and LAS two-arm coincidence measurement. In addition, the liquid hydrogen target was also used to measure the pp elastic scattering with the GR and the luminosity monitor. As shown in Fig. 4 the present data for the pp elastic cross section are consistent with the phase shift program, SAID [18], within the errors.

The $pp\gamma$ differential cross sections and analyzing powers of the present work were compared to the theoretical calculations of de Jong *et al.* [5]. In Fig.

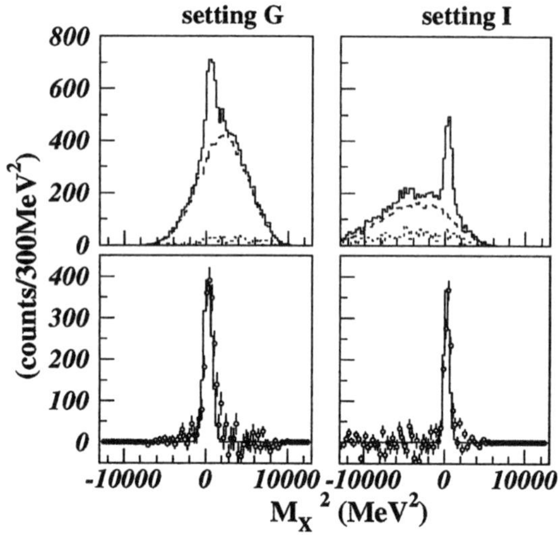

Figure 2: The squared missing mass ($M_X{}^2$) spectra reconstructed from the two observed proton momenta. The results for the G and I magnetic field settings are shown; top: The spectra as taken with the liquid hydrogen target runs (solid), contribution from the accidental coincidence events (dashed), and the "empty target" runs (dotted), bottom: The missing mass spectra after subtracting the accidental coincidence and the background events (open circles with error bars). The solid lines show the result of a Monte Carlo simulation.

3, the solid line corresponds to the calculation including the contribution from the Δ current, and the dashed line is the result of the calculation that only incorporates the nucleonic current contribution. These calculations predict that the Δ effect is seen at forward photon emission angles and the differential cross section increases by a factor two at about $\theta_\gamma = 70°$. The present data favor the calculations including the Δ current, but especially at about $\theta_\gamma = 70°$, where the effect of the Δ current seems to be large, the present result is about 70% larger than the theoretical prediction including the contribution of the Δ current. On the other hand, the present result is not much different from the theoretical predictions at the backward angles. For the analyzing powers, there are discrepancies between the data and the theoretical calculations at about

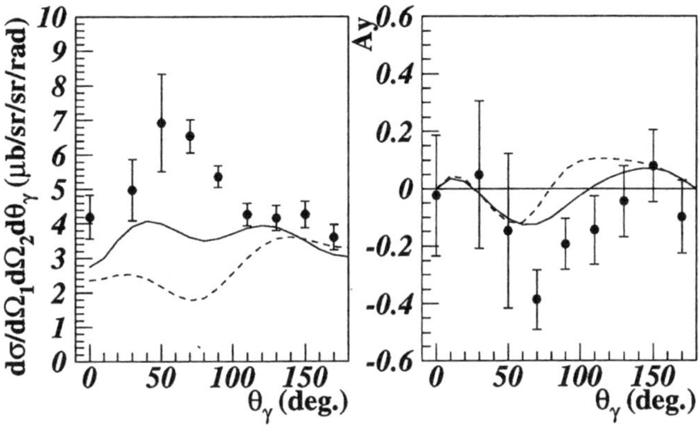

Figure 3: The differential cross sections for the $pp \to pp\gamma$ in the laboratory system. The lines indicate the results of the theoretical calculations including the Δ current (solid) and only the nucleonic current (dashed).

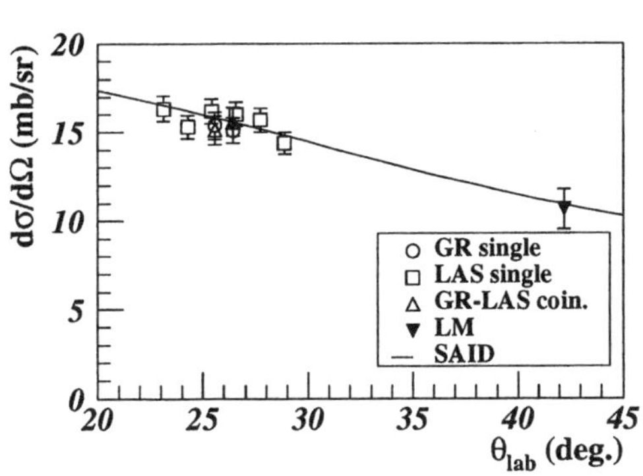

Figure 4: The differential cross sections for the *pp* elastic scattering in the laboratory system. The results of the GR single-arm measurement (open circles), the LAS single-arm measurement (open squares), the GR and LAS coincidence measurement (open triangles), and the measurement with the luminosity monitor (closed triangle) are shown. The line is the result of the *SAID*.

$\theta_\gamma = 70°$. Another result of the calculation including the meson-exchange currents and the Δ current at our kinematical conditions[7] also predict smaller differential cross section than that of the present data. This suggests that there might be contributions of some other mechanisms to the $pp\gamma$ cross section and a further theoretical study is desired.

Acknowledgments

We wish to acknowledge K. Sagara for our use of their liquid hydrogen target system. We also thank K. Nakayama, O. Scholten and H. Toki for useful discussions and theoretical calculations.

References

1. J. V. Jovanovich, in Proceedings of the Second International Conference on Nucleon-nucleon Interactions, Vancouver, 1977, (AIP, New York, 1978), p. 451
2. V. Herrmann, K. Nakayama, O. Scholten, H. Arellano Nucl. Phys. **A 582,** 568 (1995).
3. H. W. Fearing, Phys. Rev. Lett. **81,** 758 (1998).
4. J. A. Eden and M. F. Gari, Phys. Lett. **B 347,** 187 (1995); Phys. Rev. **C 53** 1102 (1996).
5. F. de Jong, K. Nakayama and T.-S. Lee, Phys. Rev. **C 51** 2334 (1995).
6. M. Jetter and H. W. Fearing, Phys. Rev. **C 51,** 1666 (1995).
7. G. H. Martinus, O. Scholten, and J. A. Tjon, Phys. Rev. **C 56,** 2945 (1997); **58** 686 (1998).
8. K. Michaelian *et al.*, Phys. Rev. **D 41,** 2689 (1990).
9. B. V. Przewoski *et al.*, Phys. Rev. **C 45,** 2001 (1992).
10. R. Bilger *et al.*, Phys. Lett. **B 429,** 195 (1998).
11. B. M. K. Nefkens *et al.*, Phys. Rev. **C 19,** 877 (1979).
12. K. Sagara *et al.*, RCNP Annual Report 1995, p. 158
13. M. Fujiwara *et al.*, in Proceedings of the 5th French-Japanese Symposium on Nuclear Physics, Dogashima, 1989, (University of Tokyo, Tokyo, 1989), p. 348.
14. N. Matsuoka *et al.*, RCNP Annual Report 1990, p. 235
15. A. Tamii *et al.*, IEEE Trans. on Nucl. Sci. **43,** 2488 (1996)
16. S. Stanislaus *et al.*, Phys. Rev. **C 41,** R1913 (1990).
17. M. Nomachi *et al.*, Nucl. Phys. **A 629,** 213c (1998).
18. R. A. Arndt, I. I. Strakovsky, and R. L. Workman Phys. Rev. **C 50,** 2731 (1994).

Photons probing dynamics in few-body systems

O. Scholten[a)] and A.Yu. Korchin[a,b)]
a) Kernfysisch Versneller Instituut, Zernikelaan 25, 9747 AA Groningen,The Netherlands
b) National Science Center "Kharkov Institute of Physics and Technology", 310108 Kharkov, Ukraine

In this presentation the use of photons is emphasized to test the dynamics in few-body systems and three different examples are shown. First, the momentum dependence of the Drell-Hearn-Gerasimov integral which is related to the polarized photo-absorbtion cross section; second, proton-proton bremsstrahlung where the photon is sensitive to off-shell effects in the proces; and third, radiative capture of proton on deuterium.

1 Drell-Hearn-Gerasimov integral

Absorption of virtual photons on the nucleon at very-high four-momentum (Q^2) has been proposed as a means to measure the spin content of the nucleon. Since helicity is conserved in the scaling limit, the cross-section difference between parallel and anti-parallel helicities for the photon and the proton should be a measure of the spin carried by the quarks in the proton. This difference, integrated over energy, is called the DHG integral I_{DHG},

$$I_{DHG}(Q^2) = \frac{M^2}{4\pi^2\alpha} \int_{Q^2/2M}^{\infty} \frac{d\nu}{\nu} \sigma^{TT} . \tag{1}$$

where $\sigma^{TT} = 1/2\,(\sigma^T_{1/2} - \sigma^T_{3/2})$. The total absorption cross section for a transverse virtual photon n a proton in a state with total helicity λ is denoted by σ^T_λ where the dependence on ν and Q^2 is suppressed. The energy of the virtual photon in the lab system is $\nu = p \cdot q/M$. At large Q^2 this integral has been expressed in terms of a sumrule by Ellis and Jaffe [1]

$$I_{DHG}(Q^2) \to \frac{2M^2}{Q^2} \int_0^1 g_1(x)dx = \frac{2M^2}{Q^2}\Gamma_1 , \tag{2}$$

where Γ_1 is the moment of g_1. Experiment gives for the proton $\Gamma_1^p \approx 0.126$ at $Q^2 = 10.7$ GeV^2 [10] while the prediction of [1] is $\Gamma_1^p = 0.185$. This discrepancy has been known as the "spin crisis".

For real photons, $Q^2 = 0$, another, rigorous sumrule has been formulated for this integral by Drell and Hearn [2] and independently by Gerasimov [3] $I_{DHG}(Q^2 = 0) = -1/4\,\kappa^2$, where κ is the anomalos magnetic moment of the nucleon. At low $Q^2 \cdot I_{DHG}$ is dominated by contributions from nucleon resonances[4,5].

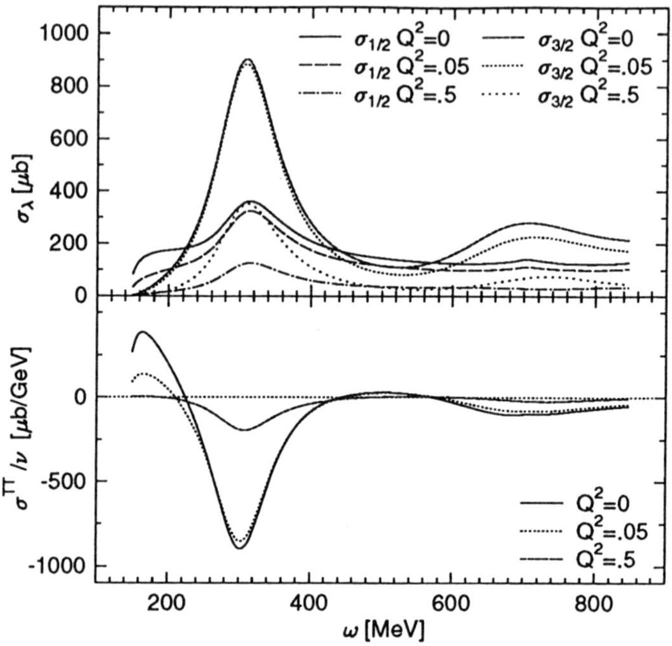

Figure 1: Energy dependence of the photo-absorption cross section for parallel ($\sigma^T_{3/2}$) and anti-parallel ($\sigma^T_{1/2}$) photon and nucleon helicities at different momentum transfer Q^2 as indicated in the figure. The lower panel shows the energy dependence of the integrand in Eq. (1). Q^2 is indicated in GeV^2.

The derivation of these sumrules is based on Lorentz and gauge invariance, crossing symmetry, unitarity and causality. We have therefore investigated[6] the sumrule in the model developed in [7,8,9], which obeys crossing symmetry, unitarity, Lorentz and gauge invariance. It is formulated in terms of meson and nucleon degrees of freedom. The model includes nucleon resonances in an effective-Lagrangian formalism and is based on the K-matrix approach. The kernel is constructed from the direct (s), exchange (u) and meson exchange (t-channel) tree-level amplitudes. In the s- and u-channels all spin -1/2 and -3/2 baryon resonances with masses below 1.7 GeV are included. The coupling parameters have been obtained from a simultaneous fit to pion-nucleon phase shifts, pion-photoproduction multipoles and cross sections for Compton scattering[9]. In the calculation also an off-shell form factor has been included which diminishes the cross section at large energies.

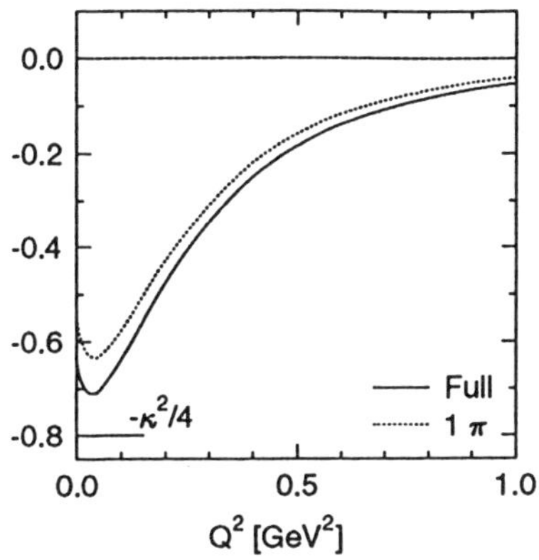

Figure 2: The momentum dependence of the DHG integral. The dotted line shows the contribution of the single-pion production channel. The sumrule value $-\kappa^2/4$ is indicated.

In Fig. (1) the cross sections for the two initial helicity states are plotted versus energy $\omega = (s - M^2)/2M = \nu - Q^2/2M$ since s-channel resonances occur at a value of ω, independent of Q^2. The large peak in the cross section at $\omega \approx 300$ MeV is due to the Δ-resonance while the D_{13}-resonance lies at $\omega \approx 700$ MeV.

At energies below the Δ-resonance the pion-photon seagull term, which ensures gauge invariance for the pion-photoproduction amplitude, gives by far the dominant contribution. It contributes to the helicity-$\frac{1}{2}$ states only and thus it gives a sizable positive contribution to the DHG integral at Q^2=0. Since this contribution to the cross section is inversely proportional to the momentum of the virtual photon, it strongly diminishes when $\sqrt{Q^2} \approx \nu$ causing the decrease (increase of the absolute magnitude) of the DHG integral seen in Fig. (2) at low Q^2. The dominant contribution to the DHG integral originates from the Δ-resonance and is negative in sign. Only at values of Q^2 of the order of the ρ-meson mass the vector-meson dominance form factor starts to cut this Δ-contribution giving rise to a general decrease of the absolute magnitude of the DHG integral seen in Fig. (2). With increasing Q^2 the absolute value of $I_{DHG}(Q^2)$ thus first increases to reach a maximum at Q^2=0.05 GeV^2 after which it strongly decreases clearly showing the sensitivity to the dynamics of the proton. At $\omega_{max} \approx 800$ MeV, the maximum energy where we can apply the present model with confidence, about 80% of the DHG sumrule value (at

$Q^2 = 0$) is reached. The single-pion production contribution to $I_{DHG}(Q^2)$ is 86% at small Q^2, comparable to the value found by Karliner[11].

2 proton-proton bremsstrahlung

In proton-proton bremsstrahlung one is testing the dynamics of an interacting proton-proton system. To be able to extract this dynamics from the data it is imperative to perform accurate theoretical calculations. In the following the present status of the theory is described and the fact that, in spite of its sophistication, still serious discrepancies with data remain. These strongly suggest that the dressing of the nucleon propagator and the nucleon vertex are important.

A fully relativistic calculation is performed, using the model of ref[12]. The contribution from negative-energy states is large, however the net effect appears to be suppressed as a consequence of a low-energy theorem.

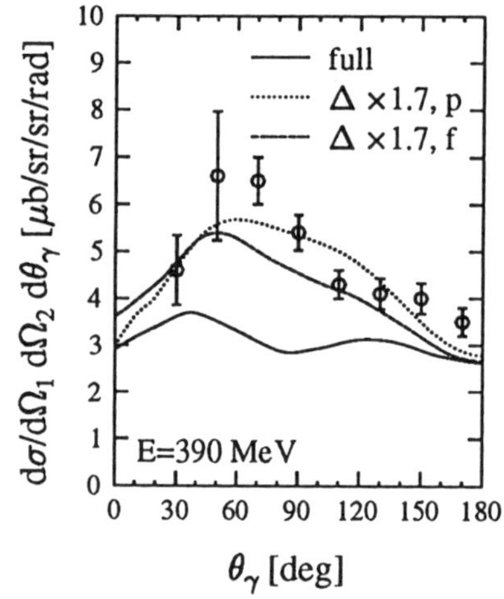

Figure 3: The proton-proton bremsstrahlung cross section as measured at RCNP[13] is compared to the result of a relativistic calculation[12] (drawn curve), a calculation in which the contribution of the Δ-isobar is enhanced by a factor 1.7 (dashed curve) and projecting the latter on positive energy states (dotted curve).

In Fig. (3) the calculation is compared to data from a recent experiment performed at RCNP at a beam energy of 390 MeV. Even though the relativistic calculation includes also contributions from non-nucleonic degrees of freedom such as the Δ-resonance and meson exchange currents, it falls considerably short of predicting the cross section. Since the calculation includes all known mechanisms that contribute, the observed difference can be attributed to off-shell effects. Off-shell effects can originate from a variety of mechanisms. One is a modification of the self-energy of the interacting nucleon and/or a modification of the nucleon-photon vertex due to dressing with meson loops or genuine quark effects. Another could be a modification of the strength of the interaction when coupling to off-mass-shell states. A third could be a different mechanism for emitting photons from the interaction region (so-called contact terms) via a mechanism similar to meson-exchange and Δ-isobar currents. As an example of the latter we have -arbitrarily- increased the contribution of the Δ-isobar current to the matrix element by a factor 1.7 shown as the dotted curve in Fig. (3) for a calculation including only positive-energy states, and as a dashed curve for a full calculation including also negative-energy states. This shows that such an enhancement could possibly be an explanation. Clearly bremsstrahlung photons in proton-proton scattering are a sensitive test of the reaction dynamics.

3 proton-deuteron radiative capture

Proton-deuteron capture is calculated in the model developed in ref.[14] which is explicitly gauge invariant, Lorentz covariant and obeys a low-energy theorem. An important input in the model is the pd^{3}He vertex function, for which recent calculations [15,16,17] of the ^{3}He wave function are used. The amplitude obtained may be called a gauge-invariant relativistic impulse approximation. An interesting aspect of the capture reaction at intermediate energies is the large value of the momenta of the involved particles. For example, at a proton energy of about 200 MeV one probes the wave function of ^{3}He at momenta of about 350 MeV.

In Fig. (4) the angular distribution and the proton analyzing power for the pd$\rightarrow \gamma^3$He at the proton lab energy $T_p = 200$ MeV are shown. The calculation (drawn curve) is in qualitative agreement with experiment. However it underpredicts the cross section at angles larger than 90°. The analyzing power A_y is identically zero since the matrix elements are all real valued.

The effect of initial state interaction can qualitatively be implemented in the calculation by multiplying the vertex function by a phase factor extracted from p-d elastic scattering [14]. The result is given by the dashed curves in Fig. (4). The analyzing power is now in reasonable agreement with the data but the cross section at backward angles is still underpredicted.

In the expression for the amplitude propagators of all particles are taken as free spin-1/2 and spin-1 propagators, which might not be appropriate for composite particles like ^{3}He or d. To improve on this we have investigated

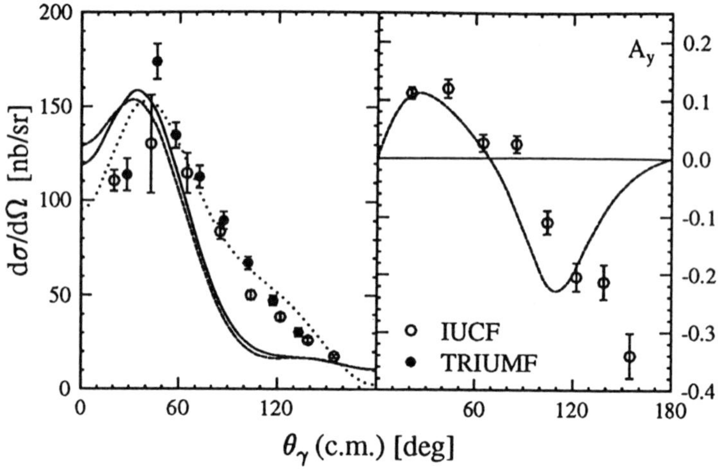

Figure 4: The proton-deuteron capture cross section at 200 MeV. The data are taken from ref.[18,19]. The curves are described in the text.

the effect of including a self-energy in the ^{3}He propagator. An excellent fit to the observed cross sections is obtained as shown by the dotted curve in Fig. (4). The magnitude of the self-energy is specified by the parameter $\alpha(p^2)$. Curiously enough, the energy dependence of $\alpha(p^2)$, as extracted form data at different energies, shows a resonance-like trend with the maximum near the pion production threshold. This suggests that $\alpha(p^2)$ might be related to the virtual production of a pion in the pd$\rightarrow^3$Heγ process.

Again it is shown that the photon acts as a rather sensitive test of the dynamics of the reaction in few-body systems. It is also shown that our picture of these "simple systems" is as yet incomplete.

Acknowledgments

We would like to thank G. Martinus, S. Kondratyuk, D. Van Neck who have closely collaborated in various aspects of this work. O.S. very gratefully acknowledge the hospitality of RCNP, Osaka, where part of this work was initiated. This work is supported by Foundation for Fundamental Research on Matter (FOM) of the Netherlands. A.Yu. Korchin thanks NWO for financial support.

1. J. Ellis and R.L. Jaffe, Phys. Rev. **D9** (1974) 1444; **D10** (1974) 1669.
2. S.D. Drell and A.C. Hearn, Phys. Rev. Lett. **16** (1966) 908.

3. S.B. Gerasimov, Sov. J. Nucl. Phys. **2** (1966) 430.
4. F.E. Close, F.J. Gilman and I. Karliner, Phys. Rev. **D6** (1972) 2533.
5. D. Drechsel, Prog. Part. Nucl. Phys. **34** (1995) 181 and references therein.
6. O. Scholten and A.Yu Korchin, preprint Nucl-Th 9905004
7. V. Pascalutsa and O. Scholten, Nucl. Phys. **A591** (1995) 658.
8. A.Yu. Korchin, O. Scholten and F. de Jong, Phys. Lett. **B402** (1997) 1.
9. A.Yu. Korchin, O. Scholten and R. Timmermans, Phys. Lett. **B438** (1998) 1.
10. J. Ashman et al. Phys. Lett. **B206** (1988) 364; Nucl. Phys. **B328** (1989) .
11. I. Karliner, Phys. Rev. **D7** (1973) 2717.
12. G.H. Martinus, O. Scholten, J.A. Tjon, Phys. Rev. **C58** (1998) 686.
13. K. Yasuda et al., to appear in Phys. Rev. Lett.
14. A. Yu. Korchin, D. Van Neck, M. Waroquier, O. Scholten, A.E.L. Dieperink, Phys. Lett. **B441** (1998) 17; A. Yu. Korchin, D. Van Neck, O. Scholten, M. Waroquier, Phys. Rev. **C59** (1999) 1890.
15. R. Schiavilla, V. R. Pandharipande and R. Wiringa, Nucl. Phys. **A 449** (1986) 219.
16. R. B. Wiringa, V. G. J. Stoks and R. Schiavilla, Phys. Rev. **C 51** (1995) 38.
17. J. L. Forest, V. R. Pandharipande, S. C. Pieper, R. B. Wiringa, R. Schiavilla and A. Arriaga, Phys. Rev. **C 54** (1996) 646; http:www.phy.anl.gov/theory/research/overlap
18. M. J. Pickar, H. J. Karwowski, J. D. Brown et al., Phys. Rev. **C 35** (1987) 37.
19. J. M. Cameron, P. Kitching, W. J. McDonald et al., Nucl. Phys. **A 424** (1984) 549.

PION PRODUCTION IN NUCLEON-NUCLEON COLLISIONS

J. ZŁOMAŃCZUK

RCNP, Osaka University, Japan. On leave of absence from Department of Radiation Sciences, Uppsala University, S-751 21 Uppsala, Sweden

K. FRANSSON, G. FÄLDT, L. GUSTAFSSON, B. HÖISTAD, J. JOHANSON, A. JOHANSSON, T. JOHANSSON, S. KULLANDER, A. KUPSC, P. MARCINIEWSKI, P. SUNDBERG

Department of Radiation Sciences, Uppsala University, S-751 21 Uppsala, Sweden

H. CALÉN, C. EKSTRÖM, R.J.M.Y. RUBER

The Svedberg Laboratory, S-751 21 Uppsala, Sweden

R. BILGER, W. BRODOWSKI, H. CLEMENT, G.J. WAGNER

Physikalisches Institut, Tübingen University, D-72076 Tübingen, Germany

K. KILIAN, W. OELERT

IKP - Forschungszentrum Jülich GmbH, D-52425 Jülich, Germany

B. MOROSOV, A. SUKHANOV, A. ZERNOV

Joint Institute for Nuclear Research Dubna, 101000 Moscow, Russia

J. STEPANIAK

Institute for Nuclear Studies, PL-00681 Warsaw, Poland

A. TUROWIECKI, Z. WILHELMI

Institute of Experimental Physics, Warsaw University, PL-0061 Warsaw, Poland

J. ZABIEROWSKI

Institute for Nuclear Studies, PL-90137 Łódz, Poland

C. WILKIN

Physics & Astronomy Dept., University College London, London WC1E 6BT, U.K.

Measurements of the $pp \to pp\pi^0$ reaction at 310, 320, 340, 360, 400 and 425 MeV, and quasi-free $pn \to pp\pi^-$ production in pd collisions at 320 MeV have been carried out at the PROMICE/WASA facility at CELSIUS. The $pp \to pp\pi^0$ differential cross sections have been parametrised and used to deduce the poorly known σ_{01} total cross section through the relation: $\sigma_{01} = 2\sigma(pn \to pp\pi^-) - \sigma(pp \to pp\pi^0)$. Preliminary results show the σ_{01} and σ_{11} to be of comparable size for centre-of-mass energies between 8 and 40 MeV.

1 Introduction

There are three independent $NN \to NN\pi$ total cross sections, σ_{if}, where the indices i and f represent the isospin of the initial and final NN pairs [1]. The $pp \to pp\pi^0$ reaction involves only σ_{11}, and is well measured in the threshold region [2,3]. σ_{10} may be extracted from data on $pp \to pn\pi^+$ ($\sigma_{10}^{np} + \sigma_{11}$) and $pp \to d\pi^+$ (σ_{10}^{d}), both extensively investigated at IUCF [4]. The third one, σ_{01}, is poorly known at low energies; it has recently been measured for excitation energies of the final pp pair up to 3 MeV [5] and 1.5 MeV [6]. A phase shift analysis of all $NN \to NN\pi$ experimental data below 1 GeV gives σ_{01} values compatible with zero [7], whereas the analysis of the NN inelastic scattering data below 600 MeV [8] makes σ_{01} small, but non-negligible.

At threshold only the final Ss state is allowed in the $pp \to pp\pi^0$ reaction, where S denotes the final NN orbital angular momentum and s that of the pion. Recent measurements have however demonstrated that higher partial waves might be important even at low energies. We have shown that at 310 MeV the Sd state strongly influences the pion angular distributions *via* interference with the Ss state [9]. Polarized-target - polarized-beam experiments at IUCF have established the contribution of the Ps state at several beam energies in a model-independent way [10].

In this contribution we report on new measurements of the quasi-free $pn \to pp\pi^-$ reaction in pd collisions at 320 MeV and of the $pp \to pp\pi^0$ reaction at 310, 320, 340, 360, 400 and 425 MeV carried out at the CELSIUS storage ring in Uppsala. Differential cross sections, $d\sigma/dq$ (q is the two-proton relative momentum in the final state) and $d\sigma/d\Omega_\pi$ are presented. An attempt has been made to describe these distributions in terms of four partial waves: Ss, Ps, Pp and Sd. The corresponding amplitudes are taken to be proportional to $q^L \times k^l$, where L and l represent the angular momentum of the pp pair and the pion, respectively and k is the c.m. pion momentum. In addition there is a strong final-state interaction between the S-state protons.

2 Experiment

The experiment was carried out using an electron-cooled proton beam and the PROMICE-WASA experimental set-up at the CELSIUS storage ring of the The Svedberg Laboratory. The detector system is described elsewhere [11] and only the main components are mentioned here. For the $pp \to pp\pi^0$ reaction, both protons were measured by the forward scintillator hodoscope and the tracker (FD), except for a small hole subtending an angle of $\pm 4.5°$, to accommodate the beam pipe. This is the principal source of detection inefficiency for the two protons, giving a geometrical acceptance of about 70% at 310 MeV.

At higher energies the proton maximum angle is greater than 20°, the largest angle measured in the FD, and the acceptance is consequently reduced. The overall efficiency is diminished additionally by ~20%, mainly through the interaction of protons in the scintillator. Since protons from the π^0 reaction stop in the hodoscope, their energies and angles can be measured with accuracies of 4.5% and 0.5° (r.m.s.) respectively. The π^0 is then reconstructed through the missing mass in the reaction and no use is made in this analysis of the information collected in parallel on the photons arising from the π^0 decay. Careful calibration of the FD elements lead to narrow, nearly background-free, missing mass peaks, with a FWHM ranging from 2.2 MeV/c^2 at 310 MeV to about 7 MeV/c^2 at 425 MeV.

In addition to the two protons detected in the FD, for quasi-free $pn \to pp\pi^-$ production a third charged particle was detected, either in the FD or in the CsI arrays covering angles from 30° to 90° on both sides of the target. The energy resolution for π^- was insufficient for a kinematical reconstruction of the missing mass and the resulting event sample was not background-free. The two major sources of background turned out to be deuteron break-up, with the neutron misidentified as a charged particle, and quasi-free $pp \to pp\pi^0$, with one of the photons converting in the exit windows of the scattering chamber. The first is straightforward to eliminate by requiring the missing mass for the two detected protons to be greater than the sum of the neutron and π^- masses. Background from photon conversion was partly reduced by checking the track direction; the remainder, about 15% of the total $pp\pi^-$ statistics, was simulated and subtracted from the final distributions.

Once a $pd \to pp\pi^-(p)$ event had been selected, it was possible to calculate the 4-momenta of the π^- and spectator proton using the momenta of the fast protons and the π^- angle. This allowed us to calculate the invariant mass of the $pp\pi^-$ system, needed to find the excitation function, and reject events with large energy for the undetected proton. This reduced the contribution from a $pd \to ppp\pi^-$ reaction where all three nucleons are involved.

The quasi-free $pp \to pp\pi^0$ reaction was measured in parallel in order to provide the absolute normalisation of the cross section and allow an extra check of the consistency of the data and the analysis.

3 Analysis

To evaluate the detector acceptance, we require a reasonable phenomenological description of the differential cross sections. In this work we have added an extra state, Sd, to the Ss, Sp, Ps and Pp states used by Handler [12] to parametrise the matrix element of the $pn \to pp\pi^-$ reaction. The two transitions which lead to the Sd state, ${}^3P_2 \to {}^1S_0d$ and ${}^3F_2 \to {}^1S_0d$, may interfere with the ${}^3P_0 \to {}^1S_0s$ and ${}^3S_1 \to {}^1S_0p$, ${}^3D_1 \to {}^1S_0p$ transitions (the latter two being forbidden for the $pp \to pp\pi^0$ reaction). Making several simplifying assumptions on the energy and angular dependence, we arrive at the following expression for the absolute squares of the matrix elements for the $pp \to pp\pi^0$ reaction:

$$\begin{aligned}
f_{Ss} &= \delta\, h^2, \\
f_{Ps} &= \beta_0\, q^2 \left[1 + \beta_1 \left(\mu_q^2 - \tfrac{1}{3}\right)\right], \\
f_{Pp} &= \alpha_0\, k^2 q^2 \left[1 + \alpha_1 \left(\mu_q^2 - \tfrac{1}{3}\right) + \alpha_2 \left(\mu_k^2 - \tfrac{1}{3}\right)\right], \\
f_{Sd} &= \epsilon_0\, h^2 k^4 \left[1 + \epsilon_1 \left(\mu_k^2 - \tfrac{1}{3}\right) + \epsilon_2 \left(\mu_k^4 - \tfrac{1}{5}\right)\right], \\
f_{SsSd} &= \zeta\, h^2 k^2 \left[\mu_k^2 - \tfrac{1}{3}\right].
\end{aligned}$$

Here the function $h^2(q^2)$ describes the two-proton final state interaction and $\mu_q = \cos(\vec{q}, \vec{N})$, $\mu_k = \cos(\vec{k}, \vec{N})$, $\mu = \cos(\vec{q}, \vec{k})$, with $\vec{N}$ being the beam direction. The parameters α, β, δ, ϵ and ζ are considered to be constant at a given beam energy and are to be found by fitting the predictions given by the matrix element to the experimental distributions.

In the case of the $pn \to pp\pi^-$ reaction, the Sp, Ps and Pp states belonging to the σ_{01} cross section have to be added, together with the interference terms: $SsSp$, $SdSp$ and $PsPp$. In our parametrisation we have neglected the Ps and $SdSp$ terms and, as a final addition to the square of the matrix element with σ_{01} present, we took

$$\begin{aligned}
f_{Sp} &= \eta_0\, h^2 k^2 \left[1 + \eta_1 \left(\mu_k^2 - \tfrac{1}{3}\right)\right], \\
f_{Pp}^{01} &= \kappa_0\, k^2 q^2 \left[1 + \kappa_1 \left(\mu_q^2 - \tfrac{1}{3}\right) + \kappa_2 \left(\mu_k^2 - \tfrac{1}{3}\right)\right], \\
f_{SsSp} &= \xi\, h^2 k \mu_k, \\
f_{PsPp} &= \rho\, k q^2 \mu_k.
\end{aligned}$$

To account for the proton-proton final state interaction, we have used scattering wave functions calculated for the Paris potential [13] at an energy-dependent radius.

4 Results and discussion

Our results shown in Figs. 1-3 have been normalised using existing data on the $pp \to pp\pi^0$ total cross section. Distributions of the acceptance-corrected two-proton relative momentum obtained for the $pp \to pp\pi^0$ reaction at six

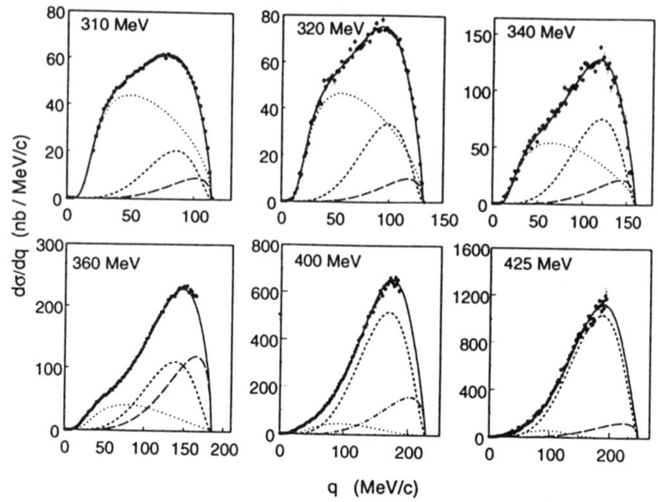

Fig. 1. Acceptance-corrected distributions of the proton-proton relative momentum for the $pp \rightarrow pp\pi^0$ reaction at six beam energies. The curves correspond to the MC-simulated contributions from *Ss* (dotted), *Ps* (long-dashed), *Pp* (dashed). The solids line represents the sum: *Ss+Ps+Pp+Sd*.

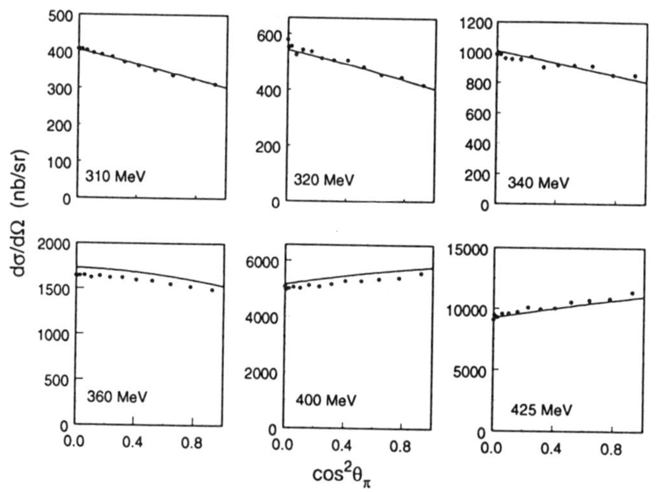

Fig. 2. Acceptance-corrected distributions in the c.m. π^0 angle for the $pp \rightarrow pp\pi^0$ reaction at six beam energies are compared to the MC predictions.

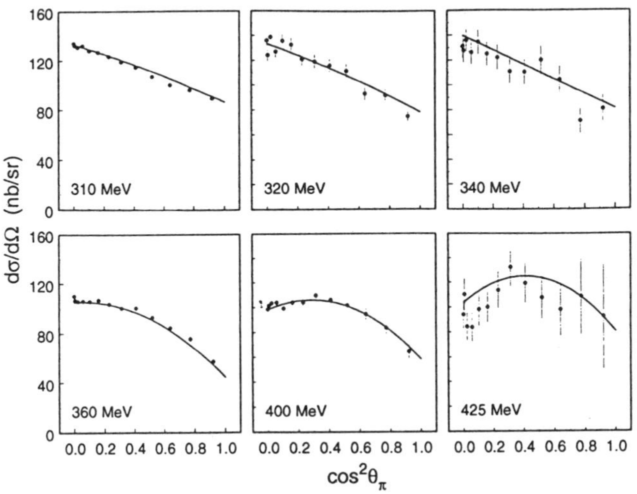

Fig. 3. Acceptance-corrected distributions in the c.m. π^0 angle for the $pp \rightarrow pp\pi^0$ reaction at six beam energies obtained for events with the proton-proton relative momentum less than 53 MeV/c. The solid curves are the MC predictions.

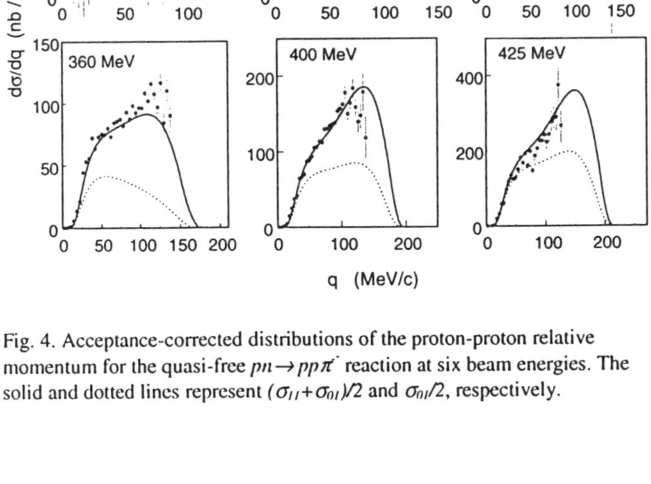

Fig. 4. Acceptance-corrected distributions of the proton-proton relative momentum for the quasi-free $pn \rightarrow pp\pi^-$ reaction at six beam energies. The solid and dotted lines represent $(\sigma_{11}+\sigma_{01})/2$ and $\sigma_{01}/2$, respectively.

beam energies are compared in Fig. 1 to the results of the Monte-Carlo (MC) simulation. In order to obtain such a good agreement, the parameters of the matrix element were fitted at each beam energy independently. In the case of the Ss state, this was achieved by varying the radius at which the pp scattering wave function was calculated. Furthermore, the strength of this state at low q had to be reduced for larger beam energies. These changes may be reflections of our attempt to use only the threshold q and k variation of the amplitudes in the fits. At 400 and 425 MeV our acceptance in q is not large enough to see if the contribution of the P states really does show a q^2 behaviour.

The acceptance-corrected c.m. pion angle distributions presented in Fig. 2 clearly show a change from a negative slope at 310 MeV to a positive one at higher energies. In the region of small q, shown in Fig. 3, there is a strong influence of the Sd state. At low energies this is mostly due to interference with the Ss state ($\cos^2\theta_\pi$ behaviour) but at higher energies f_{Sd} is significant and a $\cos^4\theta_\pi$ term shows up.

The MC model of the $pp \to pp\pi^0$ reaction may be extended to the quasi-free case if the energy dependence of the matrix element is known. Assuming that this dependence is well approximated by interpolation between our six energy points, we have calculated $d\sigma/dq$ for the quasi-free $pp \to pp\pi^0$ in pd collisions at 320 MeV using the Paris deuteron momentum-space wave function [13]. This distribution has been compared to our experimental one in order to get the normalisation constant for the quasi-free reaction. This normalisation was checked with pd elastic scattering, measured in parallel, and the two methods were found to agree within 15%.

The relative momentum distribution obtained for the quasi-free $pn \to pp\pi^-$ reaction is shown in Fig. 4, where the solid line represents $\frac{1}{2}(\sigma_{11} + \sigma_{01})$. Since σ_{11} is given by the parametrisation of the $pp \to pp\pi^0$ data, the values of σ_{01} can be deduced through the fit to the experimental points. At low Q it is seen that $\sigma_{11} > \sigma_{01}$, but the situation is reversed at higher excitation energies.

It should be stressed that the results presented here are preliminary and, especially at large values of q, there is some model dependence due to the acceptance. As a cross-check of the $pn \to pp\pi^-$ results, we shall analyse the quasi-free data obtained at 310 and 340 MeV; values deduced for different Q-bins should be independent of the beam energy.

5 Acknowledgments

We are very grateful to the TSL/ISV personnel for their continued help during the course of this work. Discussions with K. Tamura about the theoretical calculations of the $NN \to NN\pi$ reactions were much appreciated.

Financial support for this experiment and its analysis was provided by the Swedish Natural Science Research Council, the Swedish Royal Academy of Science, the Swedish Institute, the Japanese Ministry of Education, Deutsche Forschung Gesellschaft (Mu 705/3 Graduiertenkolleg), the Polish Scientific Research Committee, the Russian Academy of Science, the German Bundesministerium für Bildung und Forschung [06TU886 and DAAD], and the European Science Exchange Programme. One of us (JZł) would like to express his gratitude for the hospitality and financial support he received during his stay at RCNP of Osaka University.

References

1. M. Gell-Mann and K.M. Watson, *Ann. Rev. Nucl. Sci.* **4**, 219 (1954).
2. H.O. Meyer *et al.*, *Nucl. Phys.* A **539**, 633 (1992).
3. A. Bondar *et al.*, *Phys. Lett.* B **356**, 8 (1995).
4. J.G. Hardie *et al.*, *Phys. Rev.* C **56**, 20 (1997).
5. Y. Maeda *et al.*, *πN Newsletter* **13**, 326 (1997).
6. F. Duncan *et al.*, *Phys. Rev. Lett.* **80**, 4390 (1998).
7. B.J. VerWest and R.A. Arndt, *Phys. Rev.* C **25**, 1979 (1982).
8. J. Bystricky *et al.*, *J. Phys. (Paris)* **48**, 1901 (1987).
9. J. Złomańczuk *et al.*, *Phys. Lett.* B **436**, 251 (1998).
10. H.O. Meyer in *Proc. 8th Int. Conf. on the Structure of Baryons*, Bonn, 22-26 September, 1998.
11. H. Calén *et al.*, *Nucl. Instrum. Methods* A **379**, 57 (1996).
12. R. Handler, *Phys.Rev.* **B138**, 1230 (1965).
13. B. Loiseau and L. Mathelitsch, *Z. Pys.* **A358** (1997); M. Lacombe *et al.* *Phys. Rev.* **C21**, 861 (1980).

PION PRODUCTION MECHANISM IN NUCLEON-NUCLEON COLLISIONS

K. TAMURA
Fukui Medical University, Fukui 910-1193, Japan

Y. MAEDA, N. MATSUOKA
Research Center for Nuclear Physics, Osaka University, Osaka 567-0047, Japan

The pion production $pn \to pp\pi^-$ and $pp \to pp\pi^0$ are discussed. The theoretical analysis has been performed by using the models of heavy-meson exchange and pion-rescattering including the effect of nucleon resonances. The experimental data of the π^- production support the relative sign between the s-wave and p-wave pion-production amplitudes which is given by both of the two models. The d-wave pion production amplitude is discussed about the angular distribution of π^0 production. The spin correlation observables give information about the partial wave amplitudes.

1 Introduction

The behavior of the pion production near threshold provides us with considerable information on the nature of low-energy strong-interaction physics. Recently several measurements of the reaction $NN \to NN\pi$ near the pion-production threshold have been performed. The cross section of $pp \to pp\pi^0$ was measured very precisely[1-3]. It has been pointed out that large contribution of the s-wave pion-production amplitude is necessary to reproduce the total cross section of the reaction $pp \to pp\pi^0$. Also precise measurements for other charge channels have been performed[4-10]. Lee and Riska found that the short range effect for the two-body pion-production operator gave enhancement of the cross section by a factor 3∼5[11]. Horowitz *et al.* confirmed the importance of the short range effect by using the one-boson-exchange model (heavy-meson exchange model)[12]. Hernández and Oset took into account the off-shell properties of the πN amplitude in the pion-rescattering process (pion-rescattering model)[13]. They found constructive interference between the amplitudes of the direct pion-production process and the pion-rescattering process. Hanhart *et al.* examined the off-shell πN amplitude with a realistic meson-theoretical model[14]. Park *et al.*[15], Cohen *et al.*[16] and Sato *et al.*[17] carried out calculations in the framework of the chiral perturbation theory. These works predict the s-wave pion-production amplitude with the opposite sign to the results of the heavy-meson exchange model and the pion-rescattering model. Therefore, determination of the sign of s-wave amplitude plays an essential role to discriminate these models.

In this paper, theoretical calculations are performed by using the heavy-meson exchange model (HM) and the pion-rescattering model (RES). We report three topics about the π^- and π^0 production. 1) Comparing the experimental and theoretical results, we deduce the relative sign between s-wave and p-wave amplitudes using their interference in the $pn \to pp\pi^-$ reaction. 2) The angular distribution of π^0 production is sensitive to the short range structure of Δ excitation mechanism. 3) Recently spin correlation measurement has been done at IUCF. Large probability of s-wave amplitude is necessary to explain the experimental data.

2 Theoretical model

The parameters of the heavy-meson-exchange model are taken from Bonn potential (coupled-channel model) [18]. We adopted Hamilton model for the pion-rescattering model [13]. The effects of the Δ and $N(1440)$ resonance and the meson dissociation mechanism ($\rho\pi\pi$, $\omega\rho\pi$) are also taken into account. The partial waves for the initial and final states of the NN system are summed up to $\ell = 6$ incorporating the Coulomb interaction.

In Fig.1 are shown the calculated results of the total cross section. The hatched regions show results with the models of heavy-meson exchange and pion rescattering using the NN wave function given by the various nuclear forces (Bonn, Argonne V14, Paris, and Reid soft-core). The calculated results give good fit to the experimental data. Our results also show good explanation for $d\sigma/d\Omega$ and A_y of the reactions $pp \to pn\pi^+$ and $pp \to d\pi^+$.

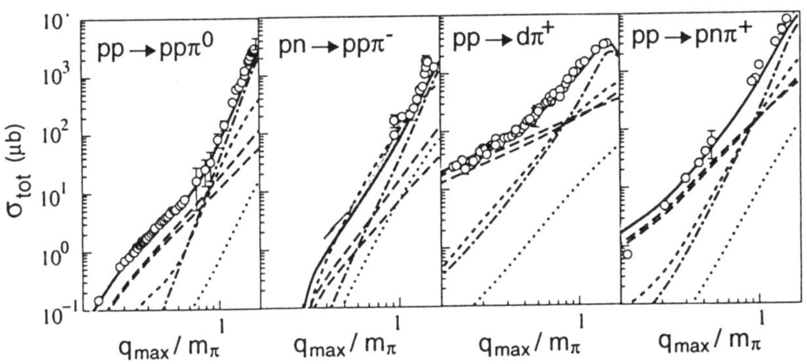

Figure 1: Total cross section of pion production. Curves show the result with Paris potential. Solid curve is the result of HM. Other lines show each mechanism: long dash (HM, RES), short dash (direct), dash-dot (Δ and N^*), dot (meson dissociation).

3 Sign of s-wave amplitude ($pn \to pp\pi^-$)

The pion-production cross sections are decomposed to four independent isospin amplitudes $T_{I_iI_f}$ (T_{11}, T_{01}, T_{10}, T_{10}^d), each represents the transition from an initial two nucleon state with isospin I_i to a final state with isospin I_f. As is well known, the amplitude T_{10} has large contribution from the p-wave pion-production, which is mainly caused by the Δ intermediate process, and gives a dominate contribution to $pp \to pn\pi^+$. In the case of the amplitude T_{01}, on the other hand, the main process of the Δ excitation is forbidden from the isospin selection rule. Thus the p-wave (T_{01}) contribution is suppressed and might be a comparable magnitude as the s-wave amplitude (T_{11}) at the threshold energy region. In this meaning the reaction $pn \to pp\pi^-$ is suitable to study the interference between s-wave and p-wave amplitudes.

In Fig.2(a) shows the angular distribution of cross section for π^- production by the heavy-meson exchange model and the pion-rescattering model as solid line and hatched region. The *p-p* relative energy of the final state was limited within 3 MeV. The experimental data point at 180° was extracted from the inclusive measurement of $d(p,pp)p\pi^-$ at RCNP. The data of TRIUMF [19] also show the asymmetric nature caused by the interference between s-wave and p-wave amplitude. The experimental data can be reproduced well by the predictions of both the heavy-meson exchange model and the pion-rescattering model. If we change the relative sign between s-wave and p-wave amplitudes, the angular distributions become the reversed shapes (dashed line). The calculated results with the opposite sign relation of the s-wave and p-wave amplitudes can not explain the experimental data.

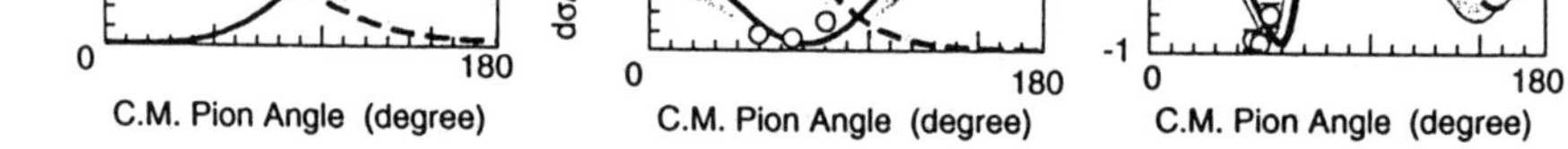

Figure 2: Angular distribution of $pn \to pp\pi^-$. Solid curve shows HM. Dashed curve is the calculation with reversed sign of s-wave amplitude.

4 d-wave pion production mechanism ($pp \to pp\pi^0$)

The angular distribution of π^0 production was measured at RCNP and TSL. The experimental data show convex structure for the angular distribution. This structure mainly comes from d-wave pion production amplitude ${}^3P_2$-${}^3F_2 \to {}^1S_0$ + d-wave. The NN state ${}^3P_2$-${}^3F_2$ has strong coupling to the ΔN state. By this reason the angular dependence of the π^0 production is sensitive to the short range structure of the Δ excitation mechanism. The pion production operator of Δ excitation contains the factor $\vec{q}^{\,2}/(\vec{q}^{\,2}+\mu^2)$. Here $\vec{q}$ and μ are the momentum and mass of the exchanged meson. If one write this factor as

$$\frac{\vec{q}^{\,2}}{\vec{q}^{\,2}+\mu^2} = 1 - \frac{\mu^2}{\vec{q}^{\,2}+\mu^2},$$

then the term 1 gives a contact interaction. Because we introduced the vertex form factor ($\Lambda_{\pi NN}$=700 MeV), the contact interaction is smeared out and gives sizable effect at short range region ($r \leq 0.5$ fm). In Fig.3, the calculations including the contact interaction (dashed line) failed to explain the convex structure of experimental data. The results without the contact interaction are shown as solid lines which give good fit to the experimental data. While we can not conclude about the necessity of the contact interaction only from this example, the angular distribution of π^0 production is very sensitive to the short range mechanism of Δ excitation mechanism.

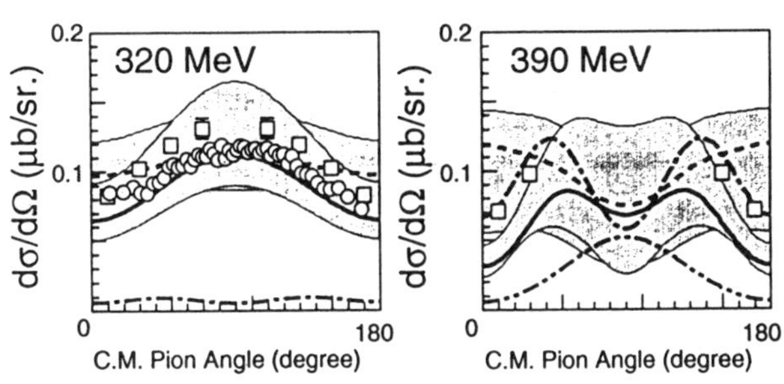

Figure 3: Angular distribution of $pp \to pp\pi^0$. Solid line is the calculation without the contact interaction. Dashed line includes the contact interaction. Dot-dashed (Double dot-dotted) line shows the contribution of Δ excitation mechanism without (with) the contact interaction. The data are from RCNP (box) and TSL (circle)[20] The data of TSL were measured at 310 MeV. The p-p relative energy of final state was limited within 3 MeV.

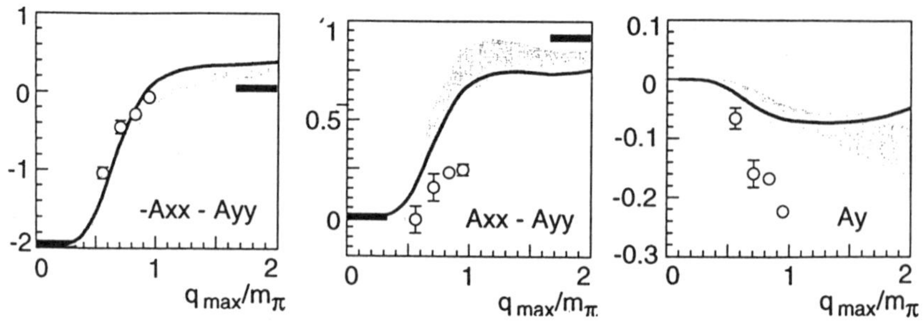

Figure 4: Spin correlation observables of $pp \to pp\pi^0$. Solid line is the calculation of HM with Paris potential. The data are from Ref. 21. The thick lines show the expected values for the s-wave and p-wave pion production.

5 Spin correlation observables ($\vec{p}\vec{p} \to pp\pi^0$)

Recently Meyer *et al.* [21] measured the spin correlation observables (A_y, A_{xx}, A_{yy}) with the beam and target polarization. The observed data was integrated over the energy and angle of the final state. By using the partial wave amplitude Pp, Ps and Ss, we can see the global structure of the observables as

$$A_{xx} + A_{yy} = 2Ss^2/\sigma\ , \qquad A_{xx} - A_{yy} = 4Pp^2/15\sigma\ ,$$
$$A_y = \pi\sqrt{5}Pp \times Ps \sin\Delta/10\sigma\ , \qquad \sigma = Ss^2 + Ps^2 + Pp^2/3.$$

Here σ means the total cross section and Δ is given by the phase shift of the initial NN states. The combination $A_{xx} + A_{yy}$ is proportional to s-wave amplitude Ss. Because the s-wave amplitude dominates at the threshold, the combination $A_{xx} + A_{yy}$ becomes the value 2 at the threshold region. And it reaches zero at the higher energy region since the pion production mechanism can be explained manly by the p-wave amplitude. Contrary to this, the combination $A_{xx} - A_{yy}$ is written by the p-wave amplitude Pp. Therefore this combination changes from 0 to 0.8. Fig.4 show the experimental data and calculated results. The combination $A_{xx} + A_{yy}$ can be explained by our model satisfactorily. Therefore the model calculations give reasonable magnitude of the Ss amplitude. But in the case of the combination $A_{xx} - A_{yy}$, our model gives overestimation. Furthermore our result of analyzing power is very small.

Analyzing power A_y is proportional to the product of p-wave amplitude Pp and s-wave amplitude Ps in which nucleons of the final state couple to P-wave. In the case of Ss amplitude, the main contribution comes from the transition ${}^3P_0 \to {}^1S_0 + s$. Because the ${}^1S_0$ state has large strength at short range region, the wave function has good overlap with the Yukawa function of the heavy meson exchange. Therefore we might expect large enhancement for Ss amplitude by the short range effects. Contrary to this, in the case of Ps amplitude, the final state becomes P-wave. Therefore we have small overlap between the wave function of P-wave and the Yukawa function at short range region. As the result the strength of Ps amplitude becomes very small. In other word, we do not have s-wave pion production mechanism which might be effective around 1 fm region in our model. If we take this problem seriously, we need another s-wave pion production mechanism with the long-range interaction.

We would like to thank T.-S.H. Lee, T. Sato, J. Haidenbauer, C. Hanhart and J. Zlomanczuk for their stimulating discussions. The experiment at RCNP has been performed under the program number E95. The numerical calculation was performed at the computer center of Fukui Medical University.

References

1. H.O. Meyer *et al.*, *Phys. Rev. Lett.* **65**, 2846 (1990).
2. S. Stanislaus *el al.*, *Phys. Rev.* C **44**, 2287 (1991).
3. A. Bondar *el al.*, *Phys. Lett.* B **356**, 8 (1995).
4. M.G. Bachman *el al.*, *Phys. Rev.* C **52**, 495 (1995).
5. W.W. Daehnick *el al.*, *Phys. Rev. Lett.* **74**, 2913 (1995).
6. D.A. Hutcheon *el al.*, *Nucl. Phys.* A **535**, 618 (1991).
7. E. Korkmaz *el al.*, *Nucl. Phys.* A **535**, 637 (1991).
8. M. Drochner *el al.*, *Phys. Rev. Lett.* **77**, 454 (1996).
9. P. Heimberg *el al.*, *Phys. Rev. Lett.* **77**, 1012 (1996).
10. B.J. VerWest and R.A. Arndt, *Phys. Rev.* C **25**, 1979 (1982), and references therein.
11. T.-S.H. Lee and D. Riska, *Phys. Rev. Lett.* **70**, 2237 (1993).
12. C.J. Horowitz *et al.*, *Phys. Rev.* C **49**, 1337 (1994).
13. E. Hernández and E. Oset, *Phys. Lett.* B **350**, 158 (1995).
14. C. Hanhart, *et al.*, *Phys. Lett.* B **358**, 21 (1995).
15. B.-Y. Park *et al.*, *Phys. Rev.* C **53**, 1519 (1996).
16. T.D. Cohen *et al.*, *Phys. Rev.* C **53**, 2661 (1996).
17. T. Sato *et al.*, *Phys. Rev.* C **56**, 1246 (1997).
18. R. Machleidt, *Adv. in Nucl. Phys.* Vol. bf 19, 189 (1989).
19. F. Duncan *et al.*, *Phys. Rev. Lett.* **80**, 4390 (1998).
20. J. Zlomanczuk *et al.*, *Phys. Lett.* B **436**, 251 (1998).
21. H.O. Meyer *et al.*, *Phys. Rev. Lett.* **81**, 3096 (1998).

PHOTODISINTEGRATION OF ^{4}He STUDIED WITH TPC USING 22-32 MeV REAL PHOTONS

T. SHIMA, T. BABA, T. KII, T. TAKAHASHI, S. NAITO, Y. NAGAI
Department of Applied Physics, Tokyo Institute of Technology,
2-12-1, O-okayama, Meguro, Tokyo 1528551, Japan

H. OHGAKI, H. TOYOKAWA
Quantum Radiation Division, Electrotechnical Laboratory,
1-1-14, Umezono, Tsukuba, Ibaraki 3058568, Japan

The photodisintegration of ^{4}He at the γ-ray energy of about 25MeV provides a useful tool to investigate the problems of the electromagnetic transition strength in the reaction and the charge symmetry in nuclear force. In this study we made a new exclusive measurement of the photodisintegration of ^{4}He using a laser-Compton-backscattered γ beam and a time projection chamber with an active target. The preliminary result indicates no anomalous violation of the charge symmetry and suppression of the transition strength compared to the expected one from the E1 sum rule.

1. Photodisintegration of ^{4}He at around 25MeV

Interests of the reactions on light nuclei are recently increasing, because they provide useful informations about the property of few-nucleon systems and the nuclear force. For this purpose there have been significant progress of both theoretical and experimental recently. For example, microscopic calculations for the break-up cross sections [1,2,3,4] and the asymptotic normalization constants [5] of the nuclei up to A =3 have been carried out, and compared with experimental data in order to obtain accurate NN potential and to study the effects of the meson-exchange current (MEC) and the final-state interaction (FSI), and so on. And it will be important to extend such study [6] to the A =4 system, i.e. ^{4}He. In addition, there are three subjects of interest concerning the photodisintegration of ^{4}He at around the incident γ-ray energy of 25MeV. Firstly, as it is well known, there is no straightforward way to solve the equation of motion for a four-body system, and therefore study of ^{4}He would be useful to develop microscopic models for few-nucleon systems. Secondly, the recent experimental data of the photodisintegration of ^{4}He suggests the transition probability is significantly smaller than the strength expected from the E1 sum rule. Recently, Efros et al. carried out a calculation of the total photoabsorption cross section including full final-state interaction effect [7]. They claimed the pronounced giant dipole resonance (GDR) does not exist in the photodisintegration of ^{4}He, and the sum-rule strength is fulfilled with the

^{4}He(γ,pn)^{2}H channel in addition to two-body break-up channels. However, as mentioned later by themselves [8], one of the previous experiment measured an inclusive cross section for the ^{4}He(γ,X)n [9], and it indicates the ^{4}He(γ,pn)d cross section is not enough large to account for the missing strength. Therefore it is quite important to know quantitatively the contribution of the three-body break-up channel. Thirdly, the ratio R of the ^{4}He(γ,p)^{3}H cross section $\sigma_{\gamma p}$ to the ^{4}He(γ,n)^{3}He cross section $\sigma_{\gamma n}$ is sensitive to the isospin-mixing in GDR state, and it is useful to search for possible charge-symmetry breaking (CSB) in the nuclear force [10]. There have already been several experimental evidences of CSB such as the differences in the NN-scattering parameters [11,12], the cross section differences in $\pi^\pm$ elastic scattering from ^{3}H and ^{3}He [13], and the Okamoto-Nolen-Schiffer mass anomaly in mirror nuclei [14,15], and so on. The origin of those phenomena would be anyway the mass difference of the up and down quarks and the electromagnetic interactions between them. But it is still interesting and important to understand the mechanisms how the above phenomena are induced by the intrinsic sources of CSB. Therefore it is worthwhile to investigate various phenomena indicating CSB. For the above reasons many experimental efforts have been made to measure the cross sections of the photodisintegration of ^{4}He [16], but the existing data of $\sigma_{\gamma p}$ and $\sigma_{\gamma n}$ contain a serious inconsistency among them, and are quite unsatisfactory for investigation of the above physics. One of the possible reason of the inconsistency could be that almost all the previous experiments were independent measurements, and therefore it would be quite necessary to make a simultaneous measurement of the photodisintegration reactions of ^{4}He, together with some reaction which can be used as a normalization of the measurement. For this purpose we carried out a new measurement by means of a laser-Compton-scattered γ beam and a time projection chamber with an active target of ^{4}He.

2. Experimental method

The experimental setup is schematically shown in Figure 1. The incident γ-ray photons were generated via the Compton backscattering of the UV photons from a Nd-YLF laser (third harmonic ; λ =351nm) with relativistic electrons circulating in the 800MeV electron storage ring TERAS of the Electrotechnical Laboratory [17]. The γ-ray photons impinged upon the ^{4}He gas target in a time projection chamber (TPC).

We made three measurements with different maximum energies of the γ-ray of E_{max} = 25MeV, 28MeV and 32MeV. The intensity and the energy distribution of the γ-ray were monitored through the measurement, using a BGO scintillation counter with 2" in diameter and 5" in length. The intensity was deter-

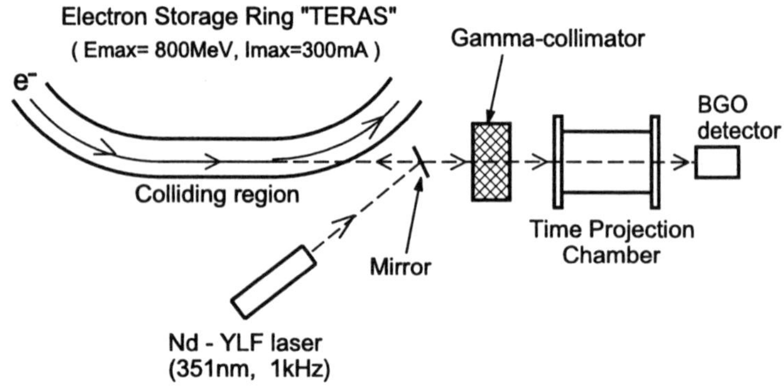

Figure 1: Experimental set up for the ^{4}He photodisintegration measurement.

mined to be $10^3\sim10^4$ photons/sec, depending on the storaged electron current. And the energy spread E_W of the γ ray was 20% (FWHM). The time width and the repitition frequency were 150ns and 1kHz, respectively. The TPC has an active volume with 60mm(width)×60mm(height)×250mm(length), and was operated with a mixture of He 80% and CH_4 20% gases with total pressure of 980Torr, and the gas serves as an active target. Therefore we could measure all the ^{4}He break-up channels simultaniously with 4π acceptance and a detection efficiency of ~100%. Also we observed three-dimensional images of the reaction products with TPC, and that enables us to separate the events of the different reaction channels and the background events from each other. Figure 2(a)~2(c), Figure 3(a) and 3(b) are examples of the observed particle tracks for the ^{4}He(γ,p)^{3}H, ^{4}He(γ,n)^{3}He, ^{4}He(γ,pn)^{2}H, ^{12}C(γ,p)^{11}B and ^{12}C(γ,n)^{11}C, respectively.

3. Result

The photodisintegration cross sections σ were given by Eq. (1) :

$$\sigma = \frac{Y}{\varepsilon \cdot n \cdot \phi_\gamma}, \tag{1}$$

where Y, n and ϕ_γ are the number of the observed reaction events, the target thickness, and the number of the incident photons, respectively. ε is the detection efficiency for the reaction, and was calculated with the Monte Carlo method. In order to cancel out most of the systematic errors, we used the

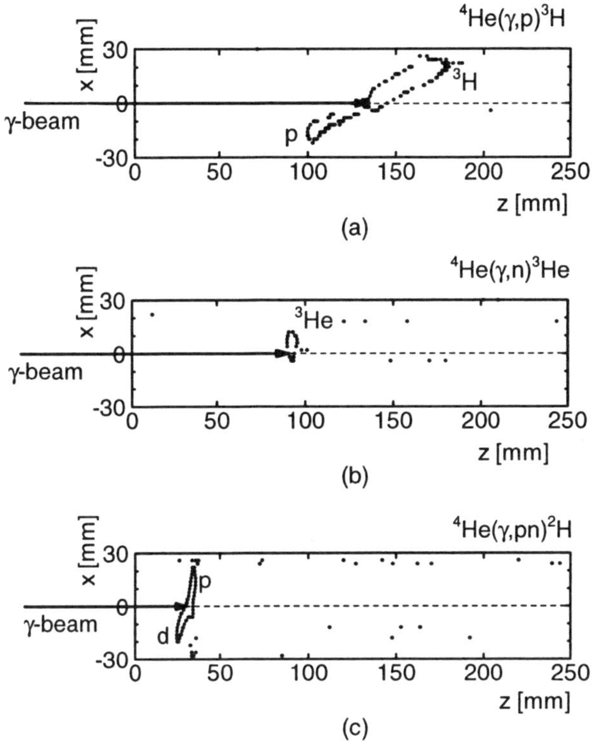

Figure 2: Projection images of the particle tracks observed for (a); ^{4}He(γ,p)^{3}H, (b); ^{4}He(γ,n)^{3}He and (c); ^{4}He(γ,pn)^{2}H.

^{12}C(γ,p)^{11}C reaction as the normalization of the measurement, where the ^{12}C(γ,p)^{11}C reaction cross section has been known experimentally within the error of about 5%[18,19,20,21]. We obtained the photodisintegration cross sections of ^{4}He relative to that of ^{12}C(γ,p)^{11}C by using Eq. (2) :

$$\overline{\sigma}(^4He) = \frac{n(^{12}C)}{n(^4He)} \cdot \frac{Y(^4He)}{Y(^{12}C(\gamma,p))} \cdot \frac{\overline{\varepsilon}(^{12}C(\gamma,p))}{\overline{\varepsilon}(^4He)} \cdot \overline{\sigma}(^{12}C(\gamma,p)) \ , \qquad (2)$$

where $Y(^4He)$ ((^{12}C)) and $n(^4He)$ ((^{12}C)) are the number of events and the target thickness for the ^{4}He photodisintegration (the ^{12}C photoproton emission), respectively. $\overline{\sigma}(^4He)$ ((^{12}C)) and $\overline{\varepsilon}(^4He)$ ((^{12}C)) are the cross section and the detection efficiency for the ^{4}He photodisintegration (the ^{12}C photoproton emission), respectively, averaged with the incident γ-ray energy distri-

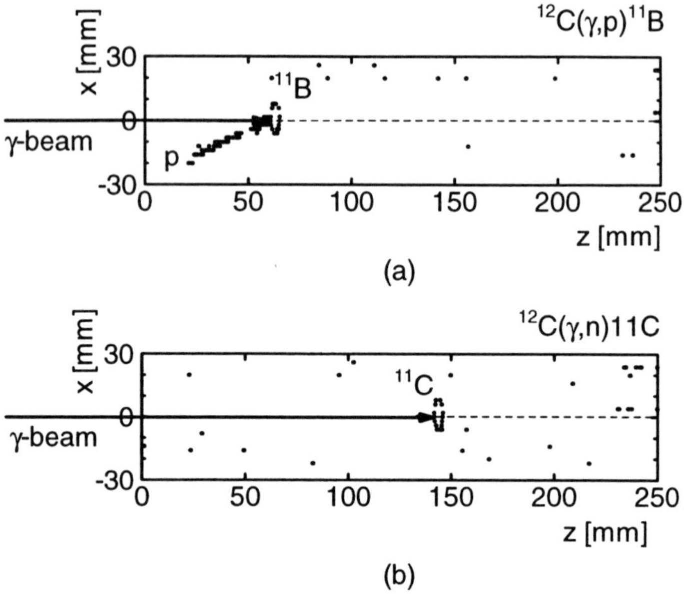

Figure 3: Projection images of the particle tracks observed for (a); $^{12}C(\gamma,p)^{11}B$ and (b); $^{12}C(\gamma,n)^{11}C$.

bution $\phi(E)$. Here $\phi(E)$ was approximately given by Eq. (3) :

$$\phi(E) = \begin{cases} \phi_0 \cdot \frac{E - E_{max} + 2E_W}{2E_W} & (E_{max} - 2E_W \leq E \leq E_{max}) \\ 0 & (E < E - 2E_W \text{ or } E_{max} < E) \end{cases}, \tag{3}$$

where ϕ_0 is the peak intensity of the γ beam at the maximum γ-ray energy E_{max}. E_W is the full width at a half maximum of the intensity, and was about 20% of E_{max}. Using Eq. (2) the ^{4}He photodisintegration cross sections were obtained as shown in Table (1). The $^4He(\gamma,pn)^2H$ cross section was as small as 0.02±0.02mb for E_{max} = 32MeV, and was negligibly small below E_{max} = 28MeV. The results of the recent calculations for the ^{4}He photodisintegration cross sections are also shown in Table (1) . In the table σ_{UH} indicates the two-body break up cross section, i.e. the sum of the $^4He(\gamma,p)^3H$ and $^4He(\gamma,n)^3He$ cross sections, calculated by Unkelbach and Hofmann [22]. And σ^{MT}_{ELO} and σ^{TN}_{ELO} are the total photoabsorption cross sections calculated by Efros et al. [7] using the Malfliet-Tjon(MT) I+II potential and the Trento (TN) potential, respectively. Details of those calculations are as follows. The calculation by Unkelbach and Hofmann is based on the refined resonating-group model with a

Table 1: Photodisintegration cross sections of ^{4}He. (Only statistical errors are shown.)

E_{max} (MeV)	$\sigma_{\gamma p}$ (mb)	$\sigma_{\gamma n}$ (mb)	σ_{total} (mb)	σ_{UH} (mb)	σ_{ELO}^{MT} (mb)	σ_{ELO}^{TN} (mb)
25	0.52±0.05	0.43±0.04	0.95±0.06	1.41	1.11	1.30
28	0.90±0.07	0.92±0.07	1.8±0.1	1.98	1.98	2.23
32	1.02±0.10	0.92±0.09	1.9±0.13	2.17	2.76	2.94

semirealistic NN potential. The contributions of the E1, M1 and E2 transition were included, and the MEC effects for the dipole transitions were calculated explicitly without using Siegert's theorem. The calculation by Efros et al. was performed via the method of the Lorentz integral transform. It included only the dipole transition, but the effect of FSI was fully taken into account. In order to take into account the MEC effect they used Siegert's theorem. The above calculations show a significant difference of the cross sections for the energy window with E_{max} = 32MeV, but the present result is not enough accurate to check the calculations because of the systematic error, which is mainly caused by the ambiguity in the energy distribution of the incident γ beam. However, it can be said that the contribution of the three-body break up channel is too small to account for the missing strength expected from the E1 sum rule. Concerning the CSB effect, the present result is consistent with the theoretical expectation assuming no exotic CSB effect exists.

4. Conclusion

In this work the photodisintegration cross sections of ^{4}He have been measured using a laser-Compton-backscattered γ beam and a TPC with an active target. The preliminary cross sections were determined as shown in Table (1), using the $^{12}C(\gamma,p)^{11}B$ reaction cross section as a normalization. The present result is consistent with the recent calculations, and the contribution of the three-body break up channel was found to be too small to account for the missing strength expected from the E1 sum rule. Also our result indicates no anomalous CSB effect in the photodisintegration of ^{4}He at around 25MeV. Now we are going to measure the energy distribution of the γ beam accurately in order to remove the main source of the systematic error.

References

1. J. Carlson *et al*, *Phys. Rev.* C **42**, 830 (1990).
2. V.D. Efros, W. Leidemann and G. Orlandini, *Phys. Lett.* B **338**, 130 (1994).
3. S. Martinelli *et al*, *Phys. Rev.* C **52**, 1778 (1995).
4. W.P. Abfalterer *et al*, *Phys. Rev. Lett.* **81**, 57 (1998).
5. J.L. Friar *et al*, *Phys. Rev.* C **37**, 2859 (1988).
6. W. Sandhas *et al*, *Nucl. Phys.* A **631**, 210c (1998).
7. V.D. Efros, W. Leidemann and G. Orlandini, *Phys. Rev. Lett.* **78**, 4015 (1997).
8. V.D. Efros, W. Leidemann and G. Orlandini, *Phys. Rev. Lett.* **80**, 1570 (1998).
9. B.L. Berman *et al*, *Phys. Rev.* C **22**, 2273 (1980).
10. F.C. Barker and A.K. Mann, *Philos. Mag.* **2**, 5 (1957).
11. O. Dumbrajs *et al*, *Nucl. Phys.* B **216**, 277 (1983).
12. S.E. Vigdor *et al*, *Phys. Rev.* C **46**, 410 (1992).
13. K.S. Dhuga *et al*, *Phys. Rev.* C **54**, 2823 (1996).
14. K. Okamoto, *Phys. Lett.* **11**, 150 (1964).
15. J.A. Nolen and J.P. Schiffer, *Annu. Rev. Nucl. Sci.* **19**, 471 (1969).
16. K.I. Hahn *et al*, *Phys. Rev.* C **51**, 1624 (1995) and references therein.
17. H. Ohgaki *et al*, *IEEE Trans. Nucl. Sci.* **38**, 386 (1991).
18. R.G. Allas *et al*, *Nucl. Phys.* **58**, 122 (1964).
19. R. Calarco *et al*, *Bull. Am. Phys. Soc.* **17**, 931 (1972).
20. S.S. Hanna, in *Proceedings of the International Conference on Photonuclear Reactions and Applications, Asilomar, 1973*, ed. B.L. Berman (Lawrence Livermore Laboratory, Livermore, CA, 1973), Vol. 1, p.417.
21. E. Kerkhove *et al*, *Phys. Rev.* C **33**, 1796 (1986).
22. M. Unkelbach and H.M. Hofmann, *Nucl. Phys.* A **549**, 550 (1992).

Lifetime Measurements of Hypernuclei at COSY

I. Zychor[a], K. Pysz[b], P. Kulessa[c,d], T. Hermes[d], Z. Rudy[c], W. Cassing[e], M. Hartmann[d], H. Ohm[d], S. Kistryn[c], W. Borgs[d], B. Kamys[c], H. R. Koch[d], R. Maier[d], D. Prasuhn[d], J. Pfeiffer[d], Y. Uozumi[f], L. Jarczyk[c], A. Strzalkowski[c], M. Matoba[f], H. Ströher[d], O. W. B. Schult[d]

[a]*Andrzej Soltan Institute for Nuclear Studies, PL-05400 Swierk, Poland*
[b]*H. Niewodniczanski Institute of Nuclear Physics, PL-31342 Cracow, Poland*
[c]*M. Smoluchowski Institute of Physics, Jagellonian University, PL-30059 Cracow, Poland,*
[d]*Institut für Kernphysik, Forschungszentrum Jülich, D-52425 Jülich, Germany*
[e]*Institut für Theoretische Physik, Universität Giessen, D-35392 Giessen, Germany*
[f]*Department of Nuclear Engineering, Kyushu University, Fukuoka 812, Japan*

ABSTRACT

At COSY-Jülich we have carried out an internal-target experiment to determine the lifetimes of very heavy hypernuclei by the recoil-distance method. For this purpose we have bombarded very thin ribbon targets of U and Bi evaporated on extremely thin carbon support foils with protons of 1.0 GeV for the background determination and of 1.5 and 1.9 GeV for the production of heavy hypernuclei. The evaluation of the data using the maximum likelihood method applied for the Poisson distribution has yielded lifetimes of $(194 \pm 50 \pm 20)$ps and $(218 \pm 26 \pm 13)$ps for the two U runs and $(153 \pm 45 \pm 20)$ps and $(161 \pm 7 \pm 14)$ps for the two Bi runs where the first errors are of statistical and the second of systematical origin. Comparison of the lifetimes of hypernuclei with mass number $A \geq 11$ reveal a very smooth mass dependence being in good agreement with the phase-space approach.

1. INTRODUCTION

The free Λ hyperon undergoes almost exclusively nonleptonic, flavour changing weak interactions and decays mesonically with 63.9 % into $p\pi^-$ and with 35,8 % into $n\pi^0$ with a lifetime of (263.2 ± 2.0)ps[1]. These nonleptonic decays are more poorly understood than the semileptonic hyperon decays[2]. For a Λ hyperon inside an atomic nucleus an additional decay channel opens up, the nonmesonic four-fermion baryon-baryon weak interaction[2]: $\Lambda + N \to n + N$. This nonmesonic weak process is unique to hypernuclear decays[3, 4]. In hypernuclei, the mesonic decay suffers Pauli suppression which rapidly increases with atomic mass A[5, 6]. Already for nuclei like ${}^{11}_{\Lambda}B$ and ${}^{12}_{\Lambda}C$ the nonmesonic decay dominates[7, 8] in spite of partial restoration[6, 9-11] of the pionic decay mode. In emulsion studies[12, 13] it has been found that it is less than ~ 1 % of the total decay rate for nuclei with $40 < A < 100$. In very heavy hypernuclei the mesonic decay width is predicted[5, 14] to be only in the order of 10^{-3} of the total decay width Γ_Λ. Thus in experiments with $A \geq 200$ the mesonic decay can be completely neglected and the decay width

$$\Gamma_\Lambda = \Gamma_-(\Lambda \to p\pi^-) + \Gamma_0(\Lambda \to n\pi^0) + \Gamma_n(n\Lambda \to nn) + \Gamma_p(p\Lambda \to pn)$$

is nonmesonic to more than 99 %:

$$\Gamma_\Lambda \sim \Gamma_n + \Gamma_p.$$

Already in 1984 Yamazaki emphasized the importance to extend Λ-lifetime measurements towards heavy hypernuclei, and he even mentioned the possibility to employ the (p, K^+) reaction[15]. A question was[2,16] whether the nonmesonic decay rate in hypernuclei scales like (A-1), in proportion to the number of ΛN pairs, or rather saturates as one goes to heavier nuclei. It was clear, that the lifetime is not easy to measure[3]. A number of theoretical studies have been devoted to calculations of partial and total decay rates[5, 14, 17-30]. Early studies of the nonmesonic decay, ΛN→nN, have been performed by Block and Dalitz[31] carrying out an analysis in terms of the spin and isospin dependence of the weak interaction. Adams[32] has used DWBA calculations with strong initial and final state interactions in the ΛN→nN decay for obtaining lifetimes of heavy hypernuclei. The theoretical treatments are usually based on meson-exchange models (e. g. refs. 19, 21), hybrid quark models (e. g. ref. 18), or phenomenological models (e. g. refs. 16, 31, 33). The meson exchange models include one-pion exchange, correlated two-pion exchange, ρ and K, K* exchange. Several authors have included the ΛNN→nNN interaction in their calculations[24, 27-29, 34]. Yet a complete understanding of the nonmesonic weak decay has not been achieved[35].
It is thus not surprizing that the predictions for the lifetimes of heavy hypernuclei range from ~64 ps to ~340 ps. In the last decade Alzetta et al.[14] obtained, depending on the depth V_N of the nuclear potential, a lifetime of ~ 98 ps for V_N ~ 50 MeV and 210 ps for V_N~60 MeV. Alberico et al.[24] found 121-192 ps, including Λnp→nnp, M. Ericson[27] obtained 183 ps including the same process. Ramos and Oset[28, 29, 34] have also included the ΛNN decay and predicted 123 ps for $^{56}_{\Lambda}$Fe and 118 ps for $^{208}_{\Lambda}$Pb, and Itonaga's prediction[30] is 141 ps.

In view of such predictions we felt it worth attempting a measurement of the lifetime of very heavy hypernuclei. At the same time we were interested in finding out if the (p, K^+) reaction could be used not only for strangeness production but also for the production of hypernuclei with sufficient cross section so that the measurement, which we were aiming at, became feasible.

Shortly before our experiment was planned, it was recognized by Yamazaki and co-workers[36] that the quasifree continuum could be a powerful source for the production of Λ hypernuclei. Even with a target as light as ^{12}C it has been shown by Ajimura, Ejiri et al.[37], that Λ hypernuclei can be produced successfully through ΛN scattering in the quasifree region. As the Λ trapping probability is rather high[38], also (p, K^+) reactions can be used for producing Λ hypernuclei through the multistep process.

2. METHOD

2.1 Calculation of the Formation Cross Section

The production of hypernuclei occurs in two steps[39, 40]: first the Λ hyperon must be produced, and in the second step it must be attached to the nucleus. In order to confirm these favourable perspectives[36-38] it was desirable to carry out a calculation of the hypernucleus production cross section for the proton bombardment of very heavy nuclei like Bi or U with beam energies of T_p=1.0–1.9 GeV.

The phase space folding model[41] was used to calculate the total cross section of the production of Λ hyperons and their momentum distribution. The predictive power of the model was checked against the experimental total K^+ production cross sections measured by Koptev et al.[42]. The fate of the produced Λ hyperons was then followed by a coupled channel BUU calculation[43, 44], where the ΛN scattering cross section was taken from the Jülich YN Interaction model[45]. In addition, the BUU calculations were followed by a Hauser-Feshbach computation in order to obtain the spectrum of hypernuclei, their recoil momenta and survival probabilities against prompt fission after emission of protons and neutrons during the cooling of the hypernuclei formed in a rather violent reaction[44, 46]. These calculations showed, that the planned experiment[47] should be feasible. For the interaction of Tp =1.9 GeV protons with a ^{209}Bi target the cross section for the production of hot hypenuclei[43] was predicted to be 330μb. This agrees well with the present measurement giving a value of (350±140)μb[48].

2.2 Approach for Lifetime Measurement

There exist two methods for measuring lifetimes in the region of ~200 ps. A delayed coincidence measurement is possible if one has extracted K^- beams[7, 8] or π^+ beams[49, 50]. With protons as projectiles the partial hypernucleus formation cross section becomes rather small compared with the total cross section, so that a direct timing measurement becomes very difficult. For very heavy nuclei, where fission induced by the hyperon decay becomes a significant decay branch, the recoil-distance method[51] can be used for the Λ lifetime determination. This method has been applied successfully by Polikanov and co-workers[52-54] for measuring the lifetime of Λ hypernuclei produced during the bombardment of U and Bi targets with an extracted $\bar{p}$ beam at LEAR, and by Noga and co-workers[55,56] for the lifetime determination of Λ hypernuclei produced when bombarding a Bi target with electrons from the Khar′kov accelerator.

A unique possibility is offered if the target can be placed in a circulating beam[57] as at COSY-Jülich[58], because here the number of measured events depends only weakly on the target thickness, and the targets can be chosen as thin as allowed by the production technique. This ensures minimal energy loss of the recoiling hypernuclei. The disadvantage is that the target has to be like a tie, suspended only from above, and without any massive backing which would ensure flatness. Furthermore, the spots of extracted beams on targets are usually much smaller[52, 53] than the region[59], where the circulating beam interacts with the target. The production of sufficiently large and very flat ribbon targets is rather difficult.

3. APPARATUS

The COSY-13 set-up consists of a detector chamber which houses two, in both directions (x and z, see Fig. 1) position sensitive, low-pressure multiwire proportional chambers (MWPC) equipped with a very thin entrance window and a target holder which permits to move the target with high precision in three directions. The detector chamber is mounted above the COSY-beam tube. The target manipulator is fixed to the COSY beam tube in x-direction[60].

A diaphragm ensured that prompt fission fragments do not overload the detectors. The narrow slit in the diaphragm permitted normalization relative to prompt fission. Tuning of the target in -z-direction with high voltage - at fixed target support
position - could be used to move the shadow edge, i. e. the border, where prompt fission fragments could no longer reach the detectors, in +z-direction. This also served to

reduce the prompt fission rate. A TV camera permitted observation of the target during the experiment. Tests and calibration measurements have been described elsewhere[60].

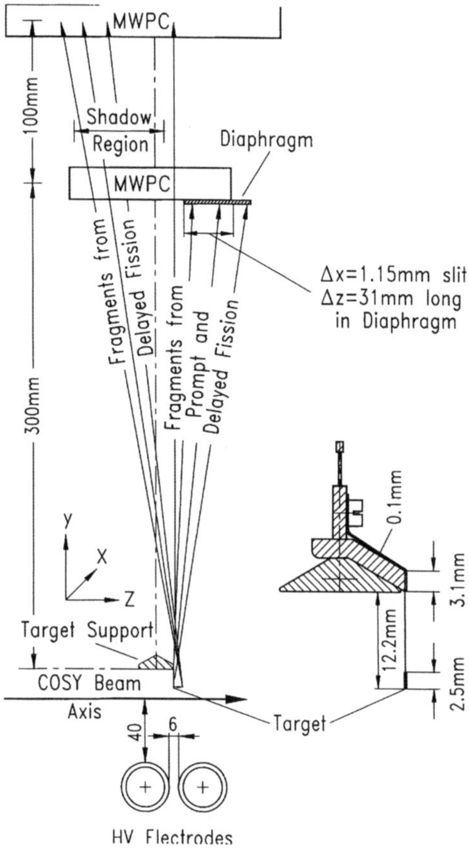

Figure 1. Sketch of the COSY-13 set-up. The target support is shown in enlarged scale to the right. The target dimensions given were those for the Bi experiment [61]. The target was stretched and could be tuned in z-direction using of high voltages on the two HV electrodes. The MWPC's are sensitive only to fission fragments. A very thin window in the bottom of the detector chamber served to separate the detector chamber vacuum from the ring vacuum[60].

4. MEASUREMENTS

In 1996 a first experiment[62] was performed with a uranium target. The background was always measured at T_p=1.0 GeV. Hypernuclei were produced at 1.5 GeV. Later in that year Bi targets were installed and hypernucleus production was observed at 1.5 and 1.9 GeV[63]. In 1997 a second run was carried out with a U target and proton energies of

1.0, 1.5 and 1.9 GeV. Starting with this run COSY was operated in a supercycle[60], where the $T_p = 1.0$ GeV COSY cycle was followed by a $T_p = 1.5$ GeV or 1.9 GeV COSY cycle. This had the effect, that the background at 1.0 GeV and hypernuclear spectra were recorded quasi simultaneously. An experiment with better statistics could be performed later in 1997 with a Bi target and projectile energies of 1.0 and 1.9 GeV[48]. During all experiments the COSY beam (up to $\sim 10^{10}$ protons in the ring) was moved to a position below the target during injection and acceleration and then bumped on to the target[60] and used up in a way that the rate was kept almost constant.

5. DATA ANALYSIS

The data which satisfied the imposed physical conditions were used for track reconstruction and only those events were accepted which came from a narrow region around the target[60]. The position spectra of these events are plotted as a function of the z coordinate of the lower MWPC in Fig. 2.

A comparison between the upper and lower parts of Fig. 2 clearly shows that bombardment of the Bi target with 1.9 GeV protons leads to a significant production of hypernuclei which recoil out of the target (in +z direction) and, due to their delayed decay, can emit fission fragments which pass by the target support and hit the lower MWPC also at small wire numbers. Due to the very large dynamic range (peak rate/rate at wire number ≤ 30) of $\sim 2 \cdot 10^8$ the sensitivity for the detection of hypernuclei is very large. The theoretical calculations[43, 44, 46,] including energy loss in the target also yielded the recoil velocity spectrum of the hypernuclei. From this velocity spectrum, the geometry of the set-up and the net position spectra the production cross sections of cold hypernuclei and the lifetimes were obtained[48,62]. For this purpose the maximum likelihood method applied the Poisson distribution was used and all data sets of the four experiments have been analyzed with the same method.

6. RESULTS

The results obtained in the first U and the last Bi run have been published[48, 62]. With the maximum likelihood method the following results were obtained:

1996	1st U run	$\tau = (194 \pm 50 \pm 20)$ ps
	1st Bi run	$\tau = (153 \pm 45 \pm 20)$ ps
1997	2nd U run	$\tau = (218 \pm 26 \pm 13)$ ps
	2nd Bi run	$\tau = (161 \pm 7 \pm 14)$ ps.

The first error is the statistical and the second is the systematical error[64]. From the first U run a Λ hyperon lifetime of (240 ± 60)ps was obtained[62] as the result of a least squares fit using a Gaussian distribution. Here the total error of 60 ps is the mean square of the statistical error of 45 ps and the systematical error of 40 ps[65]. The result of (240 ± 60)ps is fully consistent with the result of (194 ± 54)ps obtained when using the maximum likelihood method applied for the Poisson distribution. Combining these results, on finds 214 ps, but the error remains: $(\pm 50 \pm 20)$ps. To test the sensitivity of our averaged COSY-13 result: (175 ± 16)ps on the magnitude of an error, let us assume that the systematical error of the second U run is 40ps instead of 13ps. This would change the value from 175ps to 169ps.

Plotting measured[7, 8, 54, 66] lifetimes for hypernuclei with masses ≥ 11 together with the results given in this paper as function of A we observe a rather smooth mass dependence (see Fig. 3).

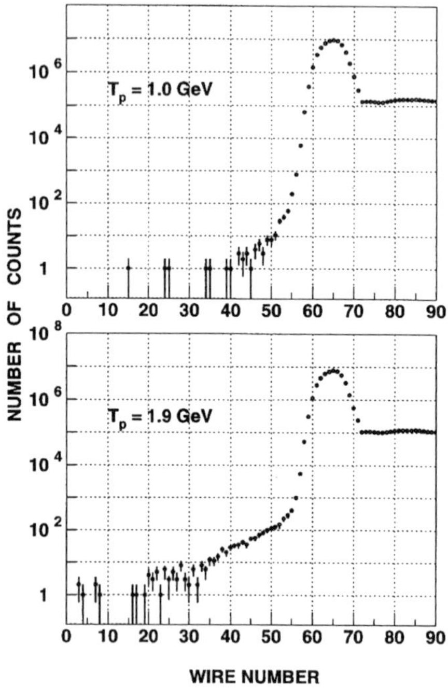

Figure 2. Position spectra of fission fragments originating from the proton bombardment of Bi and traversing the lower MWPC. The scale of the abscissa is 1 wire = mm. The upper spectrum shows the background. The sharp decrease between wire numbers ~55 and 63 is due to energy loss, absorption and small-angle scattering[60]. The peak around wire number 65 is due to prompt fission fragments passing the left edge of the diaphragm shown in Fig. 1. The plateau at wire numbers ≥72 is caused by prompt fission fragments passing the narrow slit in the diaphragm.

All measured data are well described with the phase space model[43] and a weak coupling parameter $\alpha = 1.3 \cdot 10^{-15}$, which describes the ratio of the weak interaction $\Lambda N \rightarrow NN$ relative to the strong interaction $\Lambda N \rightarrow \Lambda N$. Very recently Alberico, De Pace, Garbarino and Ramos[67] carried out a new evaluation of the Λ decay width in nuclear matter within the propagator method. Through the local density approximation they have obtained results in finite nuclei. The parameters of the model that control the short-range correlation were adjusted to reproduce the nonmesonic width of ${}^{12}_{\Lambda}C$. The lifetime prediction for ${}^{208}_{\Lambda}Pb$ is 188 ps which compares quite well with the COSY-13 result. The prediction by M. Ericson[27]: 183 ps, agrees even better with our result.

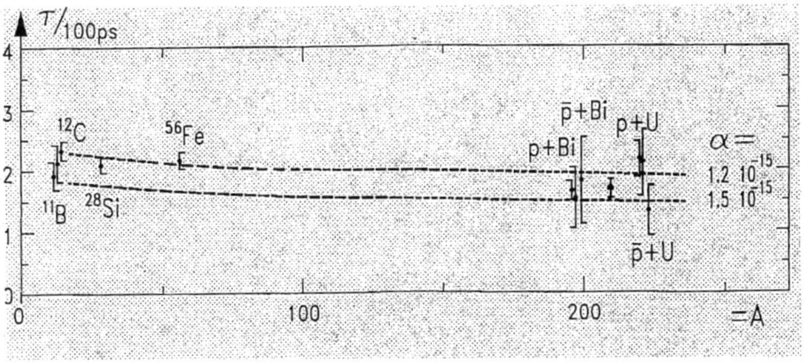

Figure 3. Lifetimes of hypernuclei with masses $A \geq 11$. The point plotted near $A \sim 210$ is the weighted mean of the $\bar{p}$ and COSY results: (167 ± 15)ps. The dashed curves are phase space predictions[43] for a weak coupling parameter $\alpha = 1.2 \cdot 10^{-15}$ and $1.5 \cdot 10^{-15}$.

It is worth noting that the experimental method for $A \leq 56$ hypernuclei was very different from that for the heavy hypernuclei. All data shown in Fig. 3 suggest that there is no strong nuclear structure effect. The observed cross sections[48,62,68] confirm the large enhancement of the hypernucleus production as predicted by Yamazaki[36] and Ejiri[37, 38]. From the measured prompt fission rate during the bombardment of Bi with protons, the target thickness, the prompt fission cross sections of Bi for 1.0 – 1.9 GeV protons[69, 70], and the number of protons in COSY we found that each proton was traversing the target on average $\sim 10^4$ times. Thus the internal target experiment, where the cycle time was ~10 s, corresponds to an experiment with an extracted beam of ~1.6 μA for 1.0 – 1.9 GeV protons.

ACKNOWLEDGEMENTS

We are most grateful to Prof. H. Ejiri for extremely fruitful discussions and his encouragement to carry out the experiment. We would also like to thank Prof. C. Guaraldo for his strong support of this work. Valuable cooperation by Dr. H.-J. Stein, regarding ion optics for COSY operation, is gratefully acknowledged. We also wish to thank the COSY-staff for cooperation during the runs. The work was supported by the DLR-International Bureau of the BMBF, Bonn, and the Polish Committee for Scientific Research (Grant No. 2 PO3B 065 12).

REFERENCES

1. Particle Physics Booklet, July 1998, extracted from the Review of Particle Physics, C. Caso et al., The European Physical Journal C3 (1998) 1
2. P. Barnes, LAMPF Workshop on (π, K) Physics, Los Alamos, NM 1990, Conference Proceedings No. 224, Particles and Fields Series 43, American Institute of Physics, (1991), p. 86, Eds. B. F. Gibson, W. R. Gibbs, M. B. Johnson
3. H. Ejiri, ibid. p. 260
4. T. Kishimoto, S. Ajimura, H. Ejiri et al., Nucl. Phys. A577 (1994) 263c
5. E. Oset, J. Nieves, Nucl. Phys. A585 (1995) 351c

6. T. Matoba, Proc. Fourth International Symposium on Weak and Electromagnetic Interactions in Nuclei (Osaka, June 1995) Eds. H. Ejiri, T. Kishimoto and T. Sato (World Scientific Singapore, 1995), p. 504
7. R. Grace, P. D. Barnes, R. A. Eisenstein et al., Phys. Rev. Lett. 55 (1985) 1055
8. J. J. Szymanski, P. D. Barnes, G. E. Diebold et al., Phys. Rev. C43 (1991) 849
9. E. Oset, L. L. Salcedo, Nucl. Phys. A 443 (1995) 704
10. K. Itonaga, T. Motoba and H. Bando, Z. Phys. A330 (1988) 209
11. T. Motoba, K. Itonaga and H. Bando, Nucl. Phys. A489 (1988) 683
12. D. H. Davis, R. Levi Setti, M. Raymund et al., Phys. Rev. Lett. 9 (1962) 464
13. J. P. Lagnaux, J. Lemonne, J. Sacton et al., Nucl. Phys. 60 (1964) 97
14. R. Alzetta, G. Liberti, G. Preparata, Nucl. Phys., A585 (1995) 307c
15. T. Yamazaki, Proc. KEK International Workshop on Nuclear Physics in GeV Region KEK Report 84 – 20 (1985) 3
16. V. N. Fetisov, Nucl. Phys. A585 (1995) 197c
17. R. H. Dalitz, Proc. Intern. Conf. on Hyperfragments, St. Cergue 1963, CERN Rept. 64-1, and Proc. Int. School of Physics „Enrico Fermi" Varenna, 1966, Course 38 (Academy Press, N. Y., 1967), p. 89
18. C. Y. Cheung, D. P. Heddle and L. S. Kisslinger, Phys. Rev. C27 (1983) 335
19. B. H. J. McKellar and B. F. Gibson, Phys. Rev. C30 (1984) 322
20. K. Takeuchi, H. Takaki and H. Bando, Prog. Theor. Phys. 73, 841 (1985) and references therein
21. J. F. Dubach, Nucl. Phys. A450 (1986) 71c
22. E. Oset and L. L. Salcedo, Nucl. Phys. A450 (1986) 371c
23. D. P. Heddle and L. S. Kisslinger, Phys. Rev. C33 (1986) 608
24. W. M. Alberico, A. Delace, M. Ericson et al., Phys. Lett. B256 (1991) 134
25. A. Ramos, C. Bennhold, E. van Meijgaard et al., Phys. Lett. B264 (1991) 233
26. T. Motoba, Nucl. Phys. A547 (1992) 115c
27. M. Ericson, Nucl. Phys. A547 (1992) 127c
28. A. Ramos, E. Oset and LL. Salcedo, Nucl. Phys. A585 (1995) 129c
29. E. Oset, C. Bennhold, A. Parreno et al., Proc. Fourth International Symposium on Weak and Electromagnetic Interactions in Nuclei (Osaka, June 1995), Eds. H. Ejiri, T. Kishimoto and T. Sato (World Scientific, Singapore, 1995), p. 350
30. K. Itonaga, T. Ueda, T. Motoba, Nucl. Phys. A639 (1998) 329c
31. M. M. Block and R. H. Dalitz, Phys. Rev. Lett. 11 (1963) 96
32. J. B. Adams, Phys. Rev. 156 (1967) 1611
33. B. Gibson, Il Nuovo Cimento 102A (1989) 367
34. A. Ramos, A. Parreno, C. Bennhold et al., Nucl. Phys. A639 (1998) 307c
35. Jung-Hwan Jun, Doohwan Kim, Hyoung Chan Bhang, Nucl. Phys. A639 (1998) 337c
36. T. Yamazaki, M. Aoki, Y. Fujita et al., Il Nuovo Cimento 102A (1989) 695
37. S. Ajimura, H. Ejiri, T. Fukuda et al., Nucl. Phys. A577 (1994) 271c
38. H. Ejiri, Proc. International Conf. on Physics with GeV – Particle Beams (Jülich, August 1994) Eds. H. Machner and K. Sistemich (World Scientific, Singapore, 1995), p. 143
39. J. Zofka, M. Wakai, M. Sano et al., Phys. Lett. B235 (1990) 25
40. H. Nifenecker and F. Malik, Proc. 29th Int. Winter Meeting on Nuclear Physics, Bormio 1991, Ed. I. Iori, Univ. Milano
41. W. Cassing, G. Batko, U. Mosel et al., Phys. Lett. B238 (1990) 25
42. V. P. Koptev, S. M. Mikirtytchiants, M. M. Nesterov et al., Sov. Phys. ITEP 67 (1988) 2177
43. Z. Rudy, W. Cassing, T. Demski et al., Z. Phys. A351 (1995) 217
44. Z. Rudy, W. Cassing, L. Jarczyk et al., Z. Physik A354 (1996) 445

45. A. Reuber, K. Holinde, J. Speth, Nucl. Phys. A570 (1994) 543
46. W. Cassing, Z. Rudy, L. Jarczyk et al., Proc. 8th Int. Conf. on Nuclear Reaction Mechanisms, Varenna, June 1997, p. 142, Ed. E. Gadioli, nucl-th/9709032
47. O. W. B. Schult, L. Jarczyk, W. Cassing et al., Nucl. Phys. A585 (1995) 247c
48. P. Kulessa, Z. Rudy, M. Hartmann et al., Phys. Lett. B427 (1998) 403
49. H. C. Bhang, S. Ajimura, K. Aoki et al., Nucl. Phys A629 (1998) 412c
50. H. C. Bhang, S. Ajimura, K. Aoki et al., Nucl. Phys. A639 (1998) 269c
51. V. Metag, E. Liukkonen, G. Sletten et al., Nucl. Instr. & Meth. 114 (1974) 445
52. J. P. Bocquet, M. Epherre-Rey-Campagnolle, G. Ericsson et al., Phys. Lett. B182 (1986) 146
53. J. P. Bocquet, M. Epherre-Rey-Campagnolle, G. Ericsson et al., Phys. Lett. B192 (1987) 312
54. T. A. Armstrong, J. P. Bocquet, G. Ericsson et al, Phys. Rev. C47 (1993) 1957
55. V. I. Noga, Yu. N. Ranyuk. N. Ya. Rutkevich et al., Yad. Fiz. 43 (1986) 1332
56. V. I. Noga, Yu. N. Ranyuk, N. Ya. Rutkevich et al., Yad. Fiz. 46 (1987) 1313
57. H. R. Koch, G. Riepe, A. Hamacher et al., Nucl. Instr. and Meth. A271 (1988) 375
58. R. Maier, Nucl. Instr. & Meth. in Phys. Res. A390 (1997) 1
59. O. W. B. Schult, R. Maier, U. Pfister et al., Physica Scripta 48 (1993) 47
60. K. Pysz, I. Zychor, T. Hermes et al., Nucl. Instr. and Meth. A420 (1999) 356
61. P. Kulessa, K. Pysz, I. Zychor et al., Nucl. Phys. A639 (1998) 283c
62. H. Ohm, T. Hermes, W. Borgs et al., Phys. Rev. C55 (1997) 3062
63. H. Ohm, W. Borgs, W. Cassing et al., Nucl. Phys. A629 (1998) 416c
64. I. Zychor, Annual Report of the Andrzej Soltan Institute for Nuclear Studies, Pl-05400 Swierk, Poland (1998)
65. O. W. B. Schult, K. Pysz, St. Kistryn et al., Proc. 35th Int. Winter Meeting on Nucl. Phys. (1997) p. 90, Ed. I. Iori, Univ. Milano, Italy
66. H. C. Bhang, S. Ajimura, K. Aoki et al., Phys. Rev. Lett 81 (1998) 4321
67. W. M. Alberico, A. De Pace, G. Garbarino, A. Ramos, Nucl. Theory, abstract nucl-th/9902023, to be published in Phys. Rev. C
68. K. Pysz, W. Borgs, W. Cassing et al., Conf. Proc. 59, Nucl. Data for Science and Technology, SIF, Bologna (1997) 354, Eds. G. Reffo, A. Ventura, C. Grandi
69. J. Hudis and S. Katcoff, Phys. Rev. C13 (1976) 1961
70. E. S. Matusevich and V. I. Regushevskii, Sov. J. Nucl. Phys. 7 (1968) 708

CHIRAL SYMMETRY AND WEAK DECAY OF HYPERNUCLEI

Makoto OKA
Dept. of Physics, Tokyo Institute of Technology
Meguro, Tokyo 152-8551, Japan
E-mail: oka@th.phys.titech.ac.jp

The weak decays of hyperons and hypernuclei are studied from the chiral symmetry viewpoint. The soft pion relations are useful in understanding the isospin properties of the weak hyperon decays. Recent development on the short-range part of the $\Lambda N \to NN$ weak transitions shows fairly good account of the weak decays of hypernuclei, though it fails to explain the n/p ratio. The π^+ decays of light hypernuclei are studied in the soft pion approach. They are related to the $\Delta I = 3/2$ amplitudes of the nonmesonic decay.

1 Introduction

While chiral symmetry is a powerful tool in understanding properties of low lying hadrons and their interactions, its role in the weak decays of hypernuclei has not been explored as far as we know. Here we suggest that chiral symmetry is significant in understanding weak mesonic decays of hyperons in the free space and also in nuclear medium. On the other hand, it was recently shown that the short range part of the weak $\Lambda N \to NN$ transition is attributed to the direct quark processes, and numerical calculation suggest that it is significant in nonmesonic weak decay of hypernuclei.[1,2]

2 Chiral Symmetry

What is the role of chiral symmetry in the hyperon decay? A useful tool to take account of the chiral symmeric dynamics of hadrons is the soft pion theorem,[3]

$$\langle \alpha \pi^a(q) | \hat{O} | \beta \rangle \stackrel{q^\mu \to 0}{\longrightarrow} -\frac{i}{f_\pi} \langle \alpha | [Q^a, \hat{O}] | \beta \rangle + (\text{pole terms}) \tag{1}$$

This can be applied to the weak pionic processes,such as,

$$\langle n \pi^0(q) | H^{PV} | \Lambda \rangle \stackrel{q^\mu \to 0}{\longrightarrow} -\frac{i}{f_\pi} \langle n | [Q_5^0, H^{PV}] | \Lambda \rangle = -\frac{i}{2f_\pi} \langle n | H^{PC} | \Lambda \rangle \tag{2}$$

Here H^{PV} is the parity violating part of the weak hamiltonian, which contains only the left-handed currents $\bar{q}_L^a \gamma^\mu q_L^a$ and the flavor singlet right handed current induced by the penguin type QCD corrections. This allows us to relate the commutation $[Q_5^0, H^{PV}]$ in the second expression of eq.(2) to H^{PC} in the final expression as

$$[Q_R^a, H_W] = 0$$
$$[Q_5^a, H_W] = -[Q_L^a, H_W] = -[I^a, H_W]$$
$$\langle n|[Q_5^a, H^{PV}]|\Lambda\rangle = \langle n|[I^a, H^{PC}]|\Lambda\rangle = -\frac{1}{2}\langle n|H^{PC}|\Lambda\rangle$$

Thus, the parity violating amplitudes, or the S-wave decay amplitudes, of pionic decays of various hyperons can be expressed in terms of the baryonic matrix elements of the parity conserving weak hamiltonian. As a result, for instance, the $\Delta I = 1/2$ dominance of $\Lambda \to N\pi$ decays follows immediately since "$\Lambda \to n$ transition" is purely $\Delta I = 1/2$. Furthermore if we conjecture that H_W is purely flavor octet, then various matrix elements, $\langle n|H^{PC}|\Sigma\rangle$, $\langle\Sigma|H^{PC}|\Xi\rangle$, are all related to $\langle n|H^{PC}|\Lambda\rangle$ and thus several relations among the pionic decay amplitudes of hyperons are obtained. Such relations are known to be satisfied fairly well for the S-wave decay amplitudes of the hyperon decays.[4]

The parity conserving decays belong to exceptions of the soft pion theorem, in which the pole terms cannot be neglected. The pole terms are such that $\Lambda \to n \to n\pi^0$ or $\Lambda \to \Sigma^0\pi^0 \to n\pi^0$, and their amplitudes are

$$\langle n\pi^0(q \to 0)|H^{PC}|\Lambda\rangle \sim \langle n\pi^0|n\rangle \frac{i}{m_\Lambda - m_n}\langle n|H^{PC}|\Lambda\rangle + \langle n|H^{PC}|\Sigma^0\rangle \frac{i}{m_n - m_\Sigma}\langle\Sigma^0\pi^0|\Lambda\rangle \quad (3)$$

Assuming the pole dominance of the parity conserving, P-wave, decay amplitudes, we find interesting relations of the $\Sigma^+ \to n\pi^+$ decay. It can be easily shown that the soft-pion amplitude vanishes,

$$\langle n\pi^+|H^{PV}|\Sigma^+\rangle_{\text{soft-pion}} = 0 \quad (4)$$

since

$$[I^-, H^{PC}] = 0 \quad (5)$$

for the $\Delta I = 1/2$ dominant H^{PC}, which has $\Delta I_3 = -1/2$. On the other hand, there is no such constraint for the pole terms of the parity conserving amplitudes, where the p, Σ^0 and Λ intermediate states with different energy denominators contribute. Therefore the soft-pion theorem suggests that the $\Sigma^+ \to n\pi^+$ decay goes only through the parity conserving P-wave channel. This is indeed what we observe experimentally, the PV amplitude 0.13 *v.s.* the PC 42.2. This example shows that the $\Delta I = 1/2$ dominance and the soft-pion relation is very well satisfied in this decay.

We later consider π^+ decays of hypernuclei in this context and see that the soft-pion theorem suggests the π^+ decays are induced only by the $\Delta I = 3/2$ part of the weak hamiltonian.

3 $\Delta I = 1/2$ Rule

In the above discussion, we have assumed that $\Delta I = 1/2$ dominance of the weak matrix element, $\langle N|H^{PC}|\Sigma\rangle$. Explanation of the "$\Delta I = 1/2$ rule" has been a long standing problem of the weak decay of the kaons and the hyperons. It was shown long time ago that the perturbative QCD corrections to the standard model weak vertex enhance the $\Delta I = 1/2$ component, while it suppresses the counterpart, $\Delta I = 3/2$ component. The mechanism can be understood easily by decomposing the weak $s + d \to u + d$ transition into the isospin-spin-color eigenstates. As the strangeness changing transition is induced only by the charged current, or the W-boson exchange, the vertex at low energy is given without the QCD corrections by

$$(\bar{u}_L^\alpha \gamma^\mu s_L^\alpha)(\bar{d}_L^\beta \gamma_\mu u_L^\beta) = (\bar{d}_L^\beta \gamma^\mu s_L^\alpha)(\bar{u}_L^\alpha \gamma_\mu u_L^\beta) \tag{6}$$

where α and β are color indices and the equality comes from the Fierz transformation. ¿From this we observe that the color+isospin combination of the final $u + d$ quarks is always symmetric, namely, ($I_f = 0$, Color $\bar{\mathbf{3}}$) or ($I_f = 1$, Color $\mathbf{6}$). In both cases, the total spin of the final quarks must be 0. When we consider gluon corrections to this vertex, we notice that the gluon exchange, or its color-magnetic component, $-(\lambda_1 \cdot \lambda_2)(\sigma_1 \cdot \sigma_2)$ term, between the final u and d is attractive for ($I_f = 0$, Color $\bar{\mathbf{3}}$),

$$\langle S = 0, C = \bar{\mathbf{3}}|(\lambda_1 \cdot \lambda_2)(\sigma_1 \cdot \sigma_2)|S = 0, C = \bar{\mathbf{3}}\rangle = -8, \tag{7}$$

while it is repulsive in the other,

$$\langle S = 0, C = \mathbf{6}|(\lambda_1 \cdot \lambda_2)(\sigma_1 \cdot \sigma_2)|S = 0, C = \mathbf{6}\rangle = +4. \tag{8}$$

Therefore the QCD correction tends to enhance the final $I_f = 0$ amplitude. The above heuristic explanation of the $\Delta I = 1/2$ enhancement can be confirmed in the renormalization group improved effective action of the strangeness-changing weak interaction.[5] It was also shown that a further enhancement of $\Delta I = 1/2$ is resulted due to the "Penguin" diagrams, which is regarded as a QCD-corrected $s \to d$ transition, and is purely $\Delta I = 1/2$.[5]

The perturbative QCD corrections are not the only source of the enhancement, but we expect further effects due to nonperturbative origin. In fact, it is known that the enhancement in the effective interaction is not large enough to explain the observed dominance of $\Delta I = 1/2$. We here concentrate on the baryonic weak interaction. Miura-Minamikawa and Pati-Woo[6] pointed out that the $\Delta I = 3/2$ part of the nonleptonic hyperon decays is suppressed due to the color symmetry of the valence quarks in the baryon. It is understood easily by considering the color structure of the 4-quark operator which belongs to the 27-dimensional irrep. of the flavor $SU(3)$:

$$O(27) = (\bar{u}_L^\alpha \gamma^\mu s_L^\alpha)(\bar{d}_L^\beta \gamma_\mu u_L^\beta) + (\bar{d}_L^\alpha \gamma^\mu s_L^\alpha)(\bar{u}_L^\beta \gamma_\mu u_L^\beta) \tag{9}$$

which is the part responsible for the $\Delta I = 3/2$ transition. ¿From the symmetry of the final u and d quarks, this operator creates two quarks with their color part being symmetric, *i.e.*, the color **6** state. As the color wave function for the valence three quarks of the baryons, either the hyperon in the initial state or the nucleon in the final state, this operator cannot be connected to two quarks inside the baryon. Thus, the only possibility is the external diagram in which both the quark and the antiquark of the meson (pion) are connected to the weak vertex directly. Such diagram is not allowed either because the external $q\bar{q}$ created by eq.(9) should be both left-handed and therefore does not make a pseudoscalar meson in the chiral limit. Gluon exchanges among the initial or final quarks do not help, while "exotic" component such as valence gluon will change the situation. Thus we observe that the $\Delta I = 3/2$ part of the $Y \to N +$ PS meson is strongly suppressed as far as we consider the valence quark picture of the baryons. It should be noted, however, that the vector mesons may couple with $\Delta I = 3/2$ verteces directly.[7]

Another possibility is the $\Delta I = 1/2$ enhancement due to the diquark components in the baryon. It is an enhancement of 0^+ $ud(I = 0)$ diquark components, which is favored by the gluon exchanges. This is in fact the same mechanism as the $\Delta I = 1/2$ enhancement in the perturbative correction discussed above. We, however, have found that such effect may be small because the naive evaluation of the H^{PC} in the harmonic oscillator valence quark model gives strong enough transition amplitude for the $\Lambda \to N\pi$ decay according to the soft-pion formula, eq.(2).

4 Direct Quark Mechanism and Weak Decay of Hypernuclei

The pionic decay of Λ is known to be suppressed in nuclear medium as the final nucleon do not have enough momentum to go above the Fermi energy. Thus the main decay mode is nonmesonic, which can be described by the simplest elementary processes, $\Lambda p \to pn$ and $\Lambda n \to nn$. These processes are viewed as weak baryonic interactions, which is unique and interesting itself as a new type of the nonleptonic weak interactions of baryons. Furthermore, this is a reaction in which the momentum transfer is so large that the quark substructures of the baryons may be significant.

Recently, we proposed the direct quark (DQ) transition mechanism to account for the short-range part of the $YN \to NN$ weak interaction.[1,8] The DQ transition potential is obtained by evaluating $su \to ud$ and $sd \to dd$ transitions among the valence quarks in two baryons.

The decay rates of the Λ in nuclear matter, and in light hypernuclei are calculated with the transition potential, which includes DQ and π and K meson exchanges. The results are compared with those without DQ, and also with experiment. We leave the details to literature,[9] while the conclusions of our study are summarized here. (1) The DQ transition is significantly large, and shows qualitative differences from the meson exchanges. (2) The one pion exchange (OPE) mechanism yields a large tensor amplitude, *i.e.*, the transition from $\Lambda p :^3 S_1$ to $np :^3 D_1$. This seems in fact too large if we employ a hard form factor as is used in the Bonn potential. We have pointed out that softer form factors for the pion-nucleon couplings are more appropriate. (3) The large tensor amplitude in OPE causes a difficulty that the nn/np ratio is too small compared to the value suggested from experiment, so-called the n/p ratio problem. We found that this problem is not solved completely only by the introduction of a soft form factor for OPE, but it is clear that the short-range mechanisms of the weak transition are important. Indeed, the DQ amplitude enhances the nn decay rate significantly and therefore improves the ratio. (4) The $\Delta I = 3/2$ contribution is significant for the $J = 0$ transition amplitudes in the DQ mechanism, while the meson exchanges are assumed purely in the $\Delta I = 1/2$ transition. It should be noted here that the mechanisms to enhance the $\Delta I = 1/2$ amplitudes in other hadronic weak interactions are not effective here and therefore the $\Delta I = 3/2$ amplitudes may be comparable to the $\Delta I = 1/2$ one. Unfortunately, it is not possible at present to determine the importance of the $\Delta I = 3/2$ contribution from the experimental data.[10]

5 π^+ decay mode and $\Delta I = 3/2$ amplitudes

Light hypernuclei may decay weakly by emitting a pion. While the free Λ decays into $p\pi^-$ or $n\pi^0$, the π^+ decay requires an assistance of a proton, *i.e.*, $\Lambda + p \to n + n + \pi^+$. Some old experimental data suggest that the ratio of π^+ and π^- emission from ${}^4_\Lambda He$ is about 5%.[11] The most natural explanation of this process is $\Lambda \to n\pi^0$ decay followed by $\pi^0 p \to \pi^+ n$ charge exchange reaction. It was evaluated for realistic hypernuclear wave functions and found to explain only 1.2% for the π^+/π^- ratio.[12] Another possibility is to consider $\Sigma^+ \to \pi^+ n$ decay after the conversion $\Lambda p \to \Sigma^+ n$ by the strong interaction such as pion or kaon exchanges. It was found, however, that the free Σ^+ decay which is dominated by P-wave amplitude, gives at most 0.2% for the π^+/π^- ratio.

In order to solve this problem, we have applied the soft-pion technique to the π^+ decay of light hypernuclei.[13] The soft-pion theorem to the process $\Lambda p \to nn\pi^+(q \to 0)$ reads

$$\lim_{q\to 0}\langle nn\pi^+(q)|H_W|\Lambda p\rangle = -\frac{i}{\sqrt{2}f_\pi}\langle nn|[Q_5^-, H_W]|\Lambda p\rangle \tag{10}$$

Again, because of

$$[Q_5^-, H_W] = -[I_-, H_W] \tag{11}$$

it discriminates the isospin properties of H_W. Similarly to the case of Σ^+ decay, we see that the $\Delta I = 1/2$ part vanishes as

$$[I_-, H_W(\Delta I = 1/2, \Delta I_z = -1/2)] = 0 \tag{12}$$

$$[I_-, H_W(\Delta I = 3/2, \Delta I_z = -1/2)] = \sqrt{3}H_W(\Delta I = 3/2, \Delta I_3 = -3/2) \tag{13}$$

We then obtain

$$\lim_{q\to 0}\langle nn\pi^+(q)|H_W|\Lambda p\rangle = \frac{i\sqrt{3}}{\sqrt{2}f_\pi}\langle nn|H_W(\Delta I = 3/2, \Delta I_3 = -3/2)|\Lambda p\rangle \tag{14}$$

Thus we conclude that the soft π^+ emission in the Λ decay in hypernuclei is caused only by the $\Delta I = 3/2$ component of the strangeness changing weak hamiltonian. In other words, the π^+ emission from hypernuclei probes the $\Delta I = 3/2$ transition of $\Lambda N \to NN$.

6 Conclusion

In this article, we have tried to demonstrate how the soft-pion approach is useful in understanding the weak hyperon transitions. It is amazing that the chiral symmetry plays so important role even in the weak processes. We, however, have further remaining problems. It is necessary to go beyond the chiral limit, so that the finite pion mass effects are to be included. Corrections due to the flavor $SU(3)$ breaking may also be important. These can be included by the chiral perturbation theory approach. In view of the qualitative success in the soft-pion approach, it is promising to apply the chiral effective theories to the hyperon and hypernuclear decay processes. Several recent papers attempt such approaches for the hyperon decays with some success.[14] It is interesting to study the hypernuclear decays in such formulations.

References

1. T. Inoue, S. Takeuchi and M. Oka, Nucl. Phys. **A577** (1994) 281c; Nucl. Phys. **A597** (1996) 563.
2. T. Inoue, M. Oka, T. Motoba and K. Itonaga, Nucl. Phys. **A633** (1998) 312.
3. S.B. Treiman, "Current Algebra and Its Applications", (Princeton Univ. Press, 1972); J.J. Sakurai, "Currents and Mesons", (Univ. Chicago Press, 1969).
4. J.F. Donoghue, E. Golowich and B. Holstein, Phys. Rev. **131** (1986) 319.
5. M.K. Gaillard and B.W. Lee, Phys. Rev. Lett. **33** (1974) 108; A.I. Vainshtein, V.I. Zakharov and M.A. Shifman, Sov. Phys. JETP **45** (1977) 670; F.J. Gillman, M.B.Wise, Phys. Rev. **D20** (1979) 2382; E.A. Paschos, T. Schneider and Y.L. Wu, Nucl. Phys. **B332** (1990) 285.
6. J.C. Pati and C.H. Woo, Phys. Rev. **D3** (1971) 2920; K. Miura and T. Minanikawa Prog. Theor. Phys. **38** (1967) 954
7. K. Maltman and M. Shmatikov, Phys. Rev. **C51** (1995) 1576.
8. K. Maltman and M. Shmatikov, Phys. Lett. **B331** (1994) 1.
9. K. Sasaki, T. Inoue and M. Oka, to be published.
10. R.A. Schumacher, Nucl. Phys. **A547** (1992) 143c R.A. Schumacher for the E788 Collaboration "Properties & Interactions of Hyperons", ed. by B. F. Gibson, P. D. Barnes and K. Nakai (World Scientific, 1994), p.85.
11. C. Mayeur, *et al.*, Nuovo Cim. **44** (1966) 698; G. Keyes, J. Sacton, J.H. Wickens and M.M. Block, Nuovo Cim. **31A** (1976) 401.
12. R.H. Dalitz and F. von Hippel, Nuovo Cim. **34** (1964) 779; F. von Hippel, Phys. Rev. **136** (1964) B455; A. Cieplý and A. Gal, Phys. Rev. **C55** (1997) 2715.
13. M. Oka, Nucl. Phys. **A647** (1999) 97.
14. E. Jenkins, Nucl. Phys. **B375** (1992) 561; B. Borasoy and B.R. Holstein, Euro. Phys. Jour., **C6** (1999) 85; Phys. Rev. **D59** (1999) 094025.

FLAVOUR CHANGING BARYON-BARYON COLLISION

Tadafumi Kishimoto
Department of Physics Osaka University, Toyonaka, Osaka 560-0043, Japan
E-mail: kisimoto@phys.sci.osaka-u.ac.jp

Weak nonmesonic decay of Λ hypernuclei gives information on the strangeness-changing weak baryon baryon interaction. Recently polarized Λ hypernuclei were produced and asymmetry emission of protons were measured. The derived asymmetry parameter showed large parity violation in the process which rather leaves a puzzle. Study of the inverse process $pn \to p\Lambda$ will clarify the situation for which an experiment is under preparation. Recent progress on the experiment is described.

1 Weak Baryon Baryon Interaction

Λ hypernuclei have two types of hadronic decay modes. One is mesonic decay (M-decay) and the other is nonmesonic decay (NM-decay). M-decay is a process in which the hypernucleus weak decays by the emission of a pion, similar to the Λ weak decay outside the nuclear medium. NM-decay is a process by which a Λ and a nucleon in a nucleus undergo a weak interaction, making two nucleons in the final state. It is the strangeness changing weak hyperon nucleon (YN) interaction and is related to the weak nucleon nucleon (NN) interaction by the $SU_3(F)$ symmetry. The weak process involves both parity-conserving and -non-conserving part. Recently, polarized ${}^{12}_{\Lambda}$C hypernuclei were produced by the ${}^{12}C(\pi^+, K^+)$ reaction, and their asymmetric NM-decay was observed [1]. The asymmetry parameter was found to be quite large (-1.0 ± 0.4) which first demonstrated that the large parity violation is involved in the process. However, the statistics was limited and calculated polarization was used since the M-decay of ${}^{12}_{\Lambda}$C was almost useless because of its small branching ratio and small asymmetry parameter [2].

A study of ${}^{5}_{\Lambda}$He by the (π^+, K^+) reaction was carried out to make up for the shortcomings of the ${}^{12}_{\Lambda}$C study. The M-decay of the ${}^{5}_{\Lambda}$He has a large asymmetry parameter (essentially equal to that of free Λ) with little theoretical ambiguity which made the polarization measurable in the experiment.

2 KEK E278 Experiment

The experiment (PS-E278) was carried out at the K6 beam line of KEK-PS. The ^{6}Li$(\pi^+, K^+p){}^{5}_{\Lambda}$He reaction at P_π=1.05 GeV/c was used to produce polarized ${}^{5}_{\Lambda}$He. The momentum of outgoing kaons was measured by the SKS spectrometer, which has a good energy resolution (2 MeV FWHM) and a large acceptance ($\sim$ 100 msr)[3]. The large solid angle especially in the horizontal direction ($-15° < \theta = dx/dz < 15°$) makes the simultaneous measurement of positive and negative scattering angles possible, which is essential to remove any spurious error in the asymmetry measurement.

Decay particles from the $^5_\Lambda$He were measured by the decay-counter system. Two sets of detector system were placed below and above the target. Each detector system consisted of a Si micro-strip detector (SSD), a multiwire proportional chamber (MWPC), a plastic scintillator hodoscope (DH), a range shower counter (RSC), and 36 NaI detectors. Pions and protons from the $^5_\Lambda$He decay were identified by dE/dx information and by their total energy. Details of the experiment have been presented elsewhere[4,5]

The up-down asymmetry (A) was obtained from the experimental coincidence yields as

$$\sqrt{\left(\frac{N^{\uparrow}(+\theta)N^{\downarrow}(-\theta)}{N^{\downarrow}(+\theta)N^{\uparrow}(-\theta)}\right)} = \frac{1+A}{1-A}. \tag{1}$$

Here, $N^{\uparrow}(+\theta)$ stands for number of counts in the up decay counter system when kaons are detected by the SKS spectrometer at $\theta = dx/dz$. where x is positive. First-order systematic errors cancel in this ratio.

The polarization of $^5_\Lambda$He is related to the observed pion asymmetry (A_π) by $A^\pi = P\alpha^\pi\varepsilon$ where α^π stands for the asymmetry parameter of the $^5_\Lambda$He M-decay, and ε represents the reduction factor due to the finite solid angle of the detector. The observed polarization is shown in table 1 together with the theoretical calculation[6,4]. The calculated polarization is consistent with the observation which demonstrates that the theoretical model well describes the polarization mechanism[4].

The obtained proton asymmetry is given in Table 1. The asymmetry parameter (α^{NM}) derived from the proton asymmetry (A_p) and the polarization (P) is $\alpha^{NM} = 0.22 \pm 0.20$. Its sign is opposite to theoretical calculations based on the meson-exchange model[7]. On the other hand, the asymmetry parameter for p-shell hypernuclei ($^{12}_\Lambda$C) was consistent with the meson-exchange model calculation[8,7]. One would think that the relative P state in the initial ΛN system plays decisive role for the p-shell hypernuclei, although most of the decay rate comes from the initial S state in the meson-exchange model[9].

Experimental results might simply reflect a statistical fluctuation. However, one thinks that the meson-exchange model is inadequate to describe the short-range part. It could also be an origin of the inconsistency that the theory predicts too much Γ_p over Γ_n. An effort to describe the short-range part of the interaction has been made by including a direct quark-exchange

Table 1: Observed proton asymmetry and asymmetry parameter obtained by Eq. ?? are shown. Measured polarization by observing asymmetry of mesonic decay is also shown.

Reaction		$\theta = 2 \sim 7°$	$\theta = 7 \sim 15°$
(π^+, K^+) $-4 < E_X < 4$	A_π	-0.128 ± 0.042	-0.203 ± 0.048
	P	-0.247 ± 0.082	-0.393 ± 0.094
	P_{cal}	-0.181	-0.368
	A_p	0.076 ± 0.058	0.027 ± 0.077
	α^{NM}	0.22 ± 0.20	

mechanism[10,11]. The model suggests a large $I_f = 1$ amplitude[11], which improves the discrepancy in the branching ratio (Γ_n/Γ_p). However, a consistent understanding of the asymmetry parameter requires both the meson-exchange and quark-exchange mechanisms including their relative phase. Further study is needed. I wish to stress that this is a clear puzzle that is quite rare recently in nuclear physics. Clear puzzles triggered progress in physics.

3 Weak ΛN interaction studied by the $pn \to p\Lambda$ reaction

In order to understand the weak YN interaction one wishes to study two body weak YN scattering as a function of relative momentum. The weak NM-decay of Λ hypernuclei gives limited information since other nucleons are not completely spectators and relative momentum is almost fixed. Study of the inverse reaction ($pn \to p\Lambda$) solves the problem. The cross section gives directly the transition amplitude of the process. The energy dependence of the cross section can be studied by which one can make a partial wave decomposition of the process.

The energy of proton beam has to be larger than 370 MeV for free neutron target. A 400 MeV proton beam is available at RCNP. Currently ^{9}Be is under consideration as the target which gives almost the largest effective neutron number for unit length. Produced Λ moves in the forward direction and decays into pions and nucleons after passing a few centimeter. This decay vertex is the genuine signal of the production of Λ, since no other processes produce pions away from the target. The cross section is related to the weak NM-decay rate by

$$\frac{1}{\tau_{\Lambda p \to pn}} =< v\sigma(\Lambda p \to pn) >_{av} \int \rho_N |u_\Lambda|^2 d^3r \tag{2}$$

where $\tau_{\Lambda p \to pn}$ is partial life of hypernuclear NM-decay, ρ is the density distribution of a nucleon, u_Λ is the wave function of a Λ. The estimated cross section by this equation is $\sim 10^{-39}cm^{2}$[12]. Similar values were obtained by recent theoretical calculations[13,14,15]. No Λ is produced by the strong interaction which inevitably accompany kaon production up to 1.6 GeV for a neutron target and 0.7 GeV for infinitely heavy nuclear target. The cross section is small though it is measurable by a carefully designed detector.

4 Spin dependent observables

It has been recently pointed out that there are many spin dependent observables in the reaction since we can control initial proton spin and measure final Λ spin [16]. The parity violation can also be studied by the analyzing power with longitudinally polarized proton beam. It is represented by

$$A = \frac{\sigma(h=1) - \sigma(h=-1)}{\sigma(h=1) + \sigma(h=-1)}, \tag{3}$$

where $h = J_p \cdot P_p/|J_p||P_p|$. The polarization of the proton beam can be as large as 0.8. This analyzing power in the proton proton scattering is very small ($10^{-7} \sim 10^{-8}$). On the other hand we can expect the effect of the order of unity for the $pn \to p\Lambda$. The large asymmetry parameter combined with large longitudinal polarization makes the systematic error easy to handle though one has to overcome very small cross section.

The T violation can be searched by observing the T odd correlation $(\vec{J_p} \times \vec{k_p}) \cdot \vec{J_\Lambda}$. In the search it is proved that there is no test that has no final state interaction effect [17]. One always has to investigate the accuracy achievable by the final state interaction. Currently the precision we expect in the reaction is inferior to other studies though it is always worth testing for the new reaction until limited by the final state interaction.

5 Experiment at RCNP

The signature of the reaction is the detection of π^- and proton from Λ decay. The $pn \to p\Lambda$ reaction takes place on a neutron moving in the ^{9}Be with Fermi momentum. Outgoing particles (π^- and p) from Λ decay are given by the Monte Carlo simulation using GEANT. Particles produced by the background process were estimated also by GEANT. The lifetime of Λ is 2.6×10^{-10}sec. The Λ has momentum typically 0.6 GeV/c thus the decay length of Λ is $\sim$4 cm. One can identify the Λ by detecting a π^- and a proton emitted from the same position that is several centimeters from the target. Typical angle and momentum distributions of π^- and proton are shown in figure 1 and 2 by which we designed our detector system.

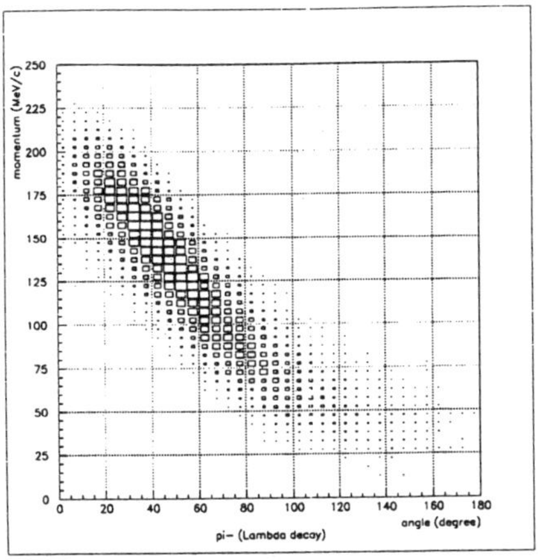

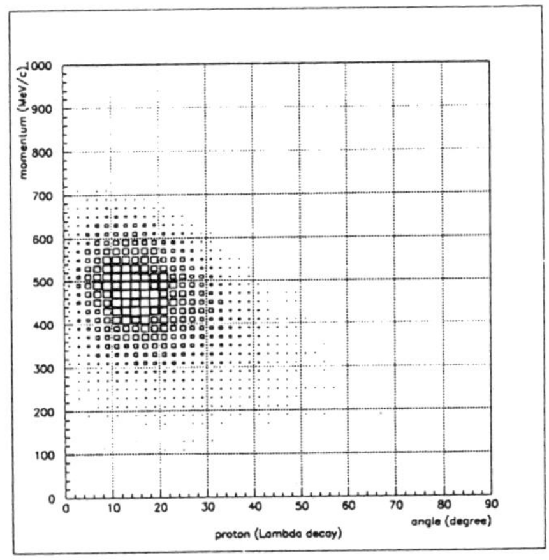

Angle and momentum distribution of pions figure 1 (left) and for proton figure 2 (right) from Λ decay are shown.

Figure 3 shows the conceptual design of our detector system. It primarily consists of a collimater, silicon microstrip detectors (SSD) at the inner most region, a CDC and trigger counters in a solenoid magnet.

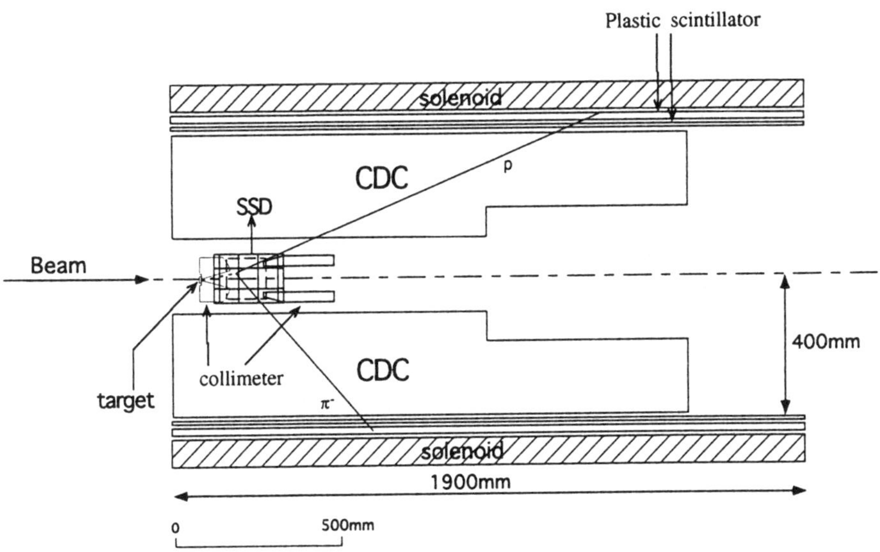

Figure 3. Schematic layout of the proposed decay counter system.

Our detector is shielded by a collimater at the target by which counting rate becomes tolerable. The small opening of the collimater reduces solid angle of Λ's though increases solid angle of detector. We thus have studied the efficiency of the signal as a function of the the opening angle. The best opening angle was found to be 15 degrees when we want to detect both a π^- and a proton. For the reality particles from the target cannot be completely shielded out. Particles leaking from the collimater have yet to be reduced for which study we are using the GEANT simulation. It is found that such leakage are well reproduced by comparing a experiment.

The signal is best identified by the vertex information where Λ decays clearly after the target. A good position resolution is obtained by the SSD. We are planning to have two layers of SSD. Strip width is 0.2 mm for z direction and 0.1 mm for azimuthal direction. We will have to read almost 10 k channels of the signal. A tiny preamplifier with multiplexer, which recently became available, makes read out of huge channels possible. Flash ADC's with multiplexer control which does data processing has been developed recently. We can now handle huge data. The outer region of the SSD will be surrounded by CDC. CDC is capable to give radial trajectory and longitudinal trajectory. It is recently completed and we are making final check.

At the outer most region we will have trigger counters which is designed sensitive only to π^-. We will use the solenoid magnet to give the momentum and charge of particles. We will give momentum of particles to finally reconstruct Λ invariant mass. It is vital to measure the charge of decay particles in a trigger level to have reasonable small trigger rate. We will thus use the solenoid magnet. Currently 3 k gauss magnet is available. A little higher field is desirable we will see real needs. A few MeV/c momentum resolution will be obtained for 100 MeV/c particles. Pions have large transverse momentum ($\sim$ 100 MeV/c). Invariant mass of Λ will be reconstructed to identify the signal. The signal to noise ratio is mostly determined by the vertex resolution though it still helps to identify the signal.

The efficiency was estimated based on the current design of the detector. The whole system will have an efficiency of 5% for Λ produced in the quasifree process on a neutron in ^{9}Be. The spectrometer efficiency (ε_{sp}) is dominantly given by Λ decay in the collimater (~ 0.15) and detection efficiency of proton.

6 Expected Results

In the present experiment we need to have a beam line that can tolerate the beam intensity. It has to provide longitudinally polarized beam to measure the parity violation. Transverse polarization is also needed to carry out the

T violation experiment. The threshold of the reaction is 370 MeV for free neutron though binding effect of a neutron in a nucleus makes the threshold a little higher effectively. Recent calculation shows that the cross section is rather large around threshold region thus beam energy of 400 MeV or less is adequate.

The expected yield of the signal is 4 events/day assuming realistic beam intensity and detector efficiency. The 30 days running times gives the ~100 events. Here we should stress that we will obtain really a new quantity in the experiment even if we can determined order-of-magnitude of the cross section. Since one can have almost 100 % polarized beam 10 events will give us asymmetry parameter with 30% error which is already better than what we have obtained in ${}^{12}_{\Lambda}C$ experiment.

References

1. S. Ajimura et al., Phys. Lett. B282 (1992) 293
2. T. Motoba K. Itonaga and H. Bandō, Nucl. Phys. A849 (1988) 683
3. T. Fukuda et al., Nucl. Inst. Meth. A 361 (1995) 485
4. S. Ajimura et al., Phys. Rev. Lett. 80 (1998) 3471
5. S. Ajimura et al., Phys. Rev. Lett. submitted
6. T. Kishimoto et al., KEK-PS proposal E278
7. A. Parreño, A. Ramos, and C. Bennhold, Phys. Rev. C **56**, 339 (1997)
8. A. Ramos, E. van Meijgaard, C. Bennhold, and B. K. Jannings, Nucl. Phys. A544 (1992) 703
9. C. Bennhold and A. Ramos, Phys. Rev. C **45**, 3017 (1992).
10. C.-Y. Cheung, D.P. Heddle, and L.S. Kisslinger, Phys. Rev. C **27**, 335 (1983); D.P. Heddle and L.S. Kisslinger, *ibid.* **33**, 608 (1986).
11. T. Inoue, S. Takeuchi, and M. Oka, Nucl. Phys. A **597**, 563 (1996); T. Inoue *et al.*, *ibid.* **633**, 312 (1998).
12. T. Kishimoto, Proc. on Weak and Electromagnetic Interactions in Nuclei (WEIN'95), 1995, Osaka, 514
13. J. Haidenbauer et al., Phys. Rev. C52 (1995)3496
14. M. Oka, Nucl. Phys. **A639**, (1998) 317c
15. A. Parreño, A. Ramos, N.G. Kelkar, C. Bennhold Phys. Rev. Cbf 59, 2122 (1999)
16. H. Nabetani, T. Ogaito, T. Sato, T. Kishimoto, Phys. Rev. C**60**, 017001, (1999)
17. H. E. Conzett, Proc. IV Int Symp. on Weak and Electromagnetic Inteactions in Nuclei"WEIN'95, June 1995, Osaka, World Scientific, (1995) 68

HADRON SPIN POLARIZATION – WHERE THE SPIN-ORBIT INTERACTION? –

K.-I. KUBO and Y. KITSUKAWA
Departoment of Physics, Tokyo Metropolitan University
1-1 Minami-osawa, Hachioji, Tokyo 192-0397, Japan

The hadron polarization induced by the high energy hadron collisions seemed to have been well established as its origin for producing the spin asymmetry is the spin-orbit interaction. However after one theoretical work which indicated a failour of the spin-orbit interaction to provide correct p_T dependence of the spin polarizations, its genuine origin to creat spin asymmetry distributions became obscure. We discuss here intuitively and numerically what kind of spin-orbit we should argue for origin of the spin polarizations and we solve the existing confusions.

1 Introduction and Motivation

For about last 5 years, we have been extensively involved in study fo the spin polarization of high energy hadron productions. Typical examples we worked on are represented in Figs. 1 and 2. The first example shown in Fig. 1 is the spin polarization of Λ particles from pp collision and from K^-p collision as a function of P_T at the various incident momenta. These two polarizations show different sign and magnitude with different slope in the P_T distributions. The second example shown in Fig. 2 is the analyzing powers of π production reactions as a function of x_F from the spin polarized pp collision at 200 GeV/c incident momentum. The obeserved analyzing powers for π^+, π^0, π^- productions are represented; they also show different sign, magnitude and slope. The solid curvers in the two figures are the results obtained by our microscopic quark recombination (QRC) model.[1] The results of calculation reproduce the experimental data very well. Basic idea of the model is represented in Fig. 3 in diagram. The two partons with momenta $\boldsymbol{p}_1$ and $\boldsymbol{p}_2$ from the colliding two hadrons H_a and H_b interact and combine to make the final hadron H_c with momenta $\boldsymbol{p}_3$ and $\boldsymbol{p}_4$. We will show some more detail of the formulation later.

What is dynamics behind the model and what is unified interaction, if exist, in such a general success of the model? For this quession, we have already answered that it is nothing but the spin-orbit interaction. [1] About two decades ago, the same answer was given by DeGrand and Miettinen.[2] They proposed a simple and useful kinematical model to estimate sign and order of magnitudes of the spin observables of hadrons produced in the high energy hadron collisions. In the model, they assumed one empirical rule "fast spin up and slow spin down" to provide spin asymmmetry depending on difference in

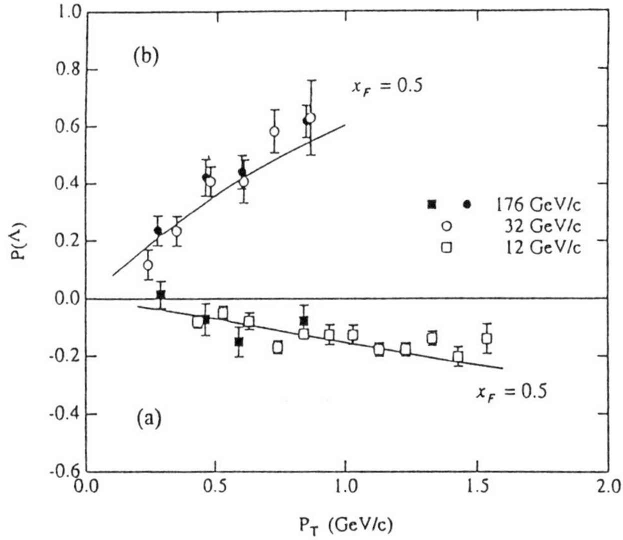

Fig. 1. The Λ spin polarization induced by the inclusive (a) pp (b) $K^- p$ collisions at the three different incident momenta.

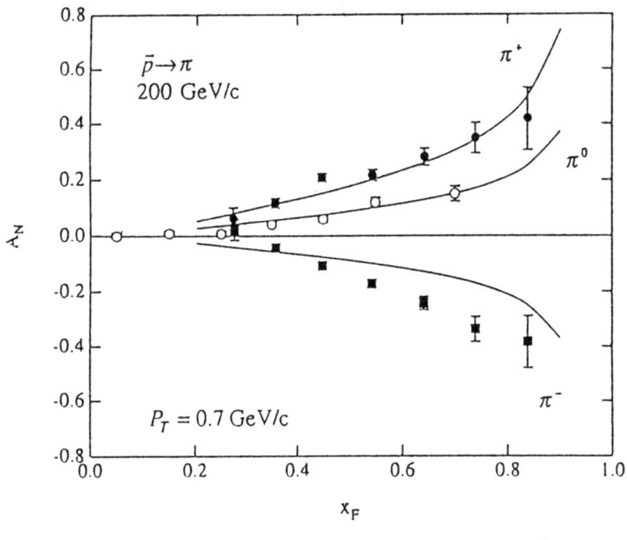

Fig. 2. The analyzing powers of the π-on production reactions induced by the spin polarized pp collision at 200 GeV/c.

velocities of the participating partons. Using the rule, they calculated the angular velocity of Thomas precession and obtained a spin-orbit type interaction potential. Then they concluded that the empirical rule they used is equivalent to account of the spin-orbit interaction and those spin asymmetries observed in the hadron productions arise from the spin-orbit interaction.

Their conclusion became, however, obscure after the work of Fujita and Matsuyama[3] and following DeGrand's comment.[4] The former paper cramed that the Λ spin polarization calculated directly from the spin-orbit interaction shows an inconsistent P_T dependence compared with that observed. The latter comment was not able to make clear cut arguement for the crame since their model does not predict spin polarization as a function of P_T, but estimate only its sign and order of magnitude.

Therefore it is quite interesting and critical for our QRC model to clarify origin of success of the model since our microscopic model can predict the spin polarization as function of P_T and x_F and we do not use the "empirical rule" mentioned above but microscopically formulate the spin observables using the Dirac spinors. As we mentioned at begining, we already indicated that our spin dependent transition amplitude obtained from the QRC formulation can be rewritten in a form including the spin-orbit interaction. As the indication was, however, made with too short words, here we discuss the same result in detail.

This is our first motivation of the present work. We will conclude that not the static or conventional one but the dynamical spin-orbit interaction is its essential origin. A significant difference between the two spin-orbit interactions will be also discussed.

The second motivation of this paper is to show where and how our spin-orbit interaction works in the hadron collisions to polarize the spin of hadrons. We will understand it in the same level as we see roles of the ordinal spin-orbit interaction in the low energy nuclear scattering and nuclear shell structure study.

2 The QRC model and the spin dependent term

The details of formulation of cross section and spin observables from the QRC model are given in Ref.1. The cross section is obtained by integration of the transition probability $|M|^2$ with the initial and final momentum distribution functions G_i of participating quark partons (see Fig. 3). The transition amplitude M is sum of the lowest-order $M^{(2)}$and the higher-order $M^{(h)}$

$$-iM = -i(M^{(2)} + M^{(h)}) \quad . \tag{1}$$

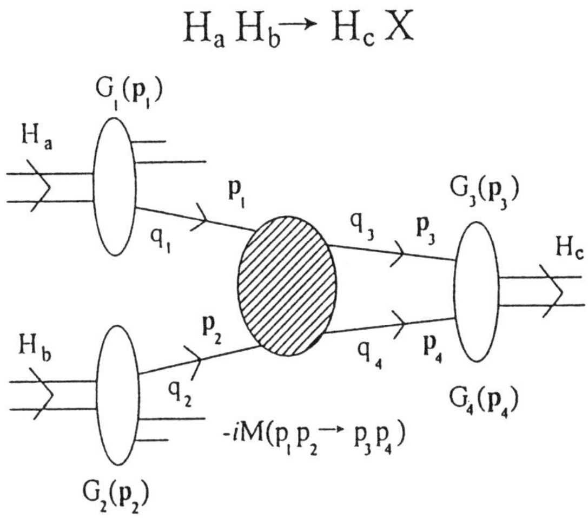

Fig. 3. The graphical description of the microscopic quark recombination (QRC) model.

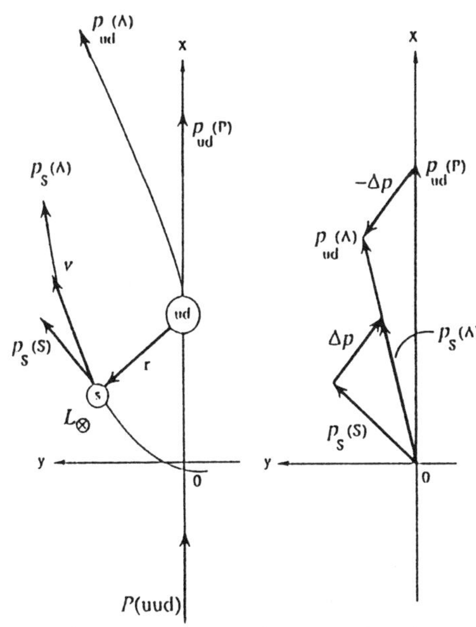

Fig. 4. The left-hand side; the pictorial presentation of recombination process of the pp→ΛX collision and the right-hand side; relations of the linear momenta appearing in the collision.

For the case of pp→ ΛX collision, the lowest-order $M^{(2)}$ is expressed as

$$-iM^{(2)}(p_4\mu_4 p_3\mu_3; p_2\mu_2 p_1\mu_1) = -g^2 V(q)\bar{u}^{(\mu_4)}(p_4)u^{(\mu_2)}(p_2) \quad , \tag{2}$$

where

$$u^{(\mu_j)}(p_j) = N \begin{pmatrix} \chi^{(\mu_j)} \\ \dfrac{\mathbf{s}_j \cdot \mathbf{p}_j}{E_j + m_j}\chi^{(\mu_j)} \end{pmatrix} \qquad \text{for} \quad j = 2, 4 \quad . \tag{3}$$

u are the Dirac four-component spinors of the spin carrier s partons in the initial sea ($j = 2$) and final lambda particle ($j = 4$).

In the model, for simplicity, we assume that the transition interaction is scalar type and constant and the higher-order term is also constant. Therefore, in the model, spin dependence arises only from property of the Dirac spinor.

The transition probability effective for the spin asymmetry appears only in the spin non-flip component [1]

$$\begin{aligned} &\left|M(\mu_4 = \mu_2 = \pm 1/2; \mu_3 = \mu_1 = 0)\right|^2 = (E_4 + m_2)(E_2 + m_2) \\ &\times \left[g^4 V(q)^2 \left\{ \left(1 - \frac{\mathbf{p}_4 \cdot \mathbf{p}_2}{(E_4 + m_2)(E_2 + m_2)}\right)^2 + \left(\frac{(\mathbf{p}_4 \times \mathbf{p}_2)_z}{(E_4 + m_2)(E_2 + m_2)}\right)^2 \right\} \right. \\ &\left. \mp 2(-g^2)V(q)\mathrm{Im}[I^{(h)}]\left(\frac{(\mathbf{p}_4 \times \mathbf{p}_2)_z}{(E_4 + m_2)(E_2 + m_2)}\right) \right] \quad . \end{aligned} \tag{4}$$

Namely, the last term in the rectangular parentheses changes its sign with change of spin projection $\mu_2 = \mu_4 = \pm 1/2$, where μ_2 and μ_4 are the spin projections of the spin carrier s parton in the initial (s_2) and final (s_4) stage. This term includes z-axis projection of an outer product form of $\boldsymbol{p}_2$ and $\boldsymbol{p}_4$, where, for the pp → ΛX production, $\boldsymbol{P}_2 (= \boldsymbol{P}_s(\mathrm{S}))$ is the initial momentum of s-parton in sea (S) and $\boldsymbol{P}_4 (= \boldsymbol{P}_s(\Lambda))$ the final momentum of s-parton in Λ particle. For the case of K⁻p → ΛX production, $\boldsymbol{P}_2 (= \boldsymbol{P}_s(\mathrm{K}^-))$ is the initial momentum of s-parton in projectile K^- and $\boldsymbol{P}_4 (= \boldsymbol{P}_s(\Lambda))$ the final momentum of s-parton in Λ particle.

3 Where the spin-orbit interaction?

(1) pp → ΛX

First we consider the pp → ΛX collision. From eq.(4), the spin dependent transition probability producing Λ is proportional to the average of the transition probability $M_\pm$-square defined for $\mu = \pm 1/2$ $(\mu = \mu_2 = \mu_4)$ in the momentum and time spaces

$$|M_{\pm}|^2 = \mp[\mathbf{P}_s(\Lambda) \times \mathbf{P}_s(\mathrm{S})]_z \quad . \tag{5}$$

The corresponding spin dependent partial cross section is given by averaging the $M_{\pm}$-square over the momentum and time spaces

$$\left\langle |M_{\pm}|^2 \right\rangle = \frac{1}{\Delta t}\int_0^{\Delta t} \left\langle |M_{\pm}(t,\mathbf{p})|^2 \right\rangle_{\mathbf{p}} = \mp\frac{1}{\Delta t}\langle [\mathbf{P}_s(\Lambda) \times \mathbf{P}_s(\mathrm{S})]_z \rangle_{t\mathbf{p}} \quad . \tag{6}$$

The incident proton with constituent uud quark partons comes in along the x-axis as shown in Fig. 4. The ud diparton picks s parton up from sea and constructs Λ. Through the collision, the incident ud parton is decelerated whereas the s parton is accelerated through recombination. The various momenta of participating partons in collision are summarized in the right side of Fig. 4 where $\boldsymbol{P}_s(\mathrm{S})$ changes to $\boldsymbol{P}_s(\Lambda)$ with momentum transfer $\Delta\boldsymbol{P}$.

It will be worthwhile for a later use to mention the following two dynamical vectors appearing in the pp $\to$ ΛX collision: in Fig. 2, (1) the direction of $\Delta\boldsymbol{P}$ is antiparallel to the radial vector $\boldsymbol{r}$ between the interacting ud and s partons and (2) the direction of orbital angular momentum $\boldsymbol{L}$ of the s parton around the ud parton faces to the back side of xy-plane. The first relation indicates that since a force $\boldsymbol{F}$ acting on the s parton from the ud parton is an attractive one

$$\mathbf{F} = \frac{d\mathbf{p}}{dt} = \frac{d\mathbf{p}}{dp}\frac{dp}{dt} = -\mathbf{r}\left(\frac{1}{r}\frac{dp}{dt}\right) < 0 \quad , \tag{7}$$

then

$$\left(\frac{1}{r}\frac{dp}{dt}\right) > 0 \quad . \tag{8}$$

Using the relation $\boldsymbol{P}_s(\mathrm{S}) = \boldsymbol{P}_s(\Lambda) + \Delta\boldsymbol{P}$, eq.(6) becomes,

$$\left\langle |M_{\pm}|^2 \right\rangle = \pm\hbar\left\langle \left(\frac{1}{r}\frac{dp}{dt}\right)(\mathbf{L})_z \right\rangle = 2\hbar\left\langle \left(\frac{1}{r}\frac{dp}{dt}\right)(\mathbf{L}\cdot\mathbf{S})_z \right\rangle \quad . \tag{9}$$

Namely we find that the spin dependent transition probability is obtained as an expectation value of a spin-orbit type interaction potential. The acting force may be nonlocal, therefore we leave $\left(\frac{1}{r}\frac{dp}{dt}\right)$ without changing into the conventional form $\left(\frac{1}{r}\frac{dV}{dr}\right)$. In the conventional interaction, V is usually a local potential energy and once we accept certain V from a static central potential, we use it for all the way through collision. In this sense we may refer such potential as the static spin-orbit interaction potential and distinguish it from the present one which keeps nonlocality due to its dynamical origin.

We calculate spin polarization of Λ and we find

$$\begin{aligned} P_{pp\to\Lambda X}(p_{\rm T}, x_{\rm F}) &\propto \left\langle |M_+|^2 \right\rangle - \left\langle |M_-|^2 \right\rangle \\ &= -(2L+1)\left\langle \left(\frac{1}{r}\frac{dp}{dt} \right) \right\rangle \leq 0 \quad . \end{aligned} \tag{10}$$

The last relation of eq.(10) results from eq.(8) and the direction of the orbital angular momentum $\boldsymbol{L}$ noted above. Sign of the Λ spin polarization thus obtained is consistent with that observed (Fig. 1).

(2) $\mathbf{K^-p \to \Lambda X}$

For the case of $K^-p \to \Lambda X$ collision, we can proceed the same calculation and we get the same expression eq.(9) for the spin dependent probability M^2, but a difference is now direction of the orbital angular momentum $\boldsymbol{L}$; as we can see from Fig. 5, it faces to the front side of xy-plane. This change from the pp $\to \Lambda X$ case is consequense of the fact that spin carrier s parton appears now as a valence parton in the incident K^- and the ud parton is picked up from sea. The s parton is then decelerated and the $\boldsymbol{L}$ becomes parallel to z-axis. Therefore the spin polarization of Λ induced by K^-p collision becomes positive value which is consistent with the observed result.

This is the way intuitively to show in our QRC model that it is the spin-orbit interaction to create the spin asymmetry distributions in the hadron productions. As we have seen above, the present spin-orbit interaction arises from a dynamical origin and its sign directly depends on the dynamical parameters. We have also seen where it appears and how it plays an essential role to predict the spin polarization induced by the high energy hadron collisions.

4 A significant feature of the LS-interaction induced dynamically

The present spin-orbit interaction obtained above shows a different feature from the conventional spin-orbit potential. This is explained for the pp $\to \Lambda X$ production case as an example. In the formation of Λ particles, every sea s parton is accelerated by the ud valence parton of which momentum is extremely large compared with that of the s parton.

Therefore, as we see in Fig. 6, the two momentum transfers $\Delta \boldsymbol{P}$ in the different trajectries face to the same direction. Hence the forces acting on the spin carrier s partons from the ud parton in near-side (far-side) collision is respectively expressed by

$$\mathbf{F}_{\rm Near(Far)} = \frac{d\mathbf{p}}{dt} = +(-)\frac{\mathbf{r}}{r}\frac{dp}{dt} \leq (\geq) 0 \quad . \tag{11}$$

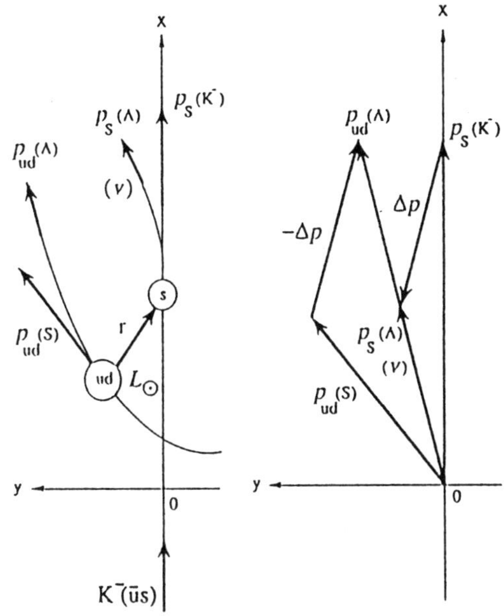

Fig. 5. The left-hand side; the pictorial presentation of recombination process of the $K^- p \to \Lambda X$ collision and the right-hand side; relations of the linear momenta appearing in the collision.

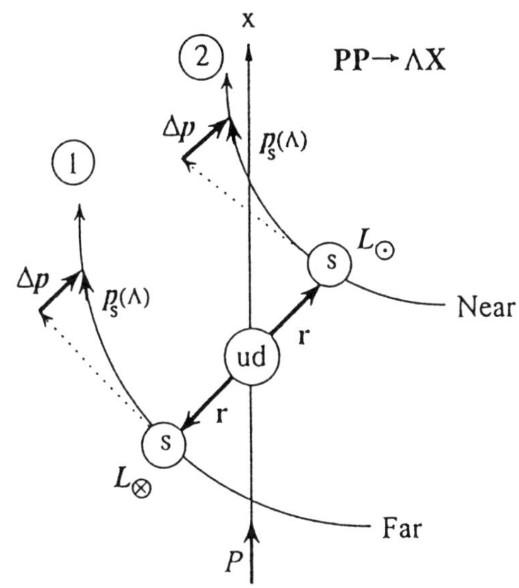

Fig. 6. The momentum transfer Δp and the orbital angular momentum L in the near-side and far-side collisions for the $pp \to \Lambda X$ collisions.

The corresponding Thomas precession angular velocity is calculated as

$$\begin{aligned}\boldsymbol{\omega}_{\mathrm{T}}^{\mathrm{Near(Far)}} &= \frac{1}{2mc^2}\left[\mathbf{F}_{\mathrm{Near(Far)}} \times \mathbf{v}\right] \\ &= +(-)\frac{\hbar}{2m^2c^2}\left(-\frac{1}{r}\frac{dp}{dt}\right)\mathbf{L} \quad . \end{aligned} \tag{12}$$

Then a spin-orbit type interaction potential is obtained

$$U_{\mathrm{T}}^{\mathrm{Near(Far)}} = \hbar\left(\boldsymbol{\omega}_{\mathrm{T}}^{\mathrm{Near(Far)}} \cdot \mathbf{S}\right) = +(-)\frac{1}{2}\left(\frac{\hbar}{mc}\right)^2\left(-\frac{1}{r}\frac{dp}{dt}\right)(\mathbf{L}\cdot\mathbf{S}) \quad . \tag{13}$$

Note that the orbital angular momentum $\boldsymbol{L}$ has opposite direction between the near-side and the far-side collisions. Therefore the spin-orbit potential, thus obtained from the dynamical origin, becomes to have the same sign in both near-side and far-side collisions. This property is quite different from the usual spin-orbit interaction arising from the static origin and appearing in the nuclear shell structure and the low energy nuclear reactions.

It is notable that the spin-orbit interaction we concern is not like one acting between a scattering particle and its center usually appearing in the conventional nuclear scattering, but the interaction between s and ud partons constituting Λ particle. If the two spin-orbit interactions corresponding to the two different trajectries had different sign, that case usually happens in the conventional spin-orbit interaction in static form, only small or no spin polarization could occur due to a large cancellation between the near-side and far-side collisions. In this sense, consideration of the present nonlocal spin-orbit interaction induced by dynamics, not the static one, is essential to describe and understand those sizable spin asymmetry distributions in the high energy hadron productions.

5 Summary

We have studied dynamics behind success of the microscopic quark recombination model and the characteristic features of the present spin-orbit interaction. The present work is summarized in the following three folds:

1. We have intuitively clarified dynamics behind the general success of the microscopic QRC model formulated for calculation of the spin observables in the high energy hadron productions; it is the spin-orbit interaction arising from the dynamical origin.

2. The spin-orbit interaction discussed here has the same sign for the both near-side and far-side collisions, which is contrast to the ordinary static

spin-orbit potential popular in the nuclear physics study. This feature is essential for producing the spin asymmetry distributions in hadron productions, otherwise small or no spin polarization occurs due to a large cancellation between the near-side and far-side collisions.

3. An obscure arguement, existed in the discussion of origin of the high energy hadron spin polarizations, has been now clearly resolved by intuitively showing that the present spin-orbit interaction gives a correct sign and by numerically showing that the presently calculated P_T dependence of Λ spin polarization is quite consistent with that observed.

It is whothwhile to note that our spin-orbit interaction has a dynamical origin and has different feature from the conventional spin-orbit interaction arising from the static origin. It was the spin-orbit potential of static origin that Fujita-Matsuyama used and cramed the incorrect P_T dependence of the Λ spin polarization.

References

1. Y. Yamamoto, K.-I. Kubo and H. Toki, Prog. Theor. Phys. **98**, 95 (1997). K.-I. Kubo, Y. Yamamoto and H. Toki, ibid **101**, 615 (1999).
2. T. A. DeGrand and H. I. Miettinen, *Phys. Rev.* D **23**, 1227 (1981). *Phys. Rev.* D **24**, 2419 (1981). *Phys. Rev.* D **31**, 661(e) (1985).
3. T. Fujita and T. Matsuyama, *Phys. Rev.* D **38**, 401 (1988). *Nucl. Phys.* A **475**, 657 (1987).
4. T. A. DeGrand *Phys. Rev.* D **38**, 403 (1988).

ISOSPIN AND SPIN-ISOSPIN MODES IN NUCLEI

R.G.T. Zegers, A.M. van den Berg, S. Brandenburg, F.R.R. Fleurot, V.M. Hannen, M.N. Harakeh, K. van der Schaaf, S.Y. van der Werf, H.W. Wilschut
Kernfysisch Versneller Instituut, Zernikelaan 25, 9747 AA Groningen, The Netherlands

J. Guillot, H. Laurent, A. Willis
Institut de Physique Nucléaire, IN2P3, Unversité de Paris-Sud, Orsay Cedex, France

J. Jänecke
Departement of Physics, University of Michigan, Ann Arbor, Michigan 48109, U.S.A.

M. Fujiwara
Research Center for Nuclear Physics, Osaka University, Suita, Osaka 567, Japan

The (^{3}He,t) reaction on Pb at $E_{^3He}$=177 MeV and the subsequent decay by proton emission were studied in order to distinguish isovector monopole strength corresponding to $2\hbar\omega$ transitions from the non-resonant continuum background. Monopole strength at excitation energies above 25 MeV was discovered and compared to the calculated strength due to the isovector giant monopole resonance and the spin-flip isovector monopole resonance. Calculations in a normal-modes framework show that all isovector monopole strength can be accounted for if the branching ratio for decay by proton emission is 20%.

1 Introduction

Spin-flip and non-spin-flip isovector resonances have been under intensive investigation for a long period. The isobaric analog state (IAS), Gamow-Teller resonance (GTR) and isovector giant dipole resonance (IVGDR) are experimentally and theoretically well understood. The isovector spin-dipole resonance (SDR) is more difficult to study since it consists of three components with spin-parity 0^-, 1^- and 2^-. Nevertheless, a reasonable picture of the strength distribution of the integrated 0^-, 1^-, 2^- multipole strength has emerged from (p,n) reaction studies[1]. Although in the NEWS 99 symposium results on the microscopic structure of the GTR and SDR in ^{208}Bi obtained via the ^{208}Pb(^{3}He,tp) reaction at 450 MeV were presented, these will not be discussed further here, since they have either been published earlier (GTR)[2,3] or will be published in the near future (SDR)[4].

For giant resonances at higher excitation energies, such as the isovector giant monopole resonance (IVGMR), its spin-flip partner, the spin-flip isovector giant monopole resonance (SIVM), and also the quadrupole resonances, the experimental investigation is seriously hampered due to the large widths of these resonances and a large, underlying, non-resonant continuum background. The high-lying isovector monopole resonances, subject of this work, play an important role in understanding nuclear structure and Coulomb effects. A good knowledge is also useful in the understanding of some astrophysical phenomena[5].

For the IVGMR success has only been achieved in π-charge-exchange experiments[6,7,8]. For the (π^-, π^0) reaction, evidence for the IVGMR was found using various targets, but for the (π^+, π^0) reaction, the evidence is less clear. Structure at high excitation energies, which might be associated with isovector monopole strength, has also been reported in the ^{90}Zr(n,p) reaction[9] and in the (^{13}C,^{13}N) reaction[10,11,12], but in these cases the observed shape and strength do not allow any conclusion with respect to the multipolarity.

Indication for the SIVM was first found in the ^{90}Zr(^{3}He,t) reaction at projectile energies of 600-900 MeV[13,14]. Also in the (p,n) reaction at 795 MeV on ^{90}Zr and Pb, strength was found[15] that is consistent with collective states of spin-parity 1^+. Observed monopole strength found in the multipole analysis of the ^{90}Zr(p,n) reaction at 295 MeV[16] was also partly associated with the SIVM.

The non-resonant continuum background, which makes interpretation of the data difficult, arises largely from quasifree processes (single-step reactions between the projectile and one of the nucleons in the target) and breakup-pickup and pickup-breakup reaction mechanisms. Few systematic studies on the continuum background exist[17,18] and usually a phenomenological description[6] is used. The subtraction of this continuum background can lead to systematic errors and, possibly, misinterpretation of the experimental results.

Various theoretical approaches have been used to describe the IVGMR and SIVM. Transition matrix elements for the IVGMR were calculated in 1972 by Auerbach[19]. Calculations in a hydro-dynamical framework were performed by Auerbach and Yeverechyahu[20]. The results from the π-charge-exchange experiments triggered more efforts, based on microscopic theories[21,22,23]. These were followed by calculations for the SIVM in the same Hartree-Fock, random-phase approximation[24]. The conclusion was that the strength distributions of the SIVM and IVGMR are very much alike. The relative contributions from the SIVM and IVGMR to monopole strength at high excitation energies depends on the kinematics of the specific reaction[25]. At high incoming energies ($E/A \geq 100$ MeV) non-spin-flip transitions are strongly quenched. Below this value both spin-flip and non-spin-flip transitions are present. In an attempt to distinguish isovector monopole strength at high excitation energies from the continuum background, the Pb(^{3}He,t)Bi reaction and the subsequent decay by proton emission was studied.

The tritons produced in quasifree and breakup-pickup processes are in coincidence with high-energy forward-peaked protons. (Note that pickup-breakup processes do not play an important role in the (^{3}He,t) reaction, since by picking up a neutron, a rather stable ^{4}He particle is created.) Since decay by proton-emission from the SIVM or IVGMR is isotropic ($\Delta L=0$), by studying coincidences between tritons at forward angles (the angular distribution of ejectiles resulting from monopole excitations peak at zero degrees) and protons at backward angles, the continuum background will be strongly reduced with respect to the monopole resonances. Of course, the result is highly dependent on the branching ratio for decay by proton-emission from the SIVM and IVGMR. The statistical decay channel is dominated by neutron emission (because of the high Coulomb barrier in heavy nuclei) and the direct decay channel by proton emission. The latter can be understood qualitatively because the IVGMR and SIVM can be described microscopically as a superposition of 1p-1h states. Data of the direct decay by proton emission of the GTR and SDR[2,3,4] indeed show a considerable branching ratio (4.9±1.3% and 13.4±3.9%, respectively) for this decay mode. Since the excitation energy for the resonances under study lead to proton-decay energies much higher than the Coulomb barrier, similar or higher branching ratios for direct decay can be expected.

2 Experiment and Results

The experiment for the IVGMR and SIVM was carried out using the Big-Bite Spectrometer[26] (BBS) at the Kernfysisch Versneller Instituut. The 177 MeV $^3He^{2+}$ beam was delivered by the AGOR cyclotron. A 7.8 mg/cm^2 thick ^{nat}Pb target was used. The BBS was set at -1° and a beam dump was placed inside the dipole-magnet chamber. In order to obtain reliable ray-trace parameters, calibrations were done at 0° using a sieve slit. Nevertheless, in order to obtain a very accurate angle definition in the excitation-energy range of interest for the IVGMR and SIVM, a special vertical-angle-defining slit was used with openings in the vertical direction chosen to coincide with the extrema of the SIVM and IVGMR (-3°, 0° and 3°)

Detection of the tritons in the focal plane was done with a detector constructed at IPN Orsay[27]. Protons were detected with the Silicon Ball[28] which, for this experiment, consisted of 15, 5 mm thick, Si(Li) detectors positioned at polar angles ranging from 95° to 160° with respect to the beam direction, and azimuthal angles between -20° and +40°, at a distance of 10 cm from the

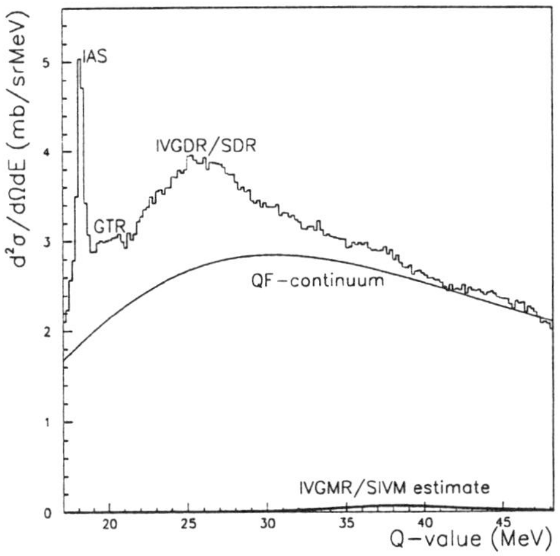

Figure 1: A singles Pb(^{3}He,t) spectrum taken at 177 MeV and $\theta = 1°$. Also indicated are the estimates for the quasifree continuum background and IVGMR/SIVM contributions; see text for more details.

target. Almost 6% of 4π was covered.

The singles spectrum is shown in fig. 1. The IAS and the combined bump from the IVGDR and SDR can easily be distinguished. In the spectrum a phenomenological estimate for the quasifree continuum [6] is drawn, as well as an estimate for the combined cross section due to the IVGMR and SIVM. This latter was obtained by performing distorted-wave Born approximation (DWBA) calculations using the code DW81 [29]. Wave functions were constructed in the normal-modes framework using the code NORMOD [30]. This procedure gives an upper limit for the total transition strength since all 1p-1h configurations that contribute to a specific resonance are added coherently. Results for the IVGMR, SIVM, IAS, as well as for the dipole resonances and the Isovector Giant Quadrupole Resonance (IVGQR) are displayed in fig. 2. Note that even at this relatively low projectile energy, the expected cross section for the SIVM is three times higher than the cross section of the IVGMR. The IVGQR, which is expected at slightly lower Q-value than the SIVM and IVGMR, has a very flat angular distribution below 4°. The estimate for the SIVM and IVGMR in fig. 1 is drawn under the assumption of a Lorentzian shape with a width (Γ) of 10 MeV.

In fig. 3 the coincident spectra are shown. In the two-dimensional spec-

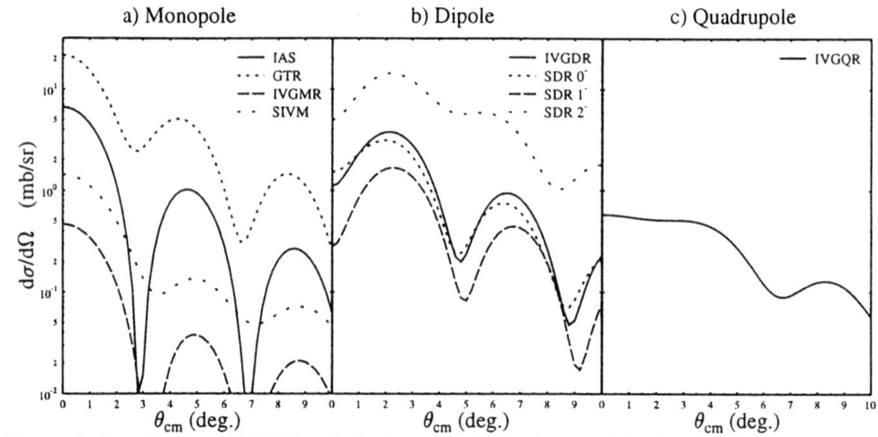

Figure 2: Results from DWBA calculations for the various multipole resonances excited in the $^{208}Pb(^{3}He,t)$ reaction at 177 MeV in the excitation-energy region of interest. See text for more details.

trum, fig. 3(a), the proton energy is plotted against the Q-value of the triton. Random events are subtracted. Around Q≈38 MeV the highest-energy protons begin to punch through the detector and to deposit less energy in the Si(Li) detectors. The most intense band in the spectrum corresponds to direct decay to low-lying states in the final nucleus, mixed with coincidences stemming from quasifree processes. The two contributions can not be distinguished kinematically since they both populate single-neutron hole states. It can be shown, by describing the quasifree scattering processes in plane-wave impulse approximation (PWIA) [31], that for the most forward proton detectors a large fraction of the coincidences in this band is due to these processes. In fig. 3(b) the projection of fig. 3(a) on the Q-axis is shown. It is found that the branching ratio for direct decay by proton emission from the IAS is 69±3%. This is within error bars equal to the result (64±4%) obtained by weighting of the branching ratios of the IAS of various lead isotopes according to their natural abundances [32].

In order to emphasise monopole strength at high Q-values, the difference spectrum between 0° and 3° was studied. The most forward detectors were excluded, to avoid interference from quasifree processes. The results are shown in fig. 4. In fig. 4(a) the spectra at 0° and 3° are shown; fig. 4(b) shows the difference. Excess cross section is present. It was checked whether this excess is isotropically distributed over the proton detectors as is expected for monopole excitations. It was found that such is indeed the case within a 95% confidence level. It must be noted that the excess contains contributions from statistical,

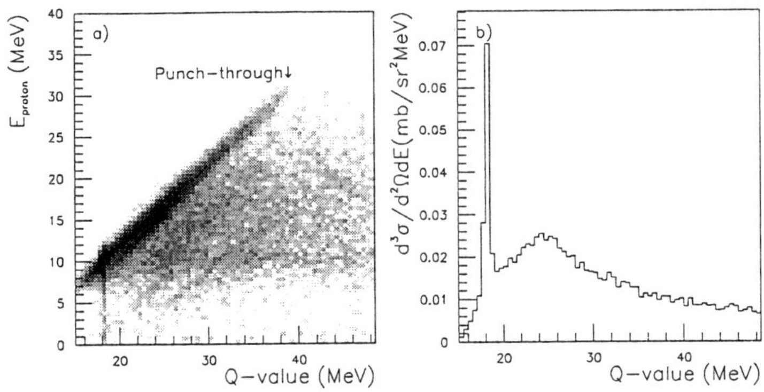

Figure 3: The two-dimensional coincidence spectrum of proton energy versus reaction Q-value (a) and its projection on the Q-axis (b) for the Pb(^{3}He,tp) reaction at 177 MeV and $\theta = 1°$.

semi-direct and direct decay by proton emission.

In order to investigate to which extent the thus-found cross section makes up for the strength due to the IVGMR and SIVM, cross-section calculations were performed for this Q-value range and the fractions of the measured cross sections relative to these were determined. The results, expressed in terms of fraction of the non-energy-weighted sum rule (NEWSR), calculated in the normal-modes framework, for the SIVM and IVGMR are shown in fig. 4(c). To produce this figure it is assumed that the SIVM and IVGMR have branching ratios of 100% for decay by proton emission. Since the calculated cross section drops strongly as a function of Q, because of the increase in momentum transfer, results above Q=45 MeV contain large error bars and are not included in the figure. The Lorentzian fits in the figures are merely to guide the eye. The slight difference in total percentage found for the NEWSR for the SIVM and IVGMR is discussed elsewhere [33].

It can be concluded that if the branching ratio for decay by proton emission from the SIVM and IVGMR is 20%, the full strength of both resonances can be accounted for. However, microscopic calculations [22,24] show a quenching of the strength with respect to the normal-modes calculations by about 40%. Assuming these values, a higher total branching ratio must be expected in order to make up for 100% of the NEWSR.

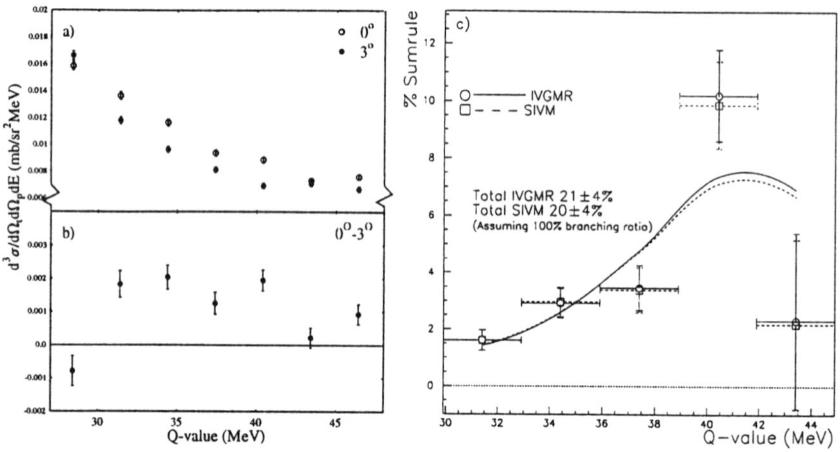

Figure 4: Coincident spectra at 0° and 3° (a) and their difference (b). (c) same as (b) after unfolding the dependence of the excitation probability of monopole strength as a function of Q-value for the separated IVGMR and SIVM contributions. The curves and fractions of NEWSR-strength exhausted are explained in the text.

3 ACKNOWLEDGEMENTS

The authors wish to thank the cyclotron crew and technical staff at KVI for their support. The research was performed as part of the research program of the "Stichting voor Fundamenteel Onderzoek der Materie" (FOM) with financial support from the "Nederlandse Organisatie voor Wetenschappelijk Onderzoek" (NWO). Support by the Scientific Affairs Division of the North Atlantic Treaty Organization (NATO), Research/Travel Grant No. CRG.971531, is gratefully acknowledged.

1. C. Gaarde et al., *Nucl. Phys.* **A369**, 258 (1981).
2. M.N. Harakeh et al., *Nucl. Phys.* **A577**, 57c (1994).
3. H. Akimune et al., *Phys. Rev.* C **52**, 604 (1995).
4. H. Akimune et al., to be published.
5. N.K. Glendenning, *Phys. Rev.* C **37**, 2733 (1988).
6. A. Erell et al., *Phys. Rev.* C **34**, 1822 (1986).
7. A. Erell et al., *Phys. Rev. Lett.* **52**, 2134 (1984).
8. F. Irom et al., *Phys. Rev.* C **34**, 2231 (1986).
9. T.D. Ford et al., *Phys. Lett.* **195B**, 311 (1987).
10. C. Bérat et al., *Nucl. Phys.* **A555**, 455 (1993).

11. I. Lhenry, *Nucl. Phys.* **A599**, 245 (1996).
12. W. von Oertzen, *Nucl. Phys.* **A482**, 357 (1988).
13. C. Ellegaard et al., *Phys. Rev. Lett.* **50**, 1745 (1983).
14. N. Auerbach et al., *Phys. Lett.* **219B**, 184 (1989).
15. D.L. Prout et al., *in Proc. of the Eighth International Symposium on Polarization Phenomena in Nuclear Physics*, AIP Conf. Proc. **339**, 458 (1995).
16. T. Wakasa et al. *Phys. Rev.* C **55**, 2909 (1997).
17. E.H.L. Aarts et al., *Phys. Lett.* **102B**, 307 (1981).
18. O. Bousshid et al., *Phys. Rev. Lett.* **45**, 980 (1980).
19. N. Auerbach, *Nucl. Phys.* **A182**, 247 (1972).
20. N. Auerbach and A. Yeverechyahu, *Ann. Phys. (N.Y.)* **95**, 35 (1975).
21. N. Auerbach and A. Klein, *Phys. Rev.* C **28**, 2075 (1983).
22. N. Auerbach and A. Klein, *Nucl. Phys.* **A395**, 77 (1983).
23. S. Adachi and N. Auerbach, *Phys. Lett.* **131B**, 11 (1983).
24. N. Auerbach and A. Klein, *Phys. Rev.* C **30**, 1032 (1984).
25. W.G. Love and M.A. Franey, *Phys. Rev.* C **24**, 1073 (1981).
26. A.M. van den Berg, *Nucl. Instr. Meth.* **B99**, 637 (1995).
27. Z. Zojceski, Ph.D. Thesis, Université Paris XI, Orsay (1997), unpublished.
28. R.G.T. Zegers et al., KVI Ann. Rep. (1998), unpublished.
29. R. Schaeffer and J. Raynal, computer code DWBA70 (1970, unpublished); extended version DW81 by J.R. Comfort (1981), unpublished.
30. S.Y. van der Werf, computer code NORMOD, unpublished.
31. C. Süskösd et al., *Nucl. Phys.* **A467**, 365 (1987).
32. J. Bordewijk, Ph.D. Thesis, Rijksuniversiteit Groningen (1993), unpublished.
33. R.G.T. Zegers et al., to be published.

Isospin- and Mirror-Symmetry Structures of Nuclei Studied through Weak, EM and Strong Interactions

Y. Fujita[a], H. Akimune[b], A. Bacher[c], G. P. A. Berg[c], T. Black[c], I. Daito[b], C. C. Foster[c], H. Ejiri[b], H. Fujimura[b], H. Fujita[a], M. Fujiwara[b], M. N. Harakeh[d], K. Harada[a], K. Hatanaka[b], T. Inomata[a], J. Jänecke[e], Y. Kanzaki[a], K. Katori[a], T. Noro[b], Y. Shimbara[a], E. J. Stephenson[c], A. Tamii[f], M. Tanaka[g], H. Ueno[a], T. Yamanaka[b] and M. Yosoi[f]

[a]*Department of Physics, Osaka Univ., Toyonaka* ,
[b]*RCNP, Osaka Univ., Ibaraki,*
[c]*IUCF, Indiana Univ., Bloomington, USA,*
[d]*KVI, Groningen, The Netherlands,*
[e]*Department of Physics, Univ. of Michigan, Ann Arbor, USA,*
[f]*Department of Physics, Kyoto Univ., Sakyo, Kyoto* ,
[g]*Kobe Tokiwa Jr. College, Nagata, Kobe,*

Abstract

Isospin symmetry structure of highly-excited states in mirror nuclei ^{27}Al and ^{27}Si having isospin $T = 1/2$ and $T_z = \pm 1/2$ has been studied by the comparison of strengths and energies of the $M1$ and Gamow-Teller (GT) transitions connected with the ground state (g.s.) of ^{27}Al. By analyzing the strengths of analogous $M1$ and GT transitions, isoscalar and orbital contributions in $M1$ transitions are discussed. The study of the symmetry structure further leads to the identification of the isospin of each pair of analogous excited-states based on the different sensitivity of the $M1$ and GT transitions to isospin transfers $\Delta T = 0$ and 1.

1. $M1$ and Gamow-Teller Transitions in Mirror Nuclei

If charge symmetry of the nuclear interaction is assumed, then isospin is a good quantum number. Every good quantum number is associated with a good symmetry structure. The aim of this paper is to look for the symmetry structure originated from quantum number isospin, i.e. "isospin symmetry structure" for the simple $T = 1/2$, odd-A mirror nuclei by comparing the strengths of analogous $M1$ and Gamow-Teller (GT) transitions. In a mirror-nuclei pair, for every state an analog state of similar structure should be found in the conjugate nucleus (see Fig. 1). Because of the analogous nature, $M1$ and GT transitions from some state or its analog to another state or its analog are all analogous. The analogous transitions are expected to have similar energies and corresponding transition strengths. Therefore, the mirror- and thus the isospin-symmetry properties of the $T = 1/2$ mirror nuclei can be investigated by combining the strength and the energy information of analog transitions in β decay, γ decay, and charge-exchange (CE) and inelastic (IE) reactions, which are caused by the weak, electromagnetic (EM), and strong interactions, respectively.

The $M1$ transitions are caused by the magnetic dipole ($M1$) interaction whose operator consists of an orbital part $g_\ell \boldsymbol{\ell}$ and a spin part $g_s \boldsymbol{s}$ [$= (1/2) g_s \boldsymbol{\sigma}$]. The $M1$ operator is further rewritten as the sum of IS and IV components (for example see

Refs. [1, 2]) as

$$B(M1) = \frac{1}{2J_i+1}\,\frac{3}{16\pi}\,\mu_N^2\Big[[M_{M1}^{\rm IS} - C_{M1}\Big(G_\ell\, M_{M1}(\ell\tau) + G_s\,\frac{1}{2}M_{M1}(\sigma\tau)\Big)\Big]^2 \quad (1)$$

$$= \frac{1}{2J_i+1}\,\frac{3}{16\pi}\,\mu_N^2\Big[M_{M1}^{\rm IS} - C_{M1}M_{M1}^{\rm IV}\Big]^2, \quad (2)$$

where μ_N is the nuclear magneton, C_{M1} is the isospin CG coefficient $(T_iT_{zi}10|T_fT_{zf})$, where $T_{zf} = T_{zi}$ holds in a $M1$ transition. The $M_{M1}(\sigma\tau)$ and $M_{M1}(\ell\tau)$ are IV-spin and IV-orbital components of the $M1$ matrix element defined by $\langle J_fT_f\|\sum_{j=1}^{A}\boldsymbol{\sigma}_j\boldsymbol{\tau}_j\|J_iT_i\rangle$, and $\langle J_fT_f\|\sum_{j=1}^{A}\boldsymbol{\ell}_j\boldsymbol{\tau}_j\|J_iT_i\rangle$, respectively. The coefficient for the $\sigma\tau$ term $G_s(= (1/2)(g_s^\pi - g_s^\nu))$, and therefore the contribution form the $\sigma\tau$ term, is the largest in a usual case [3]. The $M_{M1}^{\rm IS}$ and $M_{M1}^{\rm IV}$ are the IS and IV matrix elements, respectively, where IV term usually dominates the transition. The IS term, therefore, may interfere destructively or constructively with the IV term. In addition, the orbital term may interfere constructively or destructively with the spin term. These interference effects are strongly dependent on the configurations of the initial and final states, as was shown for $M1$ transitions in ^{28}Si [1, 4].

On the other hand, GT β decays is caused by a simple $\sigma\tau$-type operator. The situation is almost similar in hadron CE reactions at intermediate energies at 0° (in the limit of $q = 0$), because spin-isospin type interaction is dominant under such conditions [5, 6]. The B(GT) value is expressed [2, 7]

$$B({\rm GT}) = \frac{1}{2J_i+1}\,\frac{1}{2}\,C_{\rm GT}^2\,\Big[M_{\rm GT}(\sigma\tau)\Big]^2, \quad (3)$$

where $M_{\rm GT}(\sigma\tau)$ is the (IV and spin type) GT transition matrix element $\langle J_fT_f\|\sum_{j=1}^{A}(\boldsymbol{\sigma}_j\boldsymbol{\tau}_j)\|J_iT_i\rangle$ and $C_{\rm GT}$ is the isospin Clebsch-Gordan (CG) coefficient $(T_iT_{zi}1\pm1|T_fT_{zf})$.

Assuming the dominance of the $\sigma\tau$-type operator, which is mostly true for EM interaction as well as for strong interaction using hadronic probes at intermediate energy at 0°, there is a "quasi" proportional relationship between the transition strengths $B(M1)$ and B(GT), as explained in detail in Ref. [7]

$$B(M1) \approx \frac{3}{8\pi}\,(\mu_{\rm p} - \mu_{\rm n})^2\,\frac{C_{M1}^2}{C_{\rm GT}^2}\,B({\rm GT}). \quad (4)$$

Here, the numerical factor is $2.643\mu_N{}^2$ if the magnetic moments of the free nucleons are used. For the direct comparison between the values B(GT) and $B(M1)$, we define a renormalized value $B^R(M1)$ as a value $B(M1)$ divided by this factor and also by the ratio of CG coefficients.

2. Symmetry Structures in Mirror Nuclei

We studied $M1$ and GT transitions in the mirror-nuclei pair ^{27}Al and ^{27}Si connected with the g.s. of one of the two nuclei. B(M1) values for the $M1$ transitions in ^{27}Al are deduced from the data on γ decay up to $E_x = 8.5$ MeV, where the proton decay starts [8]. The analogous B(GT) values are known up to $E_x = 3.0$ MeV through

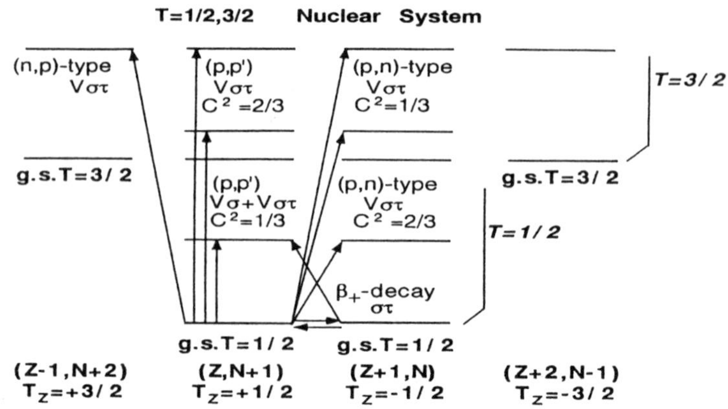

Figure 1: Schematic isospin-symmetry structures for $T_z = \pm 1/2$, odd-A mirror nuclei and the neighbouring $T_z = \pm 3/2$ nuclei. Analogous transitions connecting the ground states in the mirror nuclei with excited states in the same and the conjugate nuclei are indicated.

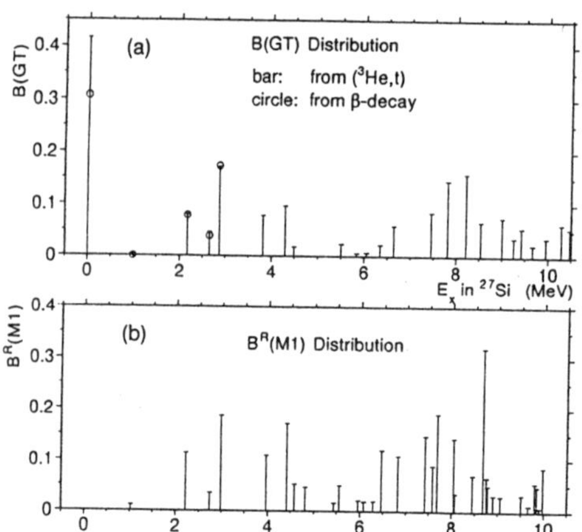

Figure 2: Comparison of (a) $B(\mathrm{GT})$ strength distribution derived from the $^{27}\mathrm{Al}(^3\mathrm{He},t)$ reaction and (b) $B^R(M1)$ distribution deduced from the measurements of γ transitions in ^{27}Al.

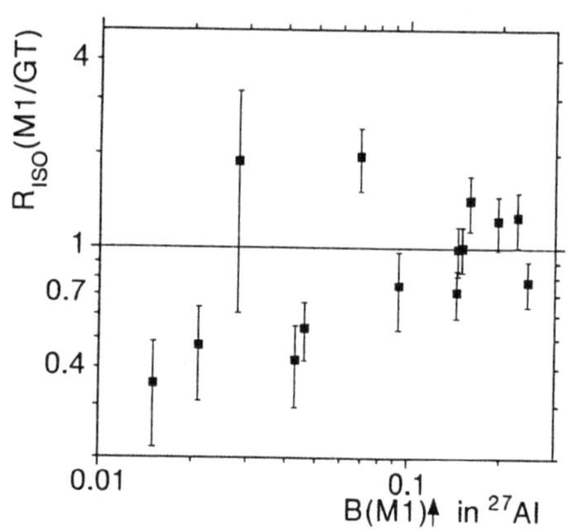

Figure 3: The ratio R_{ISO} for the $M1$ transitions in ^{27}Al as a function of $B(M1) \uparrow$ ($B(M1)$ from the g.s. to an excited state). The ratio is sensitive to the combined contribution of IS term and IV orbital term in each $M1$ transition. Values of $R_{\mathrm{ISO}} > 1 (< 1)$ suggest constructive (destructive) interference of these terms with the IV spin term. For the definition of R_{ISO}, see text.

the study of β decay from the g.s. of ^{27}Si to a few excited states of ^{27}Al. The good-resolution (p, n)-type CE reaction, (^{3}He,t) reaction, at 150 MeV/u performed at RCNP with a resolution of 90 keV is used to extend the knowledge on B(GT) values to higher excitation region based on the proportionality between the cross sections at 0° and the B(GT) values (see Ref. [7] for details).

In earlier works, the symmetry structure in the mirror nuclei ^{27}Al and ^{27}Si was studied through the comparison of excitation energies, spins, parities, and branching ratios obtained in γ-decay measurements up to $E_x = 5.5$ MeV [8]. Here the symmetry structure is studied not only from the correspondence of excitation energies, but also from the "quasi" proportionality of $B(M1)$ and B(GT) values for the analogous transitions from the g.s. of ^{27}Al. As a result of careful comparison of both excitation energies and transition strengths, the mirror-symmetry properties of the mirror-nuclei pair ^{27}Al and ^{27}Si have been established up to $E_x = 8.5$ MeV (see Fig. 2).

3. Nature of $M1$ Transitions

Although we have seen a good overall correspondence for the states in ^{27}Al and in ^{27}Si, there are noticeable differences in the strengths $B^R(M1)$ and the B(GT) of the corresponding states in a level-by-level comparison. The differences can be attributed to the different nature of the operators involved in the γ-decay and the (^{3}He, t) reaction. The CE reaction is of pure IV nature, and it is known that at 0° and at intermediate incident-energies the effective operator is to a very good approximation of the $\sigma\tau$ type [9, 5]. On the other hand, the $M1$ operator includes the IS and the IV components, and each of these contains an orbital term in addition to the spin term, as shown in Eq. (1). Thus, in an $M1$ transition between states with $T = 1/2$, not only the dominant IV spin term but also the minor IS term can make some contributions. Furthermore, the orbital contribution, although usually believed to be small, can sometimes be significant [1]. Since the interferences of the IS term and/or the IV orbital term and the IV spin term are either constructive or destructive depending on the structure of the state, it is expected that the differences in strengths by up to ~ 50 % might be expected compared to a pure GT transition [3, 10].

The IS and orbital contributions, however, cannot explain the fact that the $B^R(M1)$ values are as a whole larger than the corresponding B(GT) values as seen from the comparison of Fig. 2(a) and (b). Contributions from meson-exchange currents (MEC) are known to enhance $B(M1)$ strength over the corresponding B(GT) strength [1, 12, 11, 4]. The enhancement is traced back to larger and additive contributions of the vector MEC over the axial-vector MEC which are active in $M1$ and GT transitions, respectively [13]. Based on the reasons given in Ref. [7], we use $R_{\text{MEC}} = 1.4$.

By comparing the analogous $B(M1)$ and B(GT) strengths of the $\Delta T = 0$ transitions in $T = 1/2$ mirror nuclei, it is possible to extract the combined contribution of the IS term and the IV orbital term in the $B(M1)$ strength. Since the effect of MEC should be independent of the wave function of the individual state, the ratio of $B^R(M1)$ and B(GT) for the jth pair of corresponding states divided by R_{MEC}

$$R^j_{\text{ISO}}(M1/\text{GT}) = \frac{B^R_j(M1)}{B_j(\text{GT})} \cdot \frac{1}{R_{\text{MEC}}}, \tag{5}$$

should show the combined IS-orbital contribution to the jth $M1$ transition and indicate how the IV spin term is modified. The $R_{\rm ISO}$ should be greater than unity if the combined contributions are constructive and less for the destructive case. The results are shown in Fig. 3 for the $M1$ transitions in ^{27}Al as a function of $B(M1)\uparrow$ value. It is interesting to note that the $R_{\rm ISO}$ value tends to deviate from unity by more than a factor of two when the $B(M1)\uparrow$ is less than approximately 0.1. This shows that the combined IS-orbital contribution is rather large in weaker transitions and the "quasi" proportionality of the $B(M1)$ values for $\Delta T = 0$ $M1$ transitions and the analogous B(GT) values is lost. This finding is interpreted as follows; since the IS term and the IV orbital term are always small, the dominance of the IV spin term of the $M1$ operator is guaranteed if the transitions are at least of average strength. The IV spin term, however, can also be small. Then the relative contribution of the IS term and the IV orbital term becomes significant although the transition itself is weak.

4. Isospin Assignment Based on the Isospin Symmetry

The low-lying states of the mirror nuclei have the isospin $T = 1/2$ because of the $|T_z| = 1/2$ nature of the nuclei. As shown in Fig. 1, above certain excitation energy (about $E_x = 10$ MeV), members of the $T = 3/2$ multiplet states are also allowed. Our interest is to establish the mirror symmetry structure in the high excitation region and to identify the isospin values of excited states based on the isospin symmetry. We notice that the identification of isospin is possible from the different "sensitivity" of inelasitc and charge-exchange type transitions. Let us examine the ratio of squared CG coefficients in Eq. 4 for the transitions starting from the g.s. of $T = T_z = +1/2$ nucleus. If the transitions are to the mutually analogous $T = 1/2$ states with $T_z = \pm(1/2)$, the factor 1/2 is obtained as the ratio. On the other hand if the transitions are to the analogous $T = 3/2$ states with $T_z = \pm(1/2)$, the factor 2 is obtained as the ratio (see also Fig. 1). The $M1$ transition to a $T = 3/2$ state, therefore, is enhanced in total by a factor of 4 than the corresponding GT transition to the analog state of the $T = 3/2$ state, if transition strengths are normalized by the corresponding $M1$ and GT transitions to analogous $T = 1/2$ states.

In order to study the strengths of the $M1$ transitions in the region where no γ decay data is available, a (p,p') reaction at 0° was performed at IUCF, Indiana. Using the transmission mode of the K600 spectrometer [14], a resolution of 40 keV has been obtained for $E_p = 160$ MeV. In order to see whether or not the scheme using the different sensitivity of $M1$ and GT transitions to different T states works as a method to identify the isospin of a state, the comparison was made for the spectra of mirror nuclei pair ^{13}C and ^{13}N. The obtained (p,p') spectrum is compared with the ^{13}C(^{3}He,t)^{13}N spectrum in Fig. 4. Due to the different sensitivity, we can clearly see that the $T = 3/2$ state at 15.11 MeV is enhanced as expected from Eq. 4.

In ^{27}Al and ^{27}Si system, the level density becomes very high in the mixed T region ($E_x \geq 10$ MeV) although $\Delta L = 0$ and $\Delta S = 0$ states are strongly selected at 0°. As seen from Fig. 5, it not so easy to resolve all states in the mixed T region even with a resolution of 90 keV achieved in the (^{3}He,t) reaction, while states are almost resolved in the ^{27}Al(p,p') spectrum. What should be noted is that similar structures are seen in both spectra in spite of the different resolutions. In addition, there are prominent

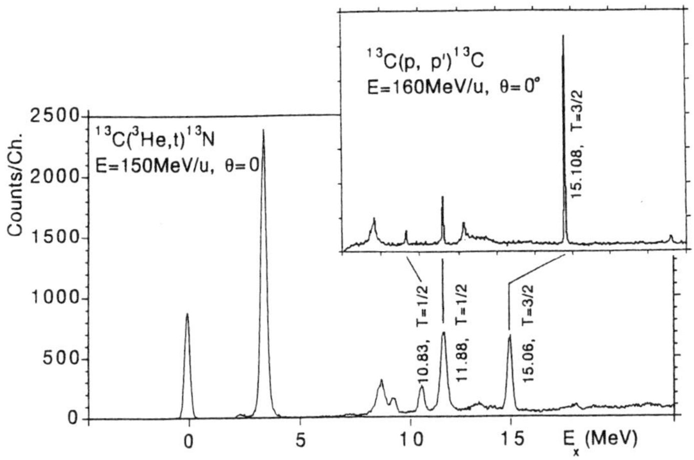

Figure 4: The comparison between the 0° $^{13}C(p,p')$ and $^{13}C(^3He,t)^{13}N$ spectra at the region of excitation energy where both known $T = 1/2$ and $T = 3/2$ states are mixed. Due to the different sensitivity for the $T_<$ ($T = 1/2$) and $T_>$ ($T = 3/2$) states, $T_>$ ($T = 3/2$) state at 15.11 MeV is enhanced by about a factor of 4 in the (p,p') spectrum.

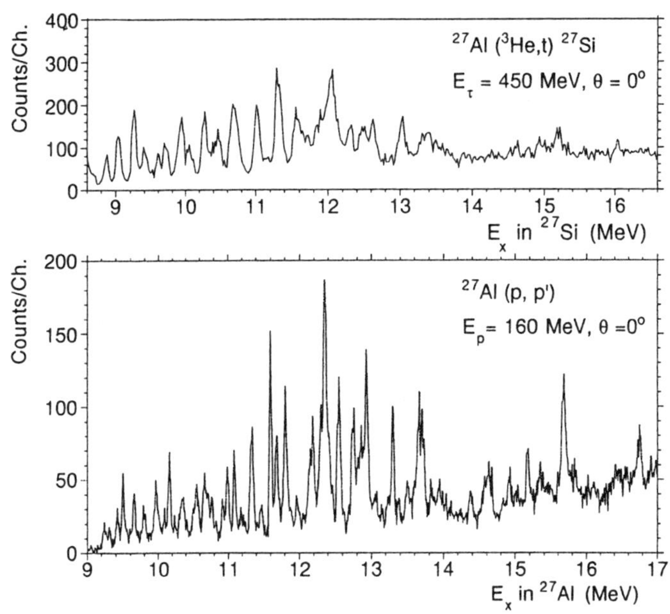

Figure 5: The comparison between the 0° $^{27}Al(^3He,t)^{27}Si$ and $^{27}Al(p,p')$ spectra measured at RCNP and at IUCF, respectively. Resolutions of 90 and 40 keV has been achieved by using a dispersion matching technique. Many states above $E_x = 11$ MeV are apparently enhanced in the (p,p') reaction, suggesting that they are $T_>$ ($T = 3/2$) states.

states at around $E_x = 11.5$ MeV region in the (p,p') spectrum. It is suspected that these states are of $T_>$ ($T = 3/2$) nature judging from the enhanced strengths in the $M1$-type (p,p') reaction.

5. Summary

Among excited states in various mass A nuclei, analogous states characterized by the same isospin T constitute isospin multiplet states. By using different selection rules and sensitivities of transitions exciting a member of the isospin multiplet states in hadron inelastic and charge-exchange reactions, β decays and γ decays, which are caused by the strong, weak, and EM interactions, respectively, it is possible to study the "isospin symmetry structure" of nuclear excitations. Through a detailed comparison of results from charge-exchange reactions and γ decays induced by EM $M1$ operator, different response of nucleus depending on different active interactions is investigated.

As we have seen, identification of isospin of excited states is not easy, because it can be known only by finding out the symmetry (or similar) structure after comparing the information on transitions exciting states in two different Z but the same A nuclei. Naturally the two transitions should be measured with a similarly high resolution to establish the symmetry (or similar) structure. At the same time, high accuracy of the angular resolution is also important to identify the transferred L of the transitions exciting the corresponding states. At RCNP, the proposal to build a new beamline combining the existing spectrometer and the cyclotron has been accepted and the construction starts in 1999. It is expected that higher resolution and better angular-resolution are realized be performing various matching conditions at the new beamline. We believe that the study on the "isospin symmetry structure" will make further progress.

References

[1] Y. Fujita *et al.*, Phys. Rev. C **55** (1997) 1137.

[2] L. Zamick and D. C. Zheng, Phys. Rev. C **37** (1988) 1675.

[3] E. K. Warburton and J. Weneser, *Isospin in Nuclear Physics*, ed. D. H. Wilkinson, (North-Holland, Amsterdam, 1969) Chap. 5.

[4] Y. Fujita, *et al.*, Nucl. Instrum. Meth. Phys. Res. A **402** (1998) 371.

[5] W. G. Love and M. A. Franey, Phys. Rev. C **24** (1981) 1073.

[6] T. N. Taddeucci *et al.*, Nucl. Phys. **A469** (1987) 125.

[7] Y. Fujita *et al.*, Phys. Rev. C **59** (1999) 90.

[8] P. M. Endt, Nucl. Phys. **A521** (1990) 1, and references therein.

[9] F. Osterfeld, Rev. Mod. Phys. **64** (1992) 491, and references therein.

[10] S. S. Hanna, *Isospin in Nuclear Physics*, ed. D. H. Wilkinson, (North-Holland, Amsterdam, 1969) Chap. 12.

[11] C. Lüttge, *et al.*, Phys. Rev. C **53** (1996) 127; and private communications.

[12] A. Richter, A. Weiss, O. Häusser, and B. A. Brown, Phys. Rev. Lett. **65** (1990) 2519.

[13] I. S. Towner and F. C. Khanna, Nucl. Phys. **A399** (1983) 334.

[14] G .P . A. Berg *et al.*, IUCF Sci. and Tech. Repost 1993-1994, p. 106.

Mixed-Symmetry Quadrupole States in Nuclei

P. von Brentano[1], N. Pietralla[1], C. Fransen[1], A. Gade[1], A. Gelberg[1], U. Kneissl[2], T. Otsuka[3], H.H. Pitz[2], V. Werner[1],

[1] *Institut für Kernphysik, Universität zu Köln, 50937 Köln, Germany*
[2] *Institut für Strahlenphysik, Universität Stuttgart, 70569 Stuttgart, Germany*
[3] *Department of Physics, University of Tokyo, Hongo, Bunkyo-ku, Tokyo 113-0033*

Abstract

We report on our recent investigations on $J^\pi = 1^+$ and 2^+ mixed-symmetry (MS) states in heavy open-shell nuclei. We discuss our investigation of the MS 2^+ state of ^{136}Ba using the photon scattering technique and we survey the properties of MS 2^+ states known from lifetime measurements on nuclei in the $A = 130$ mass region. Finally we report on our latest experiments on 1^+ and 2^+ MS states in the $N = 52$ nucleus ^{94}Mo.

1 Introduction

Nuclear electro-weak spectroscopy, which was strongly promoted for years by Hiro Ejiri in Osaka and other laboratories around the world, is a sensitive tool for the investigation of nuclear matter and symmetry concepts in quantum mechanics [1]. Symmetries and dynamical symmetry breaking play a key-role in many modern nuclear structure models, for instance in the interacting boson model (IBM), formulated by Arima and Iachello [2, 3]. The IBM provides understanding of low-lying, collective states in heavy nuclei in simple terms and has inspired experimentalists to investigate nuclear properties. The low energy excitation spectrum of a nucleus is an example for the excitation of a quantized many-body system, which is formed by two different constituents: protons and neutrons. Proton-neutron degrees of freedom are one of the most interesting topics in nuclear structure physics. The most low-energetic nuclear states are practically proton-neutron symmetric [4]. This fact makes the simplest version of the IBM, the IBM-1, a successful model for the lowest-lying levels in vibrational, rigidly deformed, transitional and γ-soft nuclei. The proton-neutron version [5] of the IBM, the IBM-2, generalizes the IBM-1 and represents the adequate formulation to consider proton-neutron degrees of freedom in terms of the IBM.

The proton-neutron symmetry of the eigenstates of an IBM-2 Hamiltonian is quantified by the F-spin quantum number [6], which represents the analogue of isospin for bosons. Proton-neutron symmetric states possess the maximum F-spin quantum number $F = F_{\text{max}} = (N_\pi + N_\nu)/2$, where $N_{\pi(\nu)}$ denotes the number of proton (neutron) bosons. IBM-2 wave functions with smaller values of the F-spin quantum number are called mixed-symmetry (MS) states because they contain at least one pair of a proton boson b_π and a neutron boson b_ν which is antisymmetric with respect to the exchange of the nucleon labels π and ν. The properties of MS states were firstly predicted [7, 8] in the early 1980ies. The most prominent experimental signature for MS states is the appearance of relatively strong magnetic dipole ($M1$) transitions to low-lying symmetric states. Typical $M1$ transition matrix elements of the order of 1 μ_N were predicted [8]. In particular, the existence of a strong orbital $M1$ excitation to a $J^\pi = 1^+$ state was expected in deformed nuclei.

In 1983, strong low-lying $M1$ excitations in deformed nuclei were discovered [9] by Richter and co-workers in electron scattering experiments on the well-deformed nucleus ^{156}Gd in Darmstadt, representing the discovery of a MS state. This discovery was confirmed [10] in photon scattering experiments performed in Stuttgart. The photon scattering technique [11] proved itself to be a very sensitive and efficient tool for the investigation of the scissors mode, as the fragmented MS $M1$ excitation is commonly called now. Consequently, the scissors mode was systematically studied [11] in even-even nuclei and in odd-mass nuclei by means of photon scattering. The data gathered on many nuclei prove the collectivity of the observed total $M1$ strength and thereby support the MS character of the fragmented $M1$ excitation [12, 13].

Table 1: Q-phonon expressions for the lowest-lying symmetric and MS states. $Q_s = Q_\pi + Q_\nu$ is the F-scalar sum of the proton and neutron boson quadrupole operators and $Q_m = Q_\pi/N_\pi - Q_\nu/N_\nu$ is the orthogonal linear combination in the ground state of the F-spin limit: $\langle 0_1^+|Q_m Q_s|0_1^+\rangle = 0$. The relations are exact in the dynamical symmetry limits U(5) and O(6) and approximate in the vicinity but not at the SU(3) limit.

$\|2_1^+\rangle$	$\propto$	$Q_s\|0_1^+\rangle$	$F = F_{\max}$
$\|2_2^+\rangle$	$\propto$	$(Q_s Q_s)^{(2)}\|0_1^+\rangle$	$F = F_{\max}$
$\|4_1^+\rangle$	$\propto$	$(Q_s Q_s)^{(4)}\|0_1^+\rangle$	$F = F_{\max}$
$\|2_{\mathrm{ms}}^+\rangle$	$\propto$	$Q_m\|0_1^+\rangle$	$F = F_{\max} - 1$
$\|1_{\mathrm{sc}}^+\rangle$	$\propto$	$(Q_s Q_m)^{(1)}\|0_1^+\rangle$	$F = F_{\max} - 1$
$\|3_{\mathrm{ms}}^+\rangle$	$\propto$	$(Q_s Q_m)^{(3)}\|0_1^+\rangle$	$F = F_{\max} - 1$

Independent support for MS assignments and information on the collectivity of the scissors mode can be deduced from $E2$ properties, which are considered to be well described by the IBM. This becomes particularly obvious in the recently developed [14] Q-phonon scheme, which formulates the relative wave functions of excited states in terms of quadrupole operators acting on the true ground state. The Q-phonon scheme was proposed [15, 14] by Otsuka and collaborators for γ-soft nuclei. The wave function generating operators can yield approximate expressions for the eigenstates of an IBM Hamiltonian even outside [16, 17, 18, 19] the analytically solvable dynamical symmetry limits. Later-on, the Q-phonon scheme was used to generate MS states, as well [20]. The Q-phonon expressions for the lowest-lying symmetric and MS states are displayed in Table 1. In the Q-phonon scheme it is obvious, that the fundamental MS state is the one-Q-phonon 2^+_{ms} state, which was predicted already in the early works [8]. Experimental information about 2^+ MS states is much sparser than for the 1^+ scissors mode. Other MS states are still unobserved.

2 Investigation of the MS 2^+ state in nuclei of the $A = 130$ mass region

In order to identify the fundamental one-Q-phonon MS 2^+ state by means of γ-spectroscopy it is necessary to measure the following four quantities: spin and parity $J^\pi = 2^+$, the ground state excitation strength $B(E2, 0_1^+ \to 2^+)$ and the $M1$ decay strength to the 2_1^+ state. Measurement of the $B(M1; 2^+ \to 2_1^+)$ value means to measure the branching ratio $I_\gamma(2^+ \to 2_1^+)/I_\gamma$, the $E2/M1$ mixing ratio δ of the $2^+ \to 2_1^+$ transition, and the lifetime τ. The fragmentation of the 2^+_{ms} state can be studied by the observation of all fragments in a complete reaction. MS states can be identified experimentally by observation of large $M1$ transition matrix elements. Therefore, the measurement of lifetimes, i.e. absolute $E2$ and $M1$ transition strengths, is very important for the identification of MS states. In the literature MS assignments are sometimes based on the measurement of small $E2/M1$ multipole mixing ratios if lifetime information is absent. We stress, however, that small mixing ratios do not necessarily imply large $M1$ transition strengths, because the corresponding $E2$ strength could be very small.

We have recently demonstrated [21] the ability of the photon scattering technique for the investigation of the 2^+_{ms} state and its fragmentation in a photon scattering experiment on the vibrational nucleus ^{136}Ba. Our experiments were carried out at the photon scattering site [11] of the Dynamitron accelerator in Stuttgart. At an excitation energy of 2129 keV we observed the 2_4^+ state, which exhibits the properties expected for the 2^+_{ms} state. It has a short lifetime $\tau = 67(7)$ fs. Its excitation strength $B(E2; 0_1^+ \to 2_4^+) = 0.045(5)\ e^2\mathrm{b}^2$ is weakly collective. It corresponds

Table 2: Summary of the relevant observables for nuclei in the $A = 130$ mass region, where $2^+_{\rm ms}$ states or its fragments have been identified from measured absolute transition strengths: We give excitation energies, lifetimes, and $E2/M1$ mixing ratios δ of $2^+_{\rm ms}$ states, and different transition strengths. If more than one fragment was observed, the reduced transition strengths have been summed up. $M1$ transition strengths of the scissors mode are obtained by the Stuttgart-Cologne photon scattering collaboration. From Ref. [29].

Nucleus	^{112}Cd	^{126}Xe	^{128}Xe	^{134}Ba	^{136}Ba	^{142}Ce	^{144}Nd
Reaction	DSAM-INS	(α,n)	(α,n)	DSAM-INS	(γ,γ')	Coulex DSAM-INS	(γ,γ')
Ref.	[28]	[30]	[25]	[24]	[21]	[26, 31]	[32, 33]
E [keV]	2156 2231	2064 (2359)	2127	2029 2088	2129	2004	2072
τ [fs]	310(35) 220(20)	351(74) 74(30)	170(70)	230(23) 85(7)	67(7)	65(7)	66 (20)
$\delta(2^+_{ms} \to 2^+_1)$	0.085(24) -0.02(3)	0.00(5) 0.8(8)	0.05(5)	-0.31(5) 0.02(5)	0.005(9)	-0.26(14)	0.08
$B(E2;2^+_{ms} \to 0^+_1)$ [W.u.]	0.14(2)			2.0(2)	2.1(2)	2.5(2)	2.2(8)
$B(E2;2^+_1 \to 0^+_1)$ [W.u.]	32	41	39	33	19	20	25
$B(M1;2^+_{ms} \to 2^+_1)[\mu_N^2]$	0.10(1)	0.06(3)	0.07(2)	0.22(4)	0.26(3)	0.23(2)	0.23(9)
$B(M1;1^+_{sc} \to 0^+_1)[\mu_N^2]$				0.19(3)	0.10(3)		0.17(1)

to 2.1 single particle units and amounts to 11% of the excitation strength of the 2^+_1 state. This is in excellent agreement with the early prediction [8]. It is three times larger than the excitation strength of the 2^+_2 state, which in the vibrator nucleus ^{136}Ba is interpreted as the 2^+ member of the isoscalar two-quadrupole phonon triplet. The $2^+_4 \to 2^+_1$ transition has a pure $M1$ character [22] with a very small $E2/M1$ mixing ratio $\delta = +0.005(9)$. From the branching ratio and the lifetime observed in our (γ,γ') experiment we deduced a decay strength $B(M1;2^+_4 \to 2^+_1) = 0.26(3)\ \mu_N^2$. This value is close to the $B(M1)$ values for the largest fragments of the 1^+ scissors mode in deformed rare earth nuclei of $B(M1;1^+ \to 0^+_1) \leq 0.5\,\mu_N^2$ [23]. Given the errors of about 10%, the $M1$ strength coincides with the total $M1$ decay strength of the fragmented $2^+_{\rm ms}$ state observed earlier [24] in the neighboring nucleus ^{134}Ba which amounts to $\sum B(M1;2^+_{3,4} \to 2^+_1) = 0.20(2)\ \mu_N^2$.

With the spin and parity quantum numbers $J^\pi = 2^+$, together with the small mixing ratio $\delta_{2^+ \to 2^+_1}$ and with the absolute $E2$ and $M1$ decay strengths to the ground state and to the 2^+_1 state, we obtained a complete set of γ-spectroscopic signatures for the $2^+_{\rm ms}$ state of ^{136}Ba. These signatures are well reproduced by an IBM-2 calculation (for details see Ref. [21]). The reduced matrix element $|\langle 2^+_1 \parallel M1 \parallel 2^+_4\rangle| = 1.14(7)\,\mu_N$ of the $2^+_4 \to 2^+_1$ $M1$ transition has, indeed, the size of about $1\,\mu_N$, which is expected for allowed $M1$ decays of MS states.

More information on lifetimes of MS 2^+ states in nuclei of the $A = 130$ mass region comes from a Coulomb excitation experiment [26] and from Doppler shift attenuation measurements after inelastic neutron scattering (DSAM-INS) [27] as frequently performed by the Budapest-Lexington group, e.g. [24, 28] and from DSAM analyses of γ-rays emitted after an (α,n) fusion evaporation reaction [25]. Table 2 summarizes information on MS 2^+ states of nuclei in the $A = 130$ mass region, which were identified from lifetime measurements. From the compilation in Table 2 we conclude the following properties of the MS 2^+ state in nuclei of the $A = 130$ mass region:

- The excitation energy is close to 2 MeV.
- The $E2$ excitation strength amounts to about $B(E2;0^+_1 \to 2^+_{\rm ms}) = 2$ single particle units,

which is weakly collective and

- the $M1$ transition strength to the 2_1^+ state amounts to $B(M1; 2_{\rm ms}^+ \rightarrow 2_1^+) = 0.05$–$0.3\ \mu_N^2$, i.e, the corresponding reduced $M1$ transition matrix element is of the order of $1\,\mu_N$.

3 $J^\pi = 1^+$ and 2^+ MS states in ^{94}Mo

Presently we are investigating low-spin states of ^{94}Mo in order to identify MS states from the measurement of absolute $E2$ and $M1$ transition strengths. We performed photon scattering experiments at the bremsstrahlung facility of the Dynamitron accelerator in Stuttgart using bremsstrahlung endpoint energies of 3.3 MeV and 4.1 MeV. At the Cologne Tandem accelerator we additionally investigated the γ spectrum of ^{94}Mo following the β-decay of ^{94}Tc nuclei. The γ spectroscopy after β-decay provides us with data on branching ratios, even for weakly intense decay channels, and $E2/M1$ multipole mixing ratios δ. Absolute ground state transition widths Γ_0 and lifetimes τ are measured in the photon scattering experiments. In combination, we can deduce a great deal of absolute $E2$ and $M1$ transition strengths of 1^+ and 2^+ states in ^{94}Mo.

Figure 1 shows a part of the measured photon scattering spectra. Spin quantum numbers were determined from the angular distribution of the photon scattering intensity. Photon scattering cross sections were measured relative to the well-known cross sections [34] of the photon flux calibration standard ^{27}Al, which was irradiated simultaneously. The ground state decay transitions of the main fragments of both low-spin MS states of ^{94}Mo are visible in the spectrum: the 2_3^+ state at 2067 keV and the 1_1^+ state at 3129 keV. Both states are short-lived and decay by strong $M1$ transitions to the 2_1^+ state and to the ground state and the 2_2^+ state, respectively. The data on the 1_1^+ state fit well into the systematics of the scissors mode extrapolated from the rare earth region [12, 13]. The 2_3^+ state at 2067 keV exhibits all the features of the MS 2^+ state enumerated and concluded above: It lies close to 2 MeV, its ground state decay $E2$ transition is weakly collective and the $M1$ transition matrix element to the 2_1^+ state has the size of about $1\ \mu_N$.

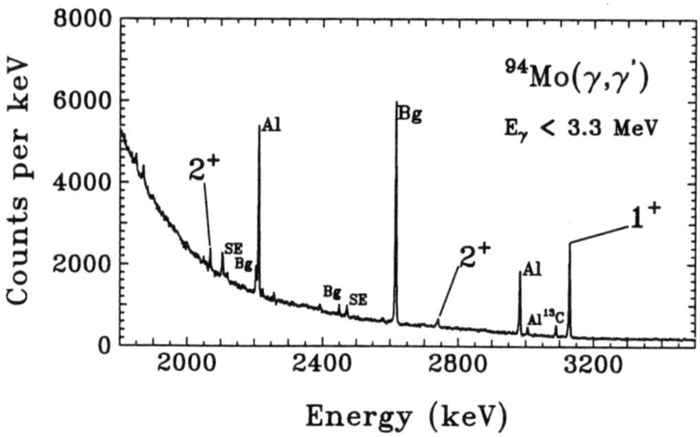

Figure 1: Part of the Photon scattering spectrum off ^{94}Mo taken at the bremsstrahlung facility of the Stuttgart Dynamitron accelerator with incident photon energies $E_\gamma < 3.3$ MeV.

Intensity ratios of γ-decay transitions in ^{94}Mo and $E2/M1$ multipole mixing ratios δ have been determined from the γ-spectroscopy study following the β-decay of ^{94m}Tc nuclei. The $J^\pi = 2^+$ isomers, ^{94m}Tc, were created by the ^{94}Mo(p,n) reaction at a beam energy of $E_p = 13$ MeV in the center of the Cologne coincidence cube spectrometer. Radioactive ^{94m}Tc nuclei with half lifes

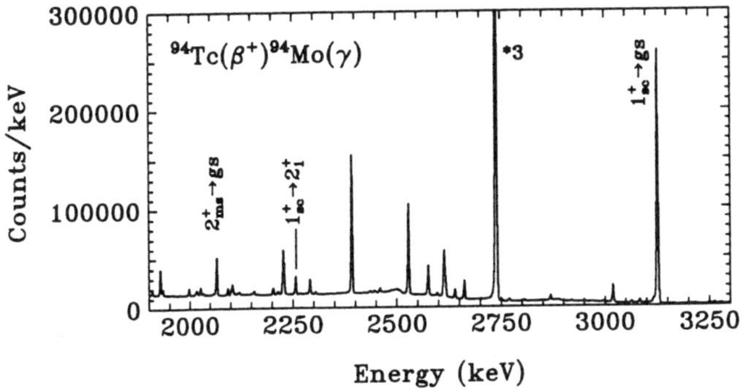

Figure 2: Part of the off-beam γ spectrum following the ^{94}Mo(p,n) reaction at a proton beam energy of 13 MeV. The data were taken at the Cologne coincidence cube spectrometer.

of $t_{1/2}(^{94m}\mathrm{Tc}) = 52$ minutes were periodically produced for 5 seconds when the proton beam hit the ^{94}Mo target. Then, the beam was switched off for another 5 seconds and γ-singles and $\gamma\gamma$-coincidence spectra of the subsequent γ radiation were taken between the beam pulses using 8 Ge detectors in the highly symmetric Cologne cube arrangement. A part of the summed up γ singles spectrum is displayed in Fig. 2. The isotropy of γ radiation following the β-decay and the high statistics and low background in the off-beam spectra enable relatively accurate determination of branching ratios even for weakly intense γ transitions.

From the measured angular correlations of the $\gamma\gamma$ coincidences we determine the $E2/M1$ multipole mixing ratio δ of many γ transitions. Multipole mixing ratios of γ transitions from MS states are particularly interesting for vibrational and γ-soft nuclei. The IBM Hamiltonian suitable for the description of such nuclei shows a specific symmetry: the d-parity [25, 35]. There exist selection rules for electromagnetic transitions with respect to the d-parity quantum number. The validity of these selection rules and the goodness of the d-parity quantum number can be tested by the measurement of F-spin allowed but d-parity forbidden $M1$ transitions. Our data enable us to extract many multipole mixing ratios and absolute $E2$ and $M1$ transition strengths. The data are presently evaluated.

4 Acknowledgments

We thank A. Fitzler, S. Kasemann, H. Maser and H. Tiesler for help with the experiments. Valuable discussions with A. Giannatiempo, J. Ginocchio, F. Iachello, R.V. Jolos, K.-H. Kim, A. Richter, and A. Zilges are gratefully acknowledged. We thank H. Ejiri for invitation to the conference on *Nuclear Electro-Weak Spectroscopy* in Osaka, Japan. Our work was supported by the *Deutsche Forschungsgemeinschaft* under contracts Br 799/6-2/8-1/9-1 and Kn 154/30.

References

[1] H. Ejiri and M.J.A. de Voigt, *Gamma-ray and electron spectroscopy in nuclear physics*, (Oxford University Press, Oxford, 1989).

[2] A. Arima, F. Iachello, Phys. Rev. Lett. **35**, 1069 (1975).

[3] F. Iachello, A. Arima, *The Interacting Boson Model* (Cambridge University Press, 1987).

[4] H. Harter, P. von Brentano, A. Gelberg, R.F. Casten, Phys.Rev. C **32**, 631 (1985).

[5] A. Arima, T. Otsuka, F. Iachello, I. Talmi, Phys. Lett. **B 66**, 205 (1977).

[6] T. Otsuka, A. Arima, F. Iachello, Nucl. Phys. **A 309**, 1 (1978).

[7] F. Iachello, Nucl. Phys. **A358**, 89c (1981).

[8] F. Iachello, Phys. Rev. Lett. **53**, 1427 (1984).

[9] D. Bohle, A. Richter, W. Steffen, A.E.L. Dieperink, N. Lo Iudice, F. Palumbo, O. Scholten, Phys. Lett. **137 B**, 27 (1984).

[10] U.E.P. Berg *et al.*, Phys. Lett. **B149**, 59 (1984).

[11] U. Kneissl, H.H. Pitz, A. Zilges, Prog. Part. Nucl. Phys. **37**, 349 (1996), and Refs. therein.

[12] N. Pietralla *et al.*, Phys. Rev. C **52**, R2317 (1995).

[13] N. Pietralla *et al.*, Phys. Rev. C **58**, 184 (1998).

[14] T. Otsuka, K.H. Kim, Phys. Rev. C **50**, R1768 (1994).

[15] G. Siems *et al.*, Phys. Lett. **320 B**, 1 (1994).

[16] N. Pietralla, P. von Brentano, R.F. Casten, T. Otsuka, N.V. Zamfir, Phys. Rev. Lett. **73**, 2962 (1994).

[17] N. Pietralla, P. von Brentano, T. Otsuka, R.F. Casten, Phys. Lett. **B 349**, 1 (1995).

[18] N. Pietralla, T. Mizusaki, P. von Brentano, R.V. Jolos, T. Otsuka, V. Werner, Phys. Rev. C **57**, 150 (1998).

[19] Yu.V. Palchikov, P. von Brentano, R.V. Jolos, Phys. Rev. C **57**, 3026 (1998).

[20] K.-H. Kim *et al.*, Proceedings of the Symposium on *Capture Gamma Ray Spectroscopy*, Budapest 1996, edited G. Molnár, (Springer, Budapest, 1998).

[21] N. Pietralla *et al.*, Phys. Rev. C **58**, 796 (1998).

[22] M.Al-Hamidi *et al.*, Rus. Jou. Nucl. Phys. **57**, 579 (1994).

[23] A. Richter, Prog. Part. Nucl. Phys. **34**, 261 (1995).

[24] B. Fazekas, T. Belgya, G. Molnár, A. Veres, R.A. Gatenby, S.W. Yates, T. Otsuka, Nucl. Phys. **A 548**, 249 (1992).

[25] I. Wiedenhöver *et al.*, Phys. Rev. C **56**, R2354 (1997).

[26] W.J. Vermeer, C.S. Lim, R.H. Spear, Phys. Rev. C **38**, 2982 (1988).

[27] T. Belgya, G. Molnár, S.W. Yates, Nucl. Phys. **A 607**, 43 (1996).

[28] P.E. Garrett, H. Lehmann, C.A. McGrath, Minfang Yeh, and S.W. Yates, Phys. Rev. C **54**, 2259 (1996).

[29] P. von Brentano *et al.*, contribution to the International Conference on *Nuclear Structure*, Gatlinburg, Tennessee, August 10 - 15, 1998.

[30] A. Gade *et al.*, in preparation.

[31] J.R. Vanhoy, J.M Anthony, B.M. Haas, B.H. Benedict, B.T. Meehan, S.F. Hicks, C.M. Davoren, C.L. Lundstedt, Phys. Rev. C **52**, 2387 (1995).

[32] W.D. Hamilton, A. Irbäck, J.P. Elliott, Phys. Rev. Lett. **53**, 2469 (1984).

[33] T. Eckert *et al.*, Phys. Rev. C **56**, 1256 (1997); **57**, 1007 (1998).

[34] N. Pietralla *et al.*, Phys. Rev. C **51**, 1021 (1995).

[35] N. Pietralla *et al.*, Phys. Rev. C **58**, 191 (1998).

Polarization Charge of Particles near Threshold due to the Coupling to Shape Oscillations [1]

Ikuko HAMAMOTO

Division of Mathematical Physics, Lund Institute of Technology at University of Lund, Lund, Sweden.

Abstract

It is shown that the isoscalar strength in the energy region just above the low-energy threshold, which is created by exciting particles to the continuum in nuclei far from β-stability lines, can be reduced by the attractive coupling to isoscalar shape oscillations. This is in contrast to the well-known fact that in β-stable nuclei low-lying isoscalar particle-hole strengths are always increased by the attractive isoscalar coupling, which leads to an appreciable amount of positive polarization charge. On the other hand, the core-polarization effect on bound particles with ℓ=0 and 1 is always positive, but vanishes as the binding energy approaches to zero.

1 Introduction

Dynamics of nuclei far from β-stability lines has become a very popular research field, in connection with the recent development of facilities of radioactive nuclear ion beams all over the world. The presence of nucleons, which have separation energies appreciably smaller than those in β-stable nuclei, leads to very interesting and unexpected phenomena. In the present contribution we study the part of polarization charge, which comes from the coupling of particles near threshold to isoscalar shape oscillations. It is known that in β-stable nuclei the coupling produces the major part of the polarization charge, which together with bare charge makes the effective charge.

The λ-pole polarization charge due to the coupling to λ-pole isoscalar (IS) shape oscillations is written as [1]

$$(e_{pol})_{\lambda,\tau\approx 0} = \frac{< j_2 \mid R\frac{dV(r)}{dr} \mid j_1 >}{< j_2 \mid r^\lambda \mid j_1 >} \frac{3}{4\pi} \frac{ZeR^\lambda}{C_\lambda} \frac{(\hbar\omega_\lambda)^2}{(\hbar\omega_\lambda)^2 - (E_2 - E_1)^2} \tag{1}$$

where the form factor of shape oscillations is expressed by $R\frac{dV}{dr}$ and the nuclear radius by R. The last factor in eq.(1) is close to unity, when either static polarization (E_2=E_1) or the coupling to giant resonances ($\hbar\omega_\lambda \gg \mid E_2 - E_1 \mid$) is considered. Then, the sign of the polarization charge in (1) is given by that of the ratio

$$\frac{< j_2 \mid R\frac{dV(r)}{dr} \mid j_1 >}{< j_2 \mid r^\lambda \mid j_1 >}. \tag{2}$$

The form-factor $R\frac{dV}{dr}$ is surface peaked for any reasonable nuclear potential, and the major contribution to $< r^\lambda >$ comes also from the surface region if relevant particles j_1 and j_2 are sufficiently bound. Thus, for harmonic oscillator model or for bound particles, which have been traditionally considered in β-stable nuclei, the sign of the ratio (2) is always

[1]The major part of the content of this talk is taken from ref.[2].

positive. This positive sign leads to the result $(e_{pol})_{\lambda,\tau\approx 0} > 0$, namely the fact [1] that the attractive coupling to IS shape oscillations produces an increase of the low-lying transition strength which is proportional to $|< j_2 \mid r^\lambda \mid j_1 >|^2$.

In section 2 we describe the variation of the coupling to shape oscillations and the resulting polarization effect, as the separation energy of relevant particles decreases from 8-10 MeV, which is a typical separation energy in β-stable nuclei, to very small values. In section 3 we illustrate the coupling of particle-hole (ph) excitations to shape oscillations, in which the particle state is in the continuum. When the angular momentum of the particle increases, the ph strength function rises more slowly as the energy above the threshold increases, due to the presence of the higher centrifugal barrier. Thus, the case of smaller particle angular momentum is of practical interest.

2 Bound particles with small separation energies

A simple model is chosen, for which we can obtain analytic expressions of matrix elements. We use the square-well potential,

$$V(r) = \begin{cases} 0 & \text{for} \quad r < R \\ U_0 > 0 & \text{for} \quad r \geq R \end{cases} \tag{3}$$

and neglect the spin of nucleons. Solving the Schrödinger equation with the above potential for E<U_0 we obtain energy eigenvalues $E_{n\ell}$ and wave functions of bound particles. Then, the analytic expressions of the matrix elements of $R\,\frac{dV}{dr}=U_0 R\delta(r-R)$ and r^2 are obtained. It is noted that for a potential well with an infinite wall ($U_0=\infty$) the result

$$< n\ell \mid R\,\frac{dV(r)}{dr} \mid n\ell > = 2E_{n\ell} \tag{4}$$

is known. In the case of a finite value of U_0 we obtain in the limit of zero binding energies of particles with orbital angular momentum ℓ, $E_{n\ell}\rightarrow U_0$,

$$< n,\ell=0 \mid R\,\frac{dV(r)}{dr} \mid n,\ell=0 > \;\propto\; (U_0 - E_{n\ell})^{1/2} \;\rightarrow\; 0 \tag{5}$$

$$< n,\ell=1 \mid R\,\frac{dV(r)}{dr} \mid n,\ell=1 > \;\rightarrow\; \frac{1}{3}\,2E_{n\ell} \tag{6}$$

$$< n,\ell=2 \mid R\,\frac{dV(r)}{dr} \mid n,\ell=2 > \;\rightarrow\; \frac{3}{5}\,2E_{n\ell} \tag{7}$$

and

$$< n,\ell=0 \mid r^2 \mid n,\ell=0 > \;\propto\; (U_0 - E_{n\ell})^{-1} \;\rightarrow\; \infty \tag{8}$$

$$< n,\ell=1 \mid r^2 \mid n,\ell=1 > \;\propto\; (U_0 - E_{n\ell})^{-1/2} \;\rightarrow\; \infty \tag{9}$$

$$< n,\ell=2 \mid r^2 \mid n,\ell=2 > \;=\; \text{finite} \tag{10}$$

Thus,

$$\frac{< j_2 \mid R\,\frac{dV(r)}{dr} \mid j_1 >}{< j_2 \mid r^2 \mid j_1 >} \;\rightarrow\; \begin{cases} 0 & \text{for} \quad \ell = 0 \text{ and } 1 \\ \text{positive finite} & \text{for} \quad \ell \geq 2 \end{cases} \tag{11}$$

That means, the quadrupole polarization charge due to the coupling of particles with ℓ to the IS quadrupole shape oscillations becomes

$$(e_{pol})_{\lambda=2,\tau\approx 0} \quad \rightarrow \quad 0 \quad \text{for} \quad \ell = 0 \quad \text{and} \quad 1 \tag{12}$$

and

$$(e_{pol})_{\lambda=2,\tau\approx 0} > 0 \quad \text{for} \quad \ell \geq 2 \tag{13}$$

in the limit of zero binding energies of particles.

3 Excitation of particles to continuum

The Shrödinger equation with the potential in (3) is solved for particle states in the continuum, $E_c > U_0$. The normalization of the continuum wave function is given by

$$\int_0^\infty R_{\ell_c}^{E_c}(r)\, R_{\ell_c}^{E'_c}(r)\, r^2 dr \;=\; \delta(E_c - E'_c) \quad . \tag{14}$$

We estimate the matrix elements between the bound particle state with quantum numbers $(n_h \ell_h)$ and the continuum particle state with energy E_c and orbital angular momentum ℓ_c. The analytic expressions of the matrix elements are found in ref.[2].

We are interested in the sign of the ratio

$$\frac{<\ell_c \mid R\,\frac{dV(r)}{dr} \mid n_h \ell_h>}{<\ell_c \mid r^2 \mid n_h \ell_h>} \qquad , \tag{15}$$

which determines whether the particle-vibration coupling increases (or decreases) the strength of the unperturbed ph excitations. Examining the analytic expressions in ref.[2] we can make the following statements : (a) For small values of ph excitation energies (namely, both the particle energy in the continuum $(E_c - U_0)/U_0$ and the separation energy of the $(n_h\, \ell_h)$ bound particle are small), the sign can be negative when the separation energy of the least-bound one-particle orbital with ℓ_c, which is an eigenstate of the potential, is small. Thus, for example, if we consider very low-energy ph excitations with $\ell_h = \ell_c$, the sign of the ratio is always negative. (b) When the separation energy of the least-bound one-particle with ℓ_c becomes larger than a certain value, the sign becomes positive even for small values of $(E_c - U_0)/U_0$. Then, the sign is positive even for small values of ph excitation energies, $(E_c - E_{n_h \ell_h})/U_0$.

Taking the example of $(n_h \ell_h\, ,\, \ell_c) = (0d, d)$, in Fig.1 we illustrate the situation in which the sign of $< \ell_c \mid R\,\frac{dV(r)}{dr}/U_0 \mid n_h \ell_h >$ is opposite to that of $< \ell_c \mid r^2/R^2 \mid n_h \ell_h >$ for low-energy threshold strength. Taking the potential parameter, which gives the $0d$ eigenstate at $E_h/U_0 = 0.95$, and choosing the energy of the continuum particle to be $E_c/U_0 = 1.05$, in Figs.(a), (b) and (c) on the l.h.s. of Fig.1 we plot radial wave functions of the hole and continuum one-particle states, $R_h(r)$ and $R_c(r)$, the products $R_c(r)\, R_h(r)\, r^2$, and $R_c(r)\, R_h(r)\, r^4$, respectively. Orthogonality of the continuum state $\mid \ell_c = 2 >$ to the hole state $\mid 0d>$ with the same angular momentum is seen from the fact that in the middle figure (b) the area above the x-axis is equal to that below the x-axis. Then, from the lower figure (c) it is clear that the sign of the matrix element $< \ell_c = 2 \mid r^2 \mid 0d>$ comes from the contribution in the region $r >> R$. Since the matrix element $< \ell_c = 2 \mid R\,\frac{dV(r)}{dr}/U_0 \mid 0d >$ receives the contribution from the value of $R_c(r)\, R_h(r)\, r^2$ at $r = R$, the sign of $< \ell_c = 2 \mid R\,\frac{dV(r)}{dr}/U_0 \mid 0d>$

must be opposite to that of $<\ell_c = 2 \mid r^2/R^2 \mid 0d>$. Taking the same quantum numbers $(n_h\ell_h\,,\,\ell_c) = (0d, d)$, on the r.h.s. of Fig.1 we illustrate the situation, in which the sign of $<\ell_c = 2 \mid R\,\frac{dV(r)}{dr}/U_0 \mid 0d>$ is the same as that of $<\ell_c = 2 \mid r^2/R^2 \mid 0d>$, when the hole state $\mid 0d>$ is deeply bound. On the r.h.s. of Fig.1 the energy of the continuum particle state is the same as that of the l.h.s., while the potential parameter is adjusted so that the $0d$ state is deeply bound at $E_h/U_0 = 0.59$. In figure (c) of the r.h.s. it is seen that the sign of the matrix element $<\ell_c = 2 \mid r^2 \mid 0d>$ is the same as that of the integrand at $r = R$.

In general, the sign of the ratio in (15) is positive when the sign of the matrix-element $<\ell_c \mid r^2/R^2 \mid n_h\ell_h>$ is the same as that of the integrand, $R_c(r)\,R_h(r)\,r^4$ at $r = R$. This occurs when the separation energy of the least-bound one-particle eigenstate with ℓ_c is larger than a certain value, which in the present example is about equal to the separation energy in β-stable nuclei.

In order to demonstrate that the physics obtained above by using the square-well potential remains unchanged for diffuse potentials, in Fig.2 we show an example numerically calculated by using a Woods-Saxon potential

$$U(r) = \frac{-U_0}{1 + exp\left(\frac{r-R}{a}\right)} \tag{16}$$

where

$$R = r_0\,A^{1/3} \quad . \tag{17}$$

We neglect the spin-orbit potential and use potential parameters : a=0.65 fm and r_0=1.25 fm. Simulating the case of Fig.1, we consider neutron excitations with ℓ_c=2 , $(n_h\,\ell_h)$=(0d) and the particle energy in the continuum E_p=+2 MeV. We adjust the mass number A so that the $0d$ state with a given energy E_B becomes an eigenstate of the potential with the depth $U_0 \approx 45.5$ MeV. The resulting mass number is 20 and 68 for E_B=−2 MeV and −20 MeV, respectively. The potential parameters used are those often used for β-stable nuclei. Since at the moment we study the coupling of ph excitations to vibrations of the core, the chosen parameters are realistic. The matrix elements of interests are

$$<p \mid r^2 \mid h> = \int_0^\infty r^4\,R_h(r)\,R_c(r)\,dr \tag{18}$$

and

$$<p \mid R\,\frac{dU(r)}{dr} \mid h> = R\int_0^\infty \frac{dU(r)}{dr}\,r^2\,R_h(r)\,R_c(r)\,dr \tag{19}$$

where $R_h(r)$ and $R_c(r)$ express radial wave functions of the hole and particle (in the continuum) state, respectively. In Fig.2 we plot the radial dependence of $R\,\frac{dU(r)}{dr}$ and the integrand of the expressions (18) and (19) by dotted, solid and dashed curves, respectively. It is clearly seen that the sign of (18) is opposite to (the same as) that of (19) for E_B=−2 MeV (−20 MeV). The change of the relative sign occurs around E_B=−12 MeV for the present parameters. The calculated result obtained by using the Woods-Saxon potential is indeed remarkably similar to that of the square-well potential.

In conclusion, we have shown that the ratio (15) can become negative for excitations of particles with small separation energies to the continuum. Consequently, the polarization charge $(e_{pol})_{\lambda,\tau\approx 0}$ can become negative. This unexpected negative sign may occur for the low-lying threshold strength in drip line nuclei, in which particles with separation energies much less than those in β-stable nuclei are excited to the continuum.

The author would like to express her deep appreciation to the numerous stimulation and encouragements extended to her for many years by professor H.Ejiri.

References

[1] A.Bohr and B.R.Mottelson, Nuclear Structure, Vol.II (Benjamin, Reading. MA, 1975).

[2] I.Hamamoto and X.Z.Zhang, Phys.Rev. **C58** (1998) 3388.

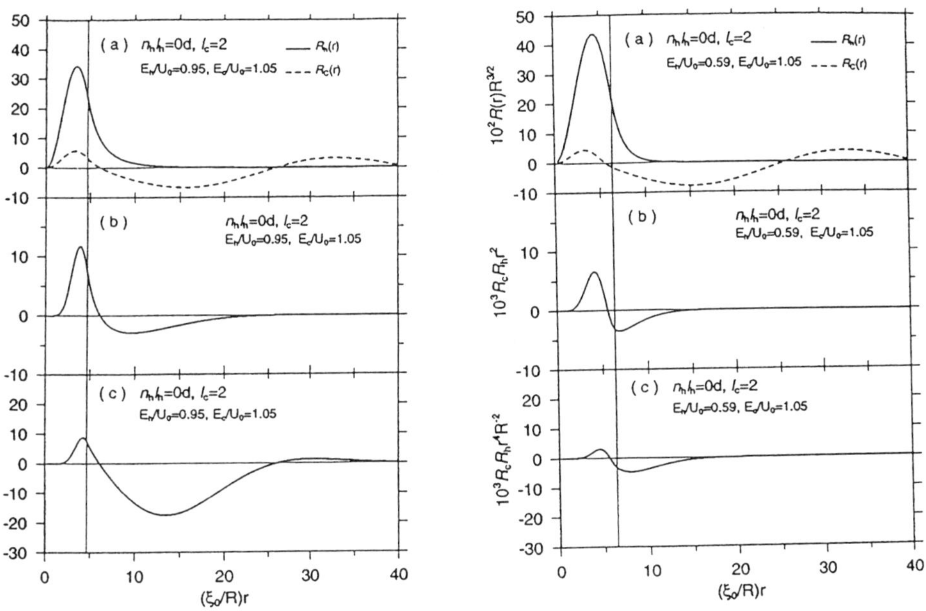

Fig.1. l.h.s. figure : Illustration of the sign difference between $<\ell_c \mid R\frac{dV(r)}{dr} \mid n_h\ell_h>$ and $<\ell_c \mid r^2 \mid n_h\ell_h>$ for low-energy threshold strength, taking an example of $(n_h\ell_h\,,\,\ell_c)$=$(0d\,,\,d)$, $E_h/U_0 = 0.95$ and $E_c/U_0 = 1.05$. The potential parameter U_0R^2 is adjusted so that the one-particle $0d$ state is an eigen state of the potential with $E_h/U_0 = 0.95$. (a) Radial wave functions of the hole state $R_h(r)$ and the continuum particle state $R_c(r)$, as a function of radial coordinate $(\xi_0/R)r$, where $\xi_0 \equiv R\sqrt{\frac{2mU_0}{\hbar^2}}$. The thin vertical line indicates the radius of the potential, $(\xi_0/R)r\,|_{r=R} = \xi_0 = 4.678$. (b) The integrand of the radial overlap integral between radial wave functions, $R_c(r)$ and $R_h(r)$. (c) The integrand of the radial integral of the matrix element $<\ell_c \mid r^2/R^2 \mid n_h\ell_h>$.
r.h.s.figure : Illustration of the same sign of $<\ell_c \mid R\frac{dV(r)}{dr} \mid n_h\ell_h>$ as that of $<\ell_c \mid r^2 \mid n_h\ell_h>$ for deeply bound hole states, taking an example of $(n_h\ell_h\,,\,\ell_c)$=$(0d\,,\,d)$, $E_h/U_0 = 0.59$, and $E_c/U_0 = 1.05$. The potential parameter U_0R^2 is adjusted so that the one-particle $0d$ state is an eigen state of the potential with $E_h/U_0 = 0.59$. In figs.(a), (b) and (c) the same quantities as on the l.h.s.figure are plotted for the present parameters. The thin vertical line indicates the radius of the potential, $(\xi_0/R)r\,|_{r=R} = \xi_0 = 6.398$.

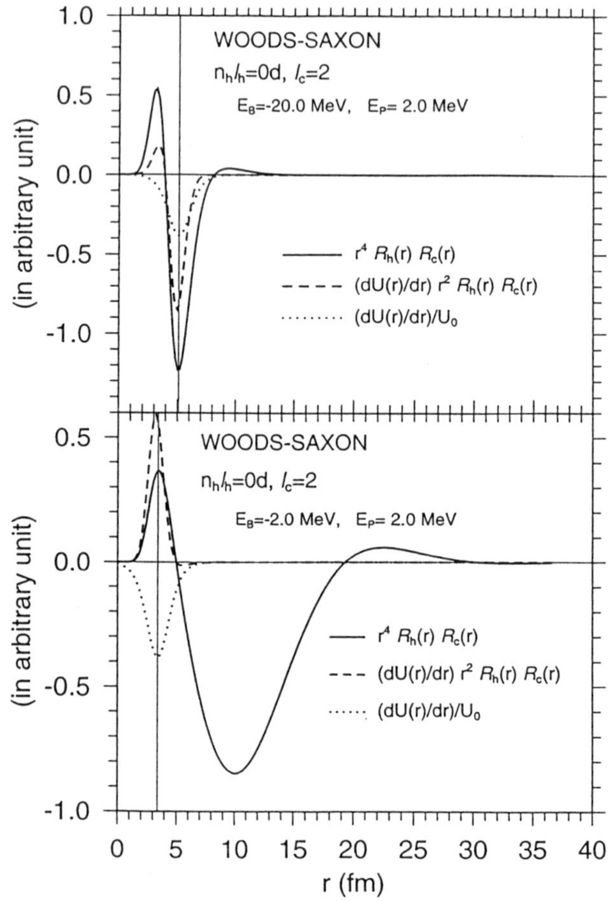

Fig.2. Comparison of the integrands of the expressions (18) and (19) for two energies of hole states, E_B=-2.0 MeV and -20.0 MeV, as a function of radial coordinate, for the case of Woods-Saxon potentials. The radial dependence of $R\,\frac{dU(r)}{dr}$ is also shown by dotted curves. Since the continuum wave function is arbitrarily normalized, the absolute magnitudes of the quantities plotted have no meaning. The particle energy in the continuum is fixed at E_p=+2.0 MeV. The upper part corresponds to A=68, while the lower to A=20. The radius of the potential is 3.39 and 5.10 fm for A=20 and 68, respectively, and is indicated by thin vertical lines. See the text for details.

Nuclear Electro Weak Spectroscopy for Symmetries in Electro Weak Nuclear-processes [a]

H. EJIRI

Research Center For Nuclear Physics, Osaka University
Ibaraki, Osaka 567-0047 Japan
ejiri@rcnp.osaka-u.ac.jp

Symmetries, which are current important subjects of nuclear particle physics, are studied by means of nuclear electro-weak spectroscopic methods. Since nuclei consist of nuclear particles in good quantum states, they are used as micro-laboratories for studying elementary particles and fundamental interaction. Recent works on the symmetries in electro-weak processes are briefly presented. Subjects discussed are left-right symmetries and neutrino mass by nuclear $\beta\beta$-γ spectroscopy, nuclear spin isospin symmetries and neutrino nuclear responses, baryon lepton symmetries by nucleon nuclear instabilites, and strange flavour asymmetry and symmetry by hyperon weak and double weak processes.

1 Introduction

Nuclei are regarded as cold crystals of particles and interactions. They are almost symmetry-broken system as a whole, but are partially symmetric quantum systems. Nuclear particles of baryons (p, n, Λ, $\cdots$) and mesons (π, ϱ, K, $\cdots$) have masses, which are much larger than the temperature of the nuclear particle system. Thus, they are well defined particles.

Nuclear states are actually good eigen states of energies E with their widths $\Gamma \ll E$. Ground states of stable nuclei are stable against electro-weak decays, while those of beta-unstable nuclei are mostly stable against electro-magnetic decays. Particle-bound excited states are stable against strong decays. Consequently, these nuclear states have well defined masses (energies).

Nuclear states have also well defined angular momenta J because of the rotation symmetry. Partial symmetries in the isospin space and the space reflection are realized in nuclei since the electromagnetic and weak interactions are much smaller than the strong interaction which bound nucleons in nuclei. Thus their isospins I and parties π are fairly well defined.

In most cases, nuclear states are described as eigen states of the nuclear Hamiltonian $H = H_s + H_r$, where H_s is the one-body nuclear potential and H_r is the residual interaction. Consequently, some nuclear states are given approximately as single particle states $\varphi_s(j,s,l)$ with relatively good shell-model

[a] The paper presented at the International Conference, NEWS99 (Nuclear Electro Weak Spectroscopy for Symmetries in Nuclear Electro Weak Nuclear-processes).

orbits of the angular momentum j, the spin s, and the orbital angular momentum l. The residual interaction gives rise to various kinds of collective/phonon states $Q^+(E, J^\pi, I)$ with relatively good quantum numbers of energy E, angular momentum J, partly π and isospin I

The nucleus is indeed a unique physics system. It consists of nucleons in good eigen states with relatively good quantum numbers of E, J^π, I, and even of j^π l. Thus the nucleus is used as an excellent micro-laboratory for studying elementary particles and fundamental interactions. Here one can select and even enhance specific processes of particle physics interests by choosing nuclear states (laboratories) with particular quantum numbers[1,2].

Electro-weak probes are quite useful for spectroscopic studies of particles in nuclear micro-laboratories.[3] Photons γ, electrons e and neutrinos ν, which are relevant to the electro-weak probes, are mass-less or very light-mass particles. Then, wave-lengths λ of these photons and leptons associated with the low-energy e$\gamma\nu$ spectroscopy are larger by 2∼3 orders of magnitudes than those of N and π and the nuclear radius R. Consequently, one can use long wave-length approximation of $\lambda \gg$R ($k_\gamma \ll \hbar/R$). In most cases, the lowest-order angular momentum is dominant.

Electro-weak probes access directry to nucleons in nuclei without much distortion because their interactions are fairly weak. It is also noted that the γ and e probes give large responses in detectors because of long-range EM interactions with atomic electrons in detectors.

2 Left-Right Symmetries Studied By Nuclear $\beta\beta - \gamma$ Spectroscopy

2.1 Left-right symmetries and nuclear double beta decays

The electro-weak standard theory of SU(2)$_L\times$U(1) is based on the pure left-handed weak interaction with mass-less neutrinos. The theory is thus completely asymmetric in left-right helicities. It is thus quite important to search for small deviations ε from the completely asymmetric weak-space toward the symmetric space. These small deviations may manifest as the possible right-handed weak current and the possible finite mass of the neutrino, which are evidences for unified theories beyond the standard theory.

The right-handed V+A current and the electron-neutrino mass have been studied by investigating longitudinal polarizations and end-point energies in single β decays. In these investigations, measured values are small deviations ε and $m_\nu c^2$from the maximal polarization of P=1 and the maximum energy release of Q_β. Thus ultra-high precision measurements are required.

Double beta decays ($\beta\beta$) are associated directly with these small quantities of ε and $m_\nu c^2$. Actually, decay rates of neutrino-less double beta decays ($0\nu\beta\beta$)

are proportional to the square of the right-handed weak current and to those of the neutrino-mass. The $\beta\beta$-decay rate itself is extremely small due to the second-order weak process. Consequently, ultra-high sensitive measurements are required for studying the ultra-rare $0\nu\beta\beta$ processes.

The $0\nu\beta\beta$ process is also a sensitive probe to study the coupling of the neutrino with the mass-less Majoron B, SUSY particles with $L-B$ violating interactions, the composite neutrino and others, which are all beyond the standard theory. The details are given in recent review articles[4–8].

Spectroscopic studies of nuclear $\beta\beta$ and γ rays are excellent ways of investigating fundamental ν-particles and fundamental weak interactions in nuclear micro-laboratories with asymmetric nuclear states as shown in Fig.1. They are found in some nuclei because of non-symmetric proton (p) and neutron (n) masses of $m_p \neq m_n$ and their interactions of $H_{pp} \neq H_{nn}$ and of the residual pairing interaction of $V_{pair} \neq 0$. In the nuclear level scheme shown in Fig.1, the single β from the even-even nucleus A to the odd-odd nucleus B is energetically forbidden, and only the $\beta\beta$ decay to the even-even nucleus C is possible. Consequently, huge background single-β rays, which would usually exist, are eliminated in this type of nuclear laboratories.

Here, we consider two-neutrino $\beta\beta$ decays of $2\nu\beta\beta$ and neutrino-less ones of $0\nu\beta\beta$.

$$2\nu\beta\beta : \mathrm{A}(nn) \rightarrow \mathrm{C}(pp) + \beta + \beta + \bar{\nu} + \bar{\nu}, \tag{1a}$$

$$0\nu\beta\beta : \mathrm{A}(nn) \rightarrow \mathrm{C}(pp) + \beta + \beta \tag{1b}$$

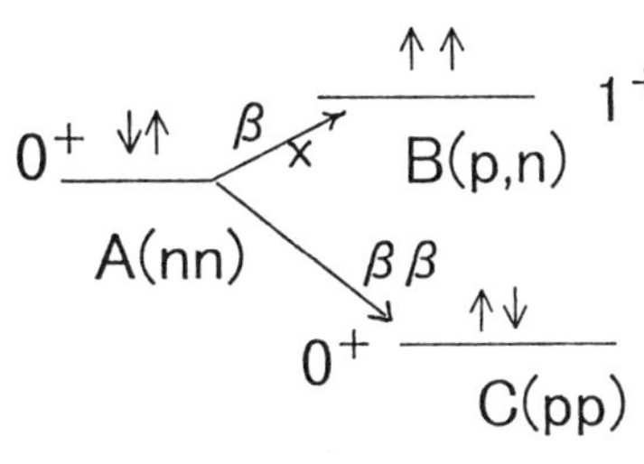

Figure 1: Schematic levels and transitions in $\beta\beta$. Spins of two nucleons involved in $\beta\beta$ are given by arrows.

The $2\nu\beta\beta$ process conserves the lepton number L, and thus it is the process within the standard theory. On the other hand the $0\nu\beta\beta$ process violates the

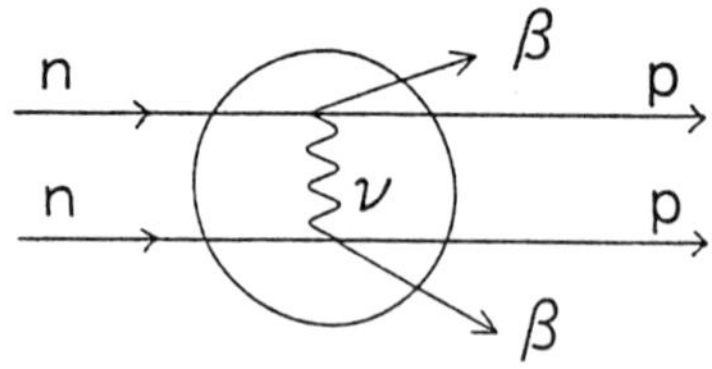

Figure 2: Two nucleon $\beta\beta$ decays with the Majorana-ν exchange between them.

lepton-number conservation law by ΔL=2. It is caused by the finite m_ν, the finite $V + A$ and other physics quantities, which are beyond the standard theory. One of major $0\nu\beta\beta$ processes is the ν exchange process between the two-nucleons in the nucleus, as shown in Fig.2. Here only tow low-energy $\beta\beta$ are emitted outside the nucleus and the virtual Majorana-ν is exchanged between the two nucleons with a small distance. Then, the phase-space factor $G^{0\nu}$ for $0\nu\beta\beta$ is larger by a factor $\sim 10^{7\sim8}$ than $G^{2\nu}$ for $2\nu\beta\beta$. Therefore, the $0\nu\beta\beta$ rate is enhanced in the nucleus by a factor $G^{0\nu}/G^{2\nu} \approx 10^{7\sim8}$.

In these respects, the nucleus used for the $0\nu\beta\beta$ spectroscopy is regarded as an excellent micro-detector (microscope) with the large enlargement factor of $G^{0\nu}/G^{2\nu} \sim 10^{7\sim8}$ and the large filtering power of $T(\beta)/T(\beta\beta) \sim 10^{23\sim26}$, which are crucial for detecting such small quantities as $m_\nu \leq 10^{-5}$ in the electron-mass unit.

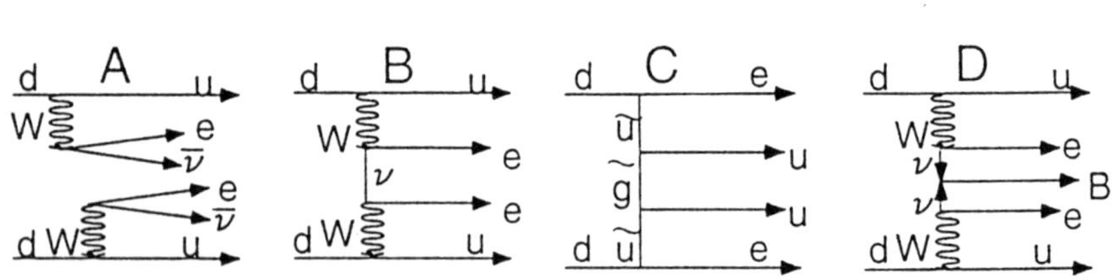

Figure 3: Schematic diagrams of $\beta\beta$[1]. u and d stand for quarks and $\tilde{u}$ and $\tilde{g}$ are for SUSY particles. W and B are the weak boson and the Majoron. A, B, C, and D show $2\nu\beta\beta$, $0\nu\beta\beta$ with the ν-exchange, $0\nu\beta\beta$ with the SUSY coupling, and $0\nu\beta\beta$ B followed by the Majoron B, respectively.

The $2\nu\beta\beta$ and $0\nu\beta\beta$ processes are shown in Fig.3. The $2\nu\beta\beta$ transition rate is written as

$$T^{2\nu} = G^{2\nu} \mid M^{2\nu} \mid^2, \tag{2a}$$

$$M^{2\nu} = M_{GT}(\sigma\tau\sigma\tau) + M_F(\tau\tau), \tag{2b}$$

where the $2\nu\beta\beta$ matrix element $M^{2\nu}$ consists of the Gamow-Teller (GT) matrix element of M_{GT} and the Fermi (F) one of M_F. The $0\nu\beta\beta$ transition rate for the Majorana-ν-exchange process (Fig.3B) is written as

$$T^{0\nu} = G^{0\nu} \mid M^{0\nu} \mid^2 (\langle m_\nu \rangle + C_\lambda \langle \lambda \rangle + C_\eta \langle \eta \rangle)^2, \tag{3a}$$

$$M^{0\nu} = M(h(r_{12}), p_1 p_2, \sigma_1 \sigma_2, \tau_1, \tau_2), \tag{3b}$$

where $M^{0\nu}$ is the matrix element for the $0\nu\beta\beta$ mode induced by the Majorana mass term $\langle m_\nu \rangle$. $M^{0\nu}$ is given as a function of the neutrino potential $h(r_{12})$, spins σ_i, isospins τ_i, and momenta p_i for the two nucleons(i=1, 2) involved in the ν-exchange. C_λ and C_η are the relative matrix elements with respect to $M^{0\nu}$ for the right-handed λ and η terms. The mass term and the right-handed λ and η terms are expressed as [4-7].

$$\langle m_\nu \rangle = \Sigma m_j U_{ej}^2, \tag{4a}$$

$$\langle \lambda \rangle = \lambda \Sigma U_{ej} V_{ej} \qquad \lambda = (M_W^L / M_W^R)^2, \tag{4b}$$

$$\langle \eta \rangle = \eta \Sigma U_{ej} V_{ej} \qquad \eta = W_L / W_R mixing, \tag{4c}$$

where M_W^L and M_W^R stand for masses of the left handed weak boson (W_L) and the right handed one (W_R), and U_{ej} and V_{ej} are the mixing amplitudes of left- and right-handed neutrinos.

The $0\nu\beta\beta$ transition rate for the SUSY particle exchange process (Fig.3C) is given as [4-7].

$$T^{0\nu}(\tilde{g}) = G f^2 M(\tilde{g}) A(M(\tilde{g}))/(M(\tilde{u}))^4, \tag{5}$$

where $M(\tilde{g})$ and $M(\tilde{u})$ are the gaugino and squark masses, respectively.

The transition rate for the $0\nu\beta\beta$ process followed by the Majoron, which is the Goldstone boson associated with the $L-B$ breaking, is given as [4-7].

$$T^{0\nu B} = G^M (M^B)^2 \langle g_M \rangle^2 \tag{6a}$$

$$\langle g_B \rangle = \Sigma g_{jk} U_{ej} \cdot V_{ek}, \tag{6b}$$

where g_B is the coupling of the electron neutrino with the Majoron B.

2.2 ELEGANT detectors for $\beta\beta-\gamma$ spectroscopy at underground laboratories

ELEGANT (ELectron GAmma-ray Neutrino Telescope) detectors have been developed for $\beta\beta-\gamma$ spectroscopic studies. EL I and EL II are test and prototype detectors. EL III consists of a pure Ge detector surrounded by the NaI detectors for γ and X-rays. The NaI detectors were used to identify the true $\beta\beta$ events and to reject such 2e and βe fake events as followed by Compton-γ rays from Compton scatterings and X-rays from internal conversion electrons. The measurement EL III gave the stringent limit of 2eV on the electron neutrino mass, and the stringent limit on the $0\nu\beta\beta$ process feeding the first 2^+ excited state[9].

The EL IV consists of multi-layer Si detector array surrounded by the NaI detectors as in EL III. The preliminary measurement with EL IV gave a first signature of the ^{100}Mo $2\nu\beta\beta$.

The EL V is a detector complex[10], as shown in Fig.4. It consists of top, central, and bottom drift chambers for β-ray trajectories, plastic scintillator arrays for β-ray energies and times, and large NaI detector arrays for γ and X rays. EL V, thus, can measure energy and angular correlations of two β rays, which are important for identifying the individual m_ν, λ, and η terms in the $0\nu\beta\beta$ process. EL V is used for any external sources of thin foils, which are set between the drift chambers. The $0\nu\beta\beta$ decays of ^{100}Mo and ^{116}Cd were studied by EL V.

EL VI consists of 25 modules of pure CaF_2 detectors surrounded by CsI detector arrays [11]. It is used to study $\beta\beta$ of ^{48}Ca in the natural Ca detectors. By using 25 modules of CaF_2 (Eu) in place of the pure CaF_2 detectors one can study cold dark matters (DM) by measuring elastic scatterings of WIMPs from ^{19}F in the CaF_2 (Eu) detectors.

Recently, EL VII with NaI detector arrays has been built to study e-capture β^+ processes by measuring two 511 keV γ-rays associated with β^+.

The EL detectors are all set in air-tight containers, and are shielded by OFHC (oxigen free high conductive cupper) and Pb breaks to reject external background γ-rays. The pure N_2 gass is introduced into the air-tight container to remove Rn in the air. All detector ensembles are set at underground laboratories to avoid cosmic rays.

The Kamioka underground laboratory with the water-equivalent(we) depth of 2700 m was used to study $\beta\beta$ decays of ^{76}Ge by EL III and those of ^{100}Mo and ^{116}Cd by EL IV and V.

Recently, a new underground laboratory, Oto Cosmo Observatory with 1500 m we has been opened, as shown in Fig. 5, at the Tentsuji tunnel, 100km south from Osaka Univ. The Rn content is most crucial for nuclear rare-decay

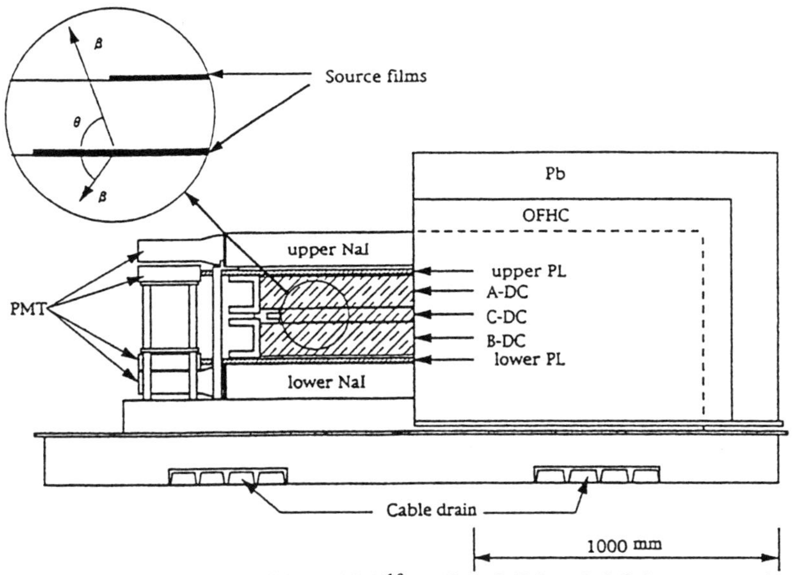

Figure 4: Schematic side view of ELEGANT V[10]. A-DC, B-DC and C-DC are upper, lower and central drift chambers, PL and PMT are plastic scintillators and photo multiplier tubes.

studies such as $\beta\beta$ and DM. A nice point of the new Oto lab. is the quite small Rn content of an order of 10 Bq/m^3, which is far below the level at the Kamioka lab. by 2~3 orders of magnitude. Other backgrounds at the Oto lab. are $4 \cdot 10^{-7}$/cm^{2}sec for cosmic muons, and $4 \cdot 10^{-5}$/cm^{2}sec for neutrons. These are satisfactry low. Actually, the Rn content inside the detector container is reduced down to 0.1 Bq/m^3 by filling it with the Rn-free N_2 gas.

2.3 The $2\nu\beta\beta$ and $0\nu\beta\beta$ results of ^{100}Mo and ^{116}Cd

The $2\nu\beta\beta$ and $0\nu\beta\beta$ decays of ^{100}Mo and ^{116}Cd were investigated by EL V at the Kamioka lab., and later those of ^{100}Mo by EL V at the Oto lab. Finite halflives of $1.15 \cdot 10^{19}$y and $2.6 \cdot 10^{19}$y were obtained for the $2\nu\beta\beta$ rates of ^{100}Mo and ^{116}Cd[10]. These are the first measurements of the finite decay rates. These halflives lead to the $2\nu\beta\beta$ matrix elements of 0.09 and 0.07, in units of m_ec^2, for ^{100}Mo and ^{116}Cd, respectively. The measured matrix elements are less than one tenth of single particle values, indicating destructive spin-isospin medium effects[12].

The new measurement on ^{100}Mo at the Oto lab., being combined with the results at the Kamioka lab. gives most stringent limits on the $0\nu\beta\beta$. The obtained halflife limits with 68% CL are as follows. $T^{0\nu}(m_\nu) > 0.65 \cdot 10^{23}y$, $\langle m_\nu \rangle$ <1.9eV. $T^{0\nu}(\lambda) > 0.49 \cdot 10^{23}y$, $\langle\lambda\rangle < 3.2 \cdot 10^{-6}$. $T^{0\nu}(\eta) > 0.64 \cdot 10^{23}y$, $\langle\eta\rangle < 2.2 \cdot 10^{-8}$. $T^{0\nu B}(B) > 0.54 \cdot 10^{22}y$, $\langle g_B\rangle < 7.3 \cdot 10^{-5}$. $T^{0\nu}(\nu^*) > 0.65 \cdot 10^{23}y$, $m(\nu^*) > 1.8 \cdot 10^4$TeV.

Here the matrix elements given in ref. 13 are used to get the limits given above. Stringent limits of the same order of magnitude have been derived from the life-time limits on ^{76}Ge, ^{136}Xe, and others by direct counting methods[8]. In fact, they depend much on the nuclear matrix elements, which are sensitive to the nuclear models to be used. Thus it is important to study $0\nu\beta\beta$ decays on several nuclei to get universal quantities of neutrinos and weak interactions.

Presently, precision measurements on the ^{100}Mo $\beta\beta$ decays are under processes with the improved EL V at Oto Cosmo Observatory. The goal is to study the $0\nu\beta\beta$ and $0\nu B\beta\beta$ processes with sensitivities of $1.4{\cdot}10^{23}$y and $0.2{\cdot}10^{23}$y, corresponding to $\langle m_\nu\rangle$ ~1.3 eV and $\langle g_B\rangle \sim 3{\cdot}10^{-5}$.

Nuclear matrix elements $M^{2\nu}$ for $2\nu\beta\beta$ have been derived on several nuclei by direct counting methods and indirect geochemical methods[6-9]. They are all quite small in comparison with simple single particle values $M_S^{2\nu}$.

It has been found that the observed $2\nu\beta\beta$ matrix element from the 0^+ initial nucleus |i⟩ to the 0^+ final nucleus |f⟩ is well reproduced by the successive single β process through the low-lying single particle-hole 1^+ state·|S⟩ in the intermediate nucleus [14]. Then, $M^{2\nu}$ is expressed as

$$M^{2\nu} \sim M_S^\nu \cdot M_{S'}^\nu / \Delta_S, \tag{7}$$

where M_S^ν and $M_{S'}^\nu$ are the single β matrix elemtents from |i⟩ and to |f⟩ through the intermediate |S⟩, and Δ_S is the denominator.

Single β matrix elements M_S^ν and $M_{S'}^\nu$ are reduced by a factor g_A^{eff}/g_A ~0.3 from the single particle value of $M(SP)$ due to the destructive interference with the GT giant resonance[12]. Thus, the 2ν matrix element $M^{2\nu}$, which is the product of M_S^ν, and $M_{S'}^\nu$ are reduced by a factor $(g_A^{eff}/g)^2$ ~0.1, in agreement with the observed reduction rate.

Recently, the $2\nu\beta\beta$ process has been analyzed in terms of the low-lying single particle state |S⟩ and the high-lying GT giant resonance |GR⟩ in the intermediate nucleus[15]. The $2\nu\beta\beta$ matrix element is mainly the GT type one of $M^{2\nu}(1^+)$ since the Fermi-type β strength is mostly absorbed into the isobaric analog state. Then $M^{2\nu}(1^+)$ is written as

$$M^{2\nu}(1^+) = \Sigma \frac{M_K^\nu \cdot M_{K'}^\nu}{\Delta_K} = \frac{M_S^\nu \cdot M_{S'}^\nu}{\Delta_S} + \frac{M_G^\nu \cdot M_{G'}^\nu}{\Delta_G}, \tag{8}$$

where M_G^ν and $M_{G'}^\nu$ are, respectively, the single β matrix element from |i⟩ and that to |f⟩ thgough |GR⟩ in the intermediate nucleus, and Δ_G is the energy

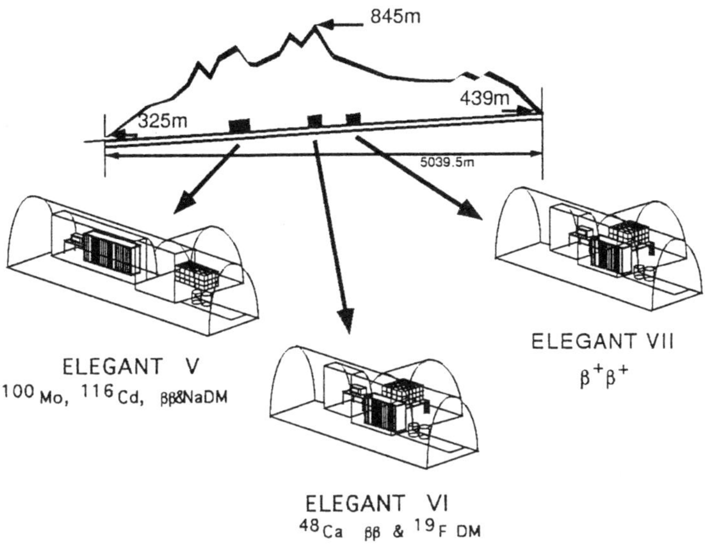

Figure 5: Oto Cosmo Observatory and ELEGANTs detectors

denominator[15]. The level scheme is as shown in Fig.6. Introducing the $\tau\sigma$ type interaction between $|S\rangle$ and $|GR\rangle$, the $|GR\rangle$ is perturbed to be given as a mixed state of $|GR\rangle = \varepsilon Q_S^0 + Q_G^0$, where Q_S^0 and Q_G^0 are the unperturbed S and GR states. Similarly, the pertubed final state is given as $|f\rangle = Q_{S'}^0 - \varepsilon' Q_{G'}^0$, where $Q_{S'}^0$ and $Q_{G'}^0$ are the unperturbed states of $|f\rangle$ and the unperturbed GT-type GR based on $|S\rangle$, respectively. Since $Q_S^0 \sim Q_{S'}^0$, $Q_G^0 \sim Q_{G'}^0$, and $\varepsilon \sim \varepsilon'$, the β transition from $|GR\rangle = \varepsilon Q_S^0 + Q_G^0$ to $|f\rangle = Q_{S'}^0 - \varepsilon' Q_S^0$ vanishes almost because of cancellation among the Q_S and Q_G terms. Furthermore, the denominator Δ_G is much larger than that of Δ_S. Consequently, the second GR term in eq. 8 may be neglected in comparison with the first SP term, in agreement with the experimental findings of eq. (7).

3 Nuclear Spin Isospin Symmetries and Neutrino Nuclear Responses

3.1 Nuclear spin isospin responses for neutrinos and charge exchange spin-flip reactions

Nuclear responses for neutrinos are very important in neutrino studies in nuclear microlaboratory. The response is represented by the nuclear matrix element M_β, as shown in Fig.7. It corresponds to the nuclear detector sensitivity.

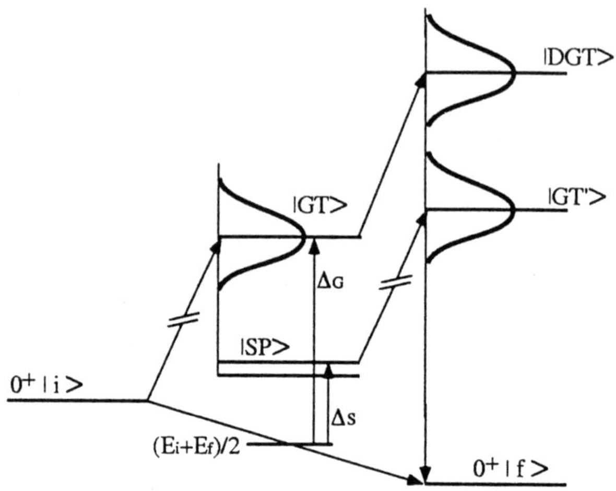

Figure 6: Levels and transition scheme for the $0^+ \to 0^+$ $\beta\beta$ decay. SP and GT stand for the 1^+ low-lying single particle-hole state and the GT-type 1^+ giant resonance, respectively. See text.

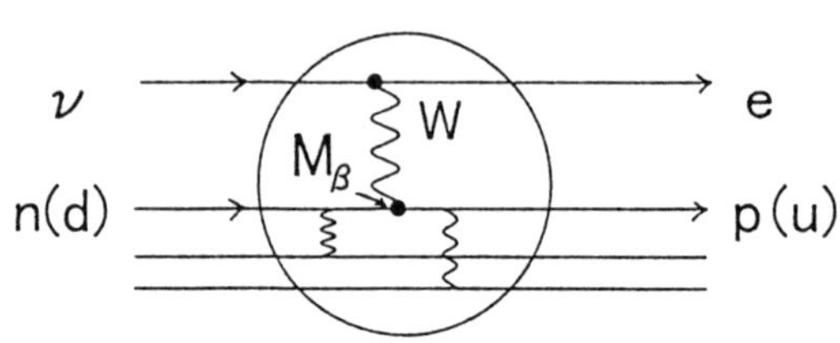

Figure 7: A nuclear weak process of ν+n(d)→e+p(u) with the weak boson W. M_β represents the nuclear response.

The nuclear response for the charged weak current is expressed as

$$M(\beta) = \tau_{\pm}(\sigma^S \times Y_l)_J, \qquad S = 0.1, \qquad l = 0, 1, 2, \cdots \tag{9}$$

where the Fremi (vector)-type and GT (axial)-type currents are given by the spin S=0 and 1, respectively. Actually, $M(\beta)$ is modified much in the nuclear medium by the strong $\tau\sigma$ correlation, which necessarily perturbs the $\tau\sigma$ symmetry in the nuclear medium [12].

The Fermi response is simply given by the isospin raising and lowering operators $\tau_{\pm}$ with S=0 and $l = J = 0$ in eq. 9. The nucleus shows a simple Fermi response with the sharp IAS (isobaric analogue state) because of the isospin symmetry in the nucleus. The commutation relation of $[H, T_-]=\Delta_c T_-$, where $T_-=\Sigma\tau_-(i)$ and Δ_c is the Coulomb energy shift, leads to the sharp IAS defined by $\mid \text{IAS}\rangle = T_- \mid 0\rangle$. IAS is a kind of a giant isospin resonance, i. e. an isospin phonon. Therefore most of Fermi-β strengths is absorbed into IAS, leaving little to the low-lying states in intermediate and final nuclei.

On the other hand, the supermultiplet ($\tau\sigma$) symmetry is not valid in the nucleus because of the strong spin isospin interaction. Then the giant GT resonance shows a broad peak, and thus a finite strength is located even at the ground state [12].

Actually, the GT nuclear matrix element for the ground state is approximately given as $M(a) = M_{SP} \cdot \varepsilon/\Delta_{GT}$, where Δ_{GT} is the GT giant resonance energy and ε is the deviation of the lowest level from the mean energy of the 1^+ particle-hole states. The ratio of ε/Δ_{GT} stands for the degree of the asmmetry.

The $\tau\sigma$ responses with $l \neq 0$ are seen even in low-lying states since the τY_l and $\tau\sigma Y_l$ giant resonances with $l \neq 0$ are rather broad because of the large deviation of the nuclear potential from the harmonic potential.

The nuclear responses for ground states in some nuclei are derived from β-decay ft values. Generally, they are shown to be obtained from charge-exchange spin-flip reaction, as shown in Fig.8. They are (p, n), (n, p), (d, ^{2}He), (^{3}He, t), (t, ^{3}He), (^{7}Li, ^{7}Be), and so on[1].

The RCNP cyclotron with E_p and K=0.4 GeV provides proton beams up to 0.4 GeV and ^{3}He beams up to 0.15 GeV/A. Recently, the MSU-RCNP collaboration has succeeded in getting triton beams with 0.38 GeV by fragmentation of α beams from the MSU cyclotron [16]. These p, ^{3}He and t beams are just adequate to study spin isospin responses for neutrinos because of the relatively large spin-isospin interaction $V_{\tau\sigma}$ at these medium energies. Since the nuclear interaction V_0 is small here, one can study the $\tau\sigma$ response without much distortion effects.

The nuclear $\tau\sigma$ responses have extensively been studied by (^{3}He, t) and (t, ^{3}He) reactions. They show good correspondences to the GT strengths of

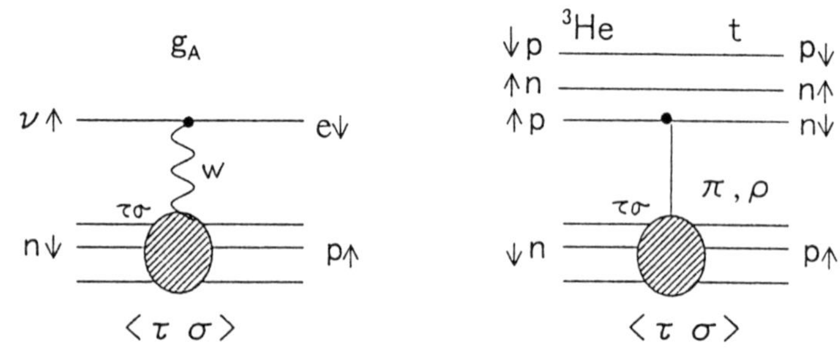

Figure 8: Nuclear responses for weak (ν, e) and strong (^{3}He, t) processes. $\langle\tau\sigma\rangle$ represents the nuclear spin isospin response.

$B(GT_-)$ and $B(GT_+)$.

The (^{7}Li, ^{7}Be) reaction is useful to study spin-flip and non spin-flip nuclear responses. The ground state of ^{7}Li has $J^\pi=\frac{1}{2}^-$, while ^{7}Be has the ground state with $J^\pi=\frac{3}{2}^-$ and the first excited state with $J^\pi=\frac{1}{2}^-$. The first excited state is identified by detecting γ-rays from the state. Thus, one can select the spin-flip and non-spin-flip processes by tagging the (^{7}Be, ^{7}Li) reaction with the decaying γ-ray.

3.2 Nuclear $\tau\sigma$ responses for $\beta\beta$-ν

Nuclear responses for $2\nu\beta\beta$ are mainly double GT $\tau\sigma\tau\sigma$ ones. They are studied by investigating both single $\beta^-(\tau_-\sigma)$ and $\beta^+(\tau_+\sigma)$ responses from $|i\rangle$ and $|f\rangle$ to intermediate states $|k\rangle$. Most nuclei used for $\beta^-\beta^-$ studies have very low-energy $Q(\beta^-)$ values. Thus it is in general difficult to get good β decay ft values.

The (^{3}He, t) reactions on ^{100}Mo and ^{116}Cd were studied at RCNP with the 0.45 GeV ^{3}He beam in order to get the $2\nu\beta\beta$ responses on these nuclei[17]. The triton was momentum-analyzed by the high-resolution spectrograph G-RAIDEN. The measured spectra are shown in Fig.9. A very tiny peak for the 1^+ ground state, broad peaks for the GT giant resonances, and a sharp peak for IAS are seen. The β matrix element M_S^ν for the ground state transition of ^{100}Mo(0^+) $\rightarrow$^{100}Te(1^+) is derived from the measured GT$_-$ strength by the (^{3}He, t) reaction. The matrix element $M_{S'}^\nu$ for the ground state transition of ^{100}Tc(1^+)$\rightarrow$ ^{100}Ru(0^+) was derived from the known ft value. Using these single β matrix elements, one gets the $2\nu\beta\beta$ matrix element through the

1^+ ground state in the intermediate nucleus of ^{100}Tc. The obtained value is $M_S^{2\nu}=M_S^{\nu} \cdot M_{S'}^{\nu}/\Delta_S=0.10$. This is in good agreement with the observed value of $M^{2\nu}=0.09$ by the EL V[10]. The $2\nu\beta\beta$ response is indeed dominated by the successive process through the ground state (single particle-hole) 1^+ state in the intermediate nucleus as shown in eq. (8).

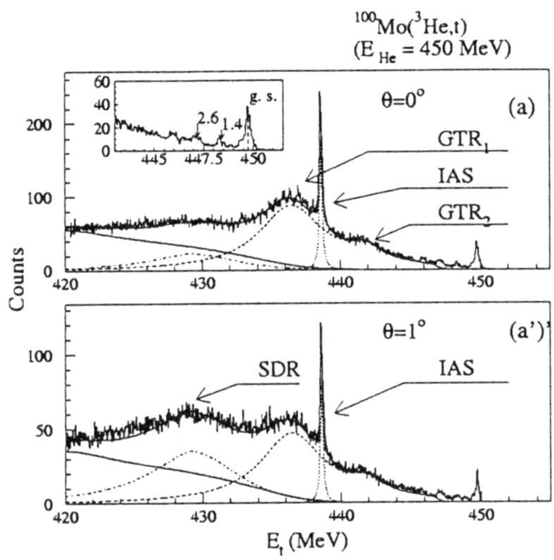

Figure 9: Energy spectra of (^{3}He, t) reactions on ^{100}Mo. GTR$_1$, GTR$_2$, IAS and g-s. stand for the higher and lower GT giant resonances, the isobaric analogue state and the ground state, respectively.

In the case of ^{116}Cd, the 1^+ single particle-hole strength is split into three low-lying 1^+ states in the intermediate nucleus of ^{116}In.

The $2\nu\beta\beta$ matrix element through the 1^+ ground state was derived as in case of ^{100}Mo by using the GT strength obtained from the (^{3}He, t) reaction and the ft value. The obtained value of $M_S^{2\nu}$ (1^+gs) is 0.03, which is 40% of the observed value[10] of $M^{2\nu}=0.07$. Obviously, the other two 1^+ states contribute to the $2\nu\beta\beta$, as studied recently for the splitting of the β strength in low-lying states.

The nuclear responses $M^{0\nu}$ for $0\nu\beta\beta$ involves many terms with angular momenta of l=0, 1, 2, $\cdots$ 6 because the virtual neutrino exchanged between two nucleons in a short distance has a large momentum. Furthermore, the $0\nu\beta\beta$

matrix element has neutrino potential terms such as $H(r_1, r_2, \tau_1\tau_2\sigma_1\sigma_2) = gf(r_1r_2)$ $\tau_1\tau_2\sigma_1\sigma_2$ with $f(r_1r_2) \approx k/ \mid r_1 - r_2 \mid$ and other momentum-dependent terms. Thus, exactly speaking, the $0\nu\beta\beta$ is not described as the sum of the successive single beta processes through intermediate states.

The neutrino potential $f(r_1r_2)$ involved in the nuclear $\beta\beta$ decay is, in fact, confined region of $r \leq r_1, r_2 \leq R$, where r and R are the nucleon hard-core radius and the nuclear one, respectively.

It has been shown by the recent work[18] that the confined coulomb-type potential is expressed approximately by a separable form,

$$f(r_1r_2) \sim \Sigma f_l h_l(r_1) h_l(r_2), \tag{10}$$

for l=0, 1, 2 and so on. Then, the matrix element with $f(r_1, r_2)$, which is associated with the neutrino-mass term, may be expressed as

$$M^{0\nu} = \Sigma f_l \langle 0_f \mid T_l^+ \mid k \rangle \langle k \mid T_l \mid 0_i \rangle. \tag{11}$$

Here $\mid 0_i \rangle$ and $\mid 0_f \rangle$ are the 0^+ initial and final states, $\mid k \rangle$ is the k th intermediate state, and $T_l = h_l(\mathrm{r})\tau\sigma$ is the β operator with the angular momentum l. Then, $M^{0\nu}$ can be given as the sum of successive β processes through the single particle-hole states and the giant resonances as

$$M^{0\nu} \sim \Sigma(M_l^+(SP)M_l^-(SP) + M_l^+(GR)M_l^-(GR)). \tag{12}$$

The second term may be small as in case of the GT resonance with l=0. Then one can express the $0\nu\beta\beta$ response as the sum over l of the successive β processes through the single particle hole states. They are well studied experimentally by means of the τ_- and τ_+ type charge-exchange reactions, as in case of $2\nu\beta\beta$.

Direct experimental studies on nuclear responses relevant to $\beta\beta$ decays are under progress by means of double charge-exchange reaction at RCNP.

3.3 Nuclear responses for solar neutrinos

Solar neutrinos (ν) are current important subjects in view of the possible neutrino oscillation. It involves finite neutrino-flavour mixings and neutrino-mass differences, which are beyond the standard theory. They have been studied by inverse beta decays of ^{71}Ga and ^{37}Cl and electron scatterings.

Nuclear response for the solar-ν is crucial for getting the solar-ν flux by the inverse β decay. They are obtained by charge-exchange reaction, as in ease of the $\beta\beta - \nu$.

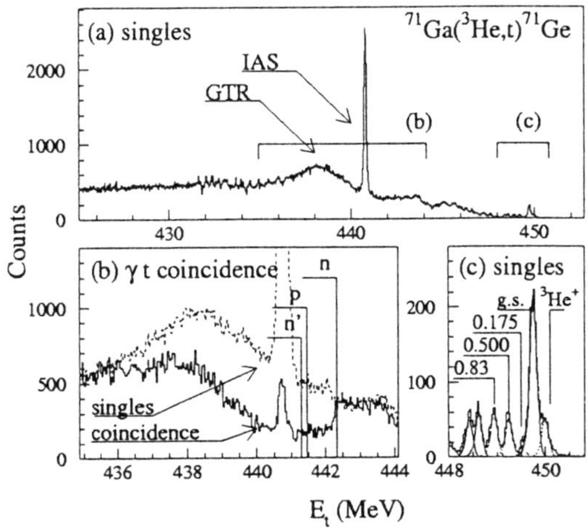

Figure 10: Energy spectra of the $^{71}Ga(^3He, t)$ reaction. The singles spectrum, the coincidence spectrum gated by decaying γ-rays, and the expanded singles spectrum for low-lying states are shown in (a), (b), and (c), respectively[19].

The $^{71}Ga(^3He, t\gamma)$ reaction[19] was studied at RCNP in order to get the ^{71}Ga response for the solar-ν. The NaI detector array was set to measure γ-rays in coincidence with the (^{3}He, t) reaction.

The measured single (^{3}He, t) and coincidence (^{3}He, tγ) spectra are shown in Fig.10. The low-lying 1^+ states, the sharp IAS and the broad GT giant resonance in ^{71}Ge are clearly excited. The GT and Fermi strengths for these states were deduced to get the solar-ν responses.

Gamma-ray branching ratios of Γ_γ/Γ_t for excited states in ^{71}Ge were obtained. Particle-unbound states up to 8.4 MeV have still large γ-branching ratios of $\Gamma_\gamma/\Gamma_t \sim$0.5. They contribute to the solar-ν absorption by the ^{71}Ga detector, where the number of the ^{71}Ge isotopes produced in the detector is counted. The solar-ν absorption rates for ^{71}Ga are 131.7 SNU with the small contribution of 1.75 SNU for the first excited state. The unbound state contribution is 0.35 SNU, which is about 3% for the ^{8}B solar-ν.

4 Baryon Lepton Symmetries by Nucleon Nuclear Instabilities

4.1 Nucleon decays in nuclei studied by nuclear $\beta - \gamma$ spectroscopy.

The baryon number B is conserved, and there are no bridge from baryon to lepton in the framework of the standard theory. The baryon lepton symmetry is completely broken there. The nucleon N forms the lightest baryon doublet with the isospin t=1/2. Thus, they are stable against electro-weak and strong processes in stable nuclei. Nucleon decays to leptons ($e^{\pm}, \mu^{\pm}, \nu$), photons (γ) and mesons ($\pi^{\pm}$, $K^{\pm}, \cdots$) have been studied mainly by observing charged decay-particles such as $e^{\pm}$, $\mu^{\pm}$, $\pi^{\pm}$, $K^{\pm}$, etc. and/or neutral particles such as γ, π^0, K^0 etc. which decay finally to charged leptons. These decay particles are thus visible particles by conventional detectors such as Cevenkov detectors, scintillators, ion-chambers, and so on. Neutron decays to neutrinos, $N \rightarrow 3\nu$, and to neutral exotic particles are invisible. In particular, rare neutrino decays are extremely hard to measure.

The nucleon decay in a nucleus leaves a nucleon hole N^{-1} there. Then, the hole deexcites by emitting γ-rays, protons, neutrons, and others, depending on the excitation energy. Thus, the nucleon decay, even the invisible one, can be studied by measuring these γ rays and/or charged particles [20]. The nuclear response for N^{-1} can be studied also by measuring their deexciting particles for the nucleon hole produced by the corresponding (p, pN) reaction.

The invisible neutron decay[21] of $n \rightarrow 3\nu$ in the ^{16}O nucleus was studied by measuring the γ ray from the inner 1s hole in ^{15}O. The energy spectrum measured by the Kamioka water Cevenkov counter shows no appreciable peak of the γ-ray from the 1s hole. The limit on the partial γ-decay lifetime is $3.0 \cdot 10^{31}$y. Using the γ decay branch of $\Gamma_\gamma / \Gamma_t \sim 10^{-4}$, the neutron halflife limit is derived as $5 \cdot 10^{26}$y. This gives the most stringent limit for the inclusive n decay, including $n \rightarrow \nu_i \nu_j \nu_k$ with i, j, k standing for any flavours.

The 1s hole in ^{15}O deexcites partially by emitting ^3He to the 15MeV 1^+ state in ^{12}C with the branching ratio of $\Gamma' / \Gamma_t \sim 10^{-3}$. This state with the isospin T=1 has a different symmetry from that of other low-lying states with T=0. Thus the α decay is forbidden, and the γ branch is as large as 80%. Then one gets the 15MeV γ branch[22] of $\Gamma_\gamma / \Gamma_t \approx (\Gamma'/\Gamma_t) \cdot (\Gamma_\gamma/\Gamma') \approx 10^{-3}$. The observed limit on the 15MeV γ ray leads to the limit of 10^{28}y on the inclusive n decay. This is much more stringent than the value obtained so far.

Nucleon decays are also studied by measuring radioactive isotopes produced by deexcitation of nucleon holes[23]. The large NaI detector array in EL V was used to study β&γ rays from radioactive I and Te isotopes produced by inclusive single- and multi-nucleon decays in ^{127}I. The measured limits lead to the lifetime limits of 10^{24}y for the inclusive multi-nucleon decays.

4.2 Non-Paulian nucleon and electron transitions and instabilities of nuclei and atoms.

Stabilities of nuclei and atoms are based on the Pauli exclusion principle for fermions. Inner shell orbits of the nuclear shell potential are filled by nucleons, and those of the atomic Coulomb potential by electrons. The nucleons and electrons in outer-shell orbits are not allowed to deexcite to the inner shell orbits due to the Pauli principle for them. Here the nucleon and the electron are considered to behave as a point-like fermion. If they, however, would have small non-fermion components, the non-Paulian transitions would be possible, as shown in Fig.11.

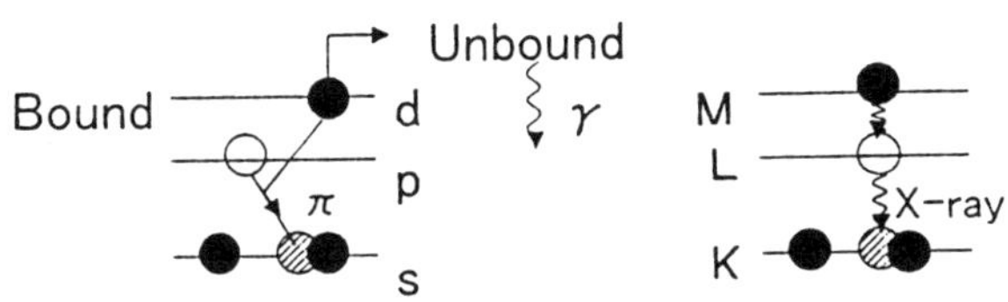

Figure 11: Non-Paulian transitions in a nucleus (left hand side) and in an atom (right hand side).

The non-Paulian nucleon and electron transitions[24] were studied by using the large NaI detector array in EL V. The limit of $\Gamma_{NP} < 10^{-25} y^{-1}$ on the non-Paulian nuclear decay leads to the upper limit of 10^{-53} on the non-Paulian branch of Γ_{NP}/Γ_t. The limit of $10^{-26} y^{-1}$ for the γ-decay to the 1s state in ^{16}O derived from the water Cerenkov data leads to the limit of $2.3 \cdot 10^{-57}$ for the non-Paulian nuclear decay[21].

Stability of the atom is studied by measuring K X-rays from higher electron shells to the inner K shell[25]. The upper limit of 10^{-23}y^{-1} on the K X-ray in the I atom was deduced by using the NaI detector array of EL V. It leads to the upper limit of 10^{-46} on the non-Paulian X-ray branch of Γ_{NP}/Γ_X. Assuming that the non-Paulian X-ray transition is possible for the composite electron by the order of the electron size[26], the measured X-ray limit leads to the limit of $1.1 \cdot 10^{-17}$cm on the size parameter r_0. This corresponds to 2 TeV in energy, which is the same order of the magnitude as the limit derived from the accelerator experiments.

5 Strange Flavour Asymmetry and Symmetry by Hyperon Weak and Double Weak Processes

5.1 Asymmetries in flavour-change weak processes

Weak decays of hyperons in hypernuclei provide unique opportunities for studying parity-violating weak processes at the strange and u-d quark sectors. Non-mesonic weak decays of Λ+p$\rightarrow$p+n with a large momentum transfer of 0.4 GeV/c were studied by measuring asymmetries of the decay protons from Λ-hypernuclei. Spin polarized $\vec{\Lambda}$ in p-shell light nuclei were produced by the (π^+, K^+) reaction on ^{12}C with the 1.06 GeV/c π^+ beam from the 12 GeV PS at KEK. The K^+ spectrometer and the weak-decay proton detector are shown in Fig.12. Since π and K mesons are partially absorbed in the nucleus, the near-side scattering with the small pass length for the π^+ and K^+ mesons are favored than the far-side scattering, as shown in Fig.13. This asymmetry leads to the asymmetry in the transferred angular momentum, resulting in the polarization[28]. The Λ spin itself produced by the π^++n$\rightarrow K^+ + \Lambda$ is polarized with respect to the reaction plane. These two effects give rize to the polarization of the Λ in the hypernucleus[28,29].

Asymmetries for the non-mesonic weak decays are expressed as

$$W(\theta) = A_0(1 + a_1 P_\Lambda k cos\theta), \tag{13}$$

where θ is the angle of the decaying proton with respect to the polarization axis, P_Λ is the Λ-spin polarization, k is the attenuation factor due to the rescattering effect, and a_1 is the asymmetry coefficient. The finite polarizations for Λ in hypernuclei and the large asymmetry coefficient of a_1=–1 for the non-mesonic weak decay were found for the first time[30], as shown in Fig.14. The large asymmetry indicates contributions of heavy mesons and quarks to the non-mesonic weak process with the large momentum transfer.

5.2 Exotic strange baryon and light quark symmetry

The exotic H dybaryon[31,32] with J^π=0^+ and I=0 has been studied theoretically and experimentally. This is an uds-quark symmetry dybaryon with B=2 and S=–2, as shown in Fig.15. The light stable H with the mass m_H below the two nucleon mass of $2m_N$ were searched for by investigating double weak decays of p+n$\rightarrow$H+$e^+ + \nu$ and n+n$\rightarrow$H in nuclei[33]. The high sensitive detector EL III and EL V were used. Stringent upper limits of $10^{-20} \sim 10^{-24} y^{-1}$ on double weak processes exclude the light H with m_H <1.875 GeV ($\sim 2m_N$).

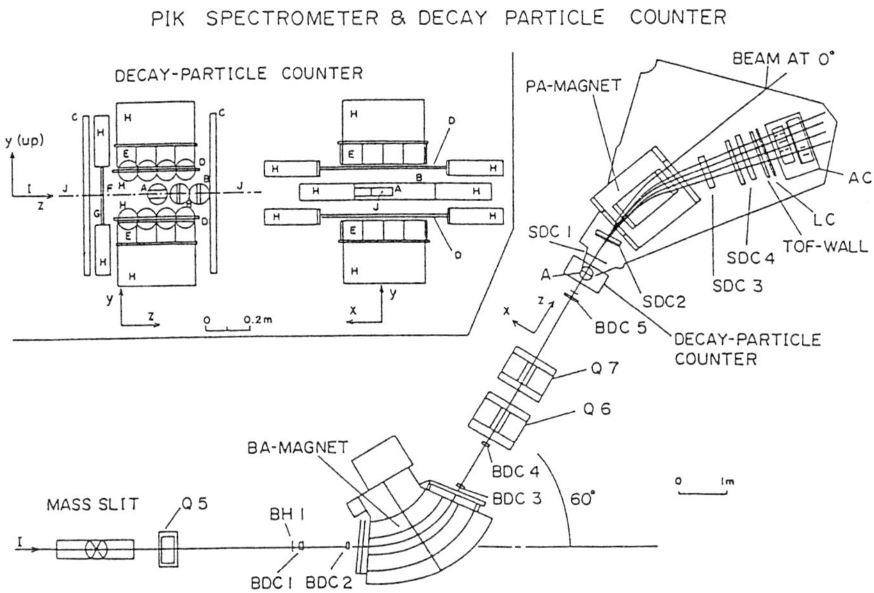

Figure 12: Experimental set-up with the K^+ spectrometer and the decay p detector of non-mesonic decays of Λ-hypernuclei produced by (π^+, K^+) reactions[27].

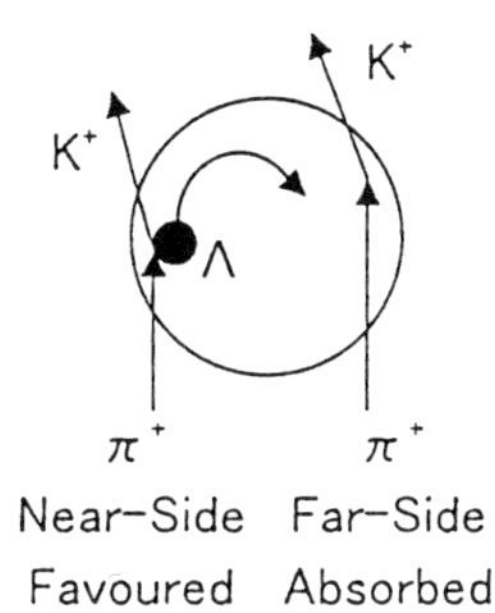

Figure 13: Near-side and far-side scatterings of the (π^+, K^+) reaction and the polarization of the transferred orbital angular momentum[28].

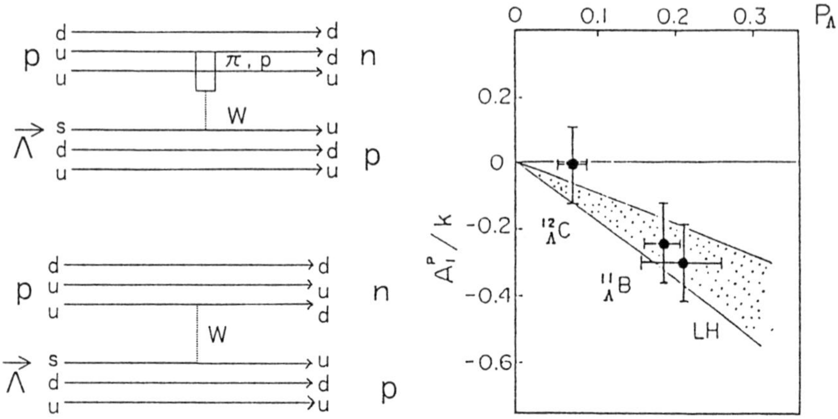

Figure 14: Non-mesonic weak decay diagrams for Λ, and asymmetries $a_1 = A_1/k$ against Λ-spin polarizations[30].

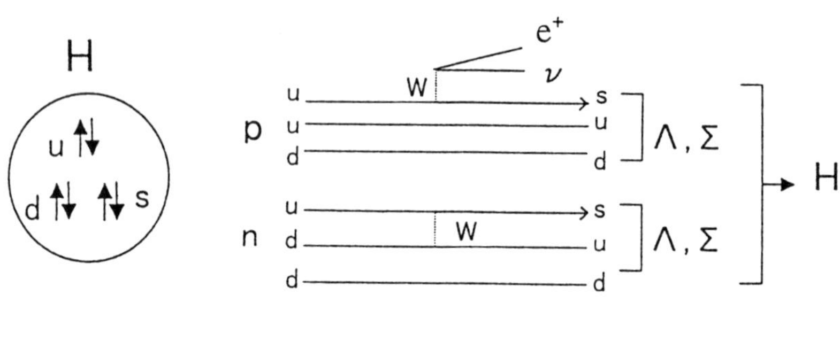

Figure 15: Quark flavour and spin configuration of H dyhyperon (left hand side) and a double weak diagram of p+p→H+e^++ν (right hand side).

6 Concluding Remarks

1. Symmetries and asymmetries are key subjects of nuclear particle physics. Symmetries and asymmetries are investigated by precise spectroscopic studies of small deviations from symmetries and also from complete asymmetries in the nuclear particle systems.

2. Nuclei are quantum systems of nucleons in good quantum states of masses, energies, spins, flavours, parites, isospins and so on. Then, nuclei are excellent microlaboratories for studying elementary particles and fundamental interactions.

3. Neutrino-less double beta decays ($0\nu\beta\beta$) are quite sensitive to the Majorana neutrino mass (m_ν), the right-handed weak currents (λ, η), the coupling with the Majoron field, the coupling with SUSY particles, the composite neutrinos and others, which are beyond the standard theory. Here $\beta\beta$-nuclei are used as nuclear microscopes to enhance the $0\nu\beta\beta$ rate and to reject background single β-rays. Two neutrino double beta decays ($2\nu\beta\beta$), which conserve the lepton number, are within the standard theory. They give the nuclear responses $M^{2\nu}$, which are used to get the nuclear spin isospin interactions relevant to the $0\nu\beta\beta$ nuclear responses $M^{0\nu}$.

4. ELEGANTs (ELEctron GAmma-ray Neutrino Telescope) have been used to study the $0\nu\beta\beta$ and $2\nu\beta\beta$ processes at the Kamioka and Oto underground laboratories. The measurements give finite $2\nu\beta\beta$ halflives and their matix elements $M^{2\nu}$ for ^{100}Mo and ^{116}Cd, and stringents limits on $\langle m_\nu \rangle$ <1.9eV, $\langle g_B \rangle$ <$7{\cdot}10^{-5}$, $\langle V+A \rangle$ <$10^{-6} \sim 10^{-8}$, and on others.

5. The $2\nu\beta\beta$ process in medium and heavy nuclei are well represented by the successive single β processes through the low-lying single particle-hole (SP) 1^+ state in the intermediate nucleus. The GT giant resonance (GR) in the intermediate nucleus does not contribute to the $2\nu\beta\beta$ because of the cancellation between the GR and SP processes.

6. Nuclear spin isospin responses for $\beta\beta - \nu$ and solar-ν have been well studied by using the corresponding charge-exchange spin-flip reactions such as (^{3}He, t), (t, ^{3}He), (d, ^{2}He), (^{7}Li, ^{7}Be), and others.

7. The measured response for $2\nu\beta\beta$ of ^{100}Mo agrees with the dominance of the single particle-hole state in the intermediate nucleus. The unbound

GT states up to 1MeV above the neutron threshold energy in ^{71}Ge deexcite partially to the ^{71}Ge by emitting γ-rays, and thus contribute to the production of ^{71}Ge isotopes in the ^{71}Ga detector.

8. Invisible neutron decays of n→ 3ν and others are studied by measuring deexcitations of the nucleon hole produced by the nucleon decay in the nucleus. The limits on γ deexcitations associated with neutron holes in ^{16}O set quite stringent limits of $10^{26} \sim 10^{28}$ on the inclusive nucleon decay, including n→ 3ν. Limits on non-Paulian γ and X rays are derived from the measured limits on these γ and X-rays from stable nuclei in the ELEGANT detectors. They give limits on stabilities of $10^{26\sim23}$y for stable nuclei and atoms.

9. Spin-polarized Λ-hypernuclei were produced by (π^+, K$^+$) reaction on ^{12}C in order to study non-mesonic weak decays of Λ+p→n+p in the hypernuclei. Large asymmetries of the proton with respect to the Λ-spin polarization were observed, indicating contributions of heavy mesons and even six quarks in the weak process with the large momentum transfer of q ~0.4GeV/c.

10. Possible existence of the light H dyhyperon with J^π=0$^+$, I=0, S=-2, which is the uds quark symmetry dybaryon, was investigated by studying double weak decays of n+n→H and n+p→H+e$^+$+ν in nuclei of the ELEGANT detectors. The measured upper limits of $10^{-24} \sim 10^{-26}y^{-1}$ for these decay rates exclude the light H dyhyperon with mass $m_H \leq 2m_n$.

11. RCNP (Research Center for Nuclear Physics), Osaka University consists of the cyclotron laboratory with E_p and K=0.42 GeV, the Oto underground laboratory with high sensitive ELEGANT detectors and the laser electron photon laboratory with multi-GeV polarized photons. They provide excellent opportunities for studying symmetries in nuclei. The underground lab is used to study small deviations from symmetries and complete asymmetries by investigating rare nuclear processes. The laser electron photon lab. is used to study symmetries and QCD. The cyclotron lab. is used to study nuclear responses for neutrinos and other nuclear properties associated with the symmetry studies in nuclear micro-laboratory.

Acknowledgments

The author would like to express his hearty thank to all of his collaborators, especially to those at Dept. Phys. and RCNP, Osaka Univ., for their nice collaborations. He also wish to thank his colleagues, friends and students for their cooperations.

References

1. H. Ejiri, Int. J. Modern Phys. E. Vol.6, No.1 (1997) 1, and refs. therein.
2. H. Ejiri, in Nucleon Hadron Many Body System, eds, H. Ejiri and H. Toki, (Oxford Science Pub.) 1999
3. H. Ejiri and M. J. de Voigt, Gamma-ray and Electron Spectroscopy in Nuclear Physics (Oxford Science Pub.) 1989
4. W. S. Haxton and G. J. Stephenson Jr., Prog. Part. Nucl. Phys. 12 (1984) 409
5. M. Doi, T. Kotani, and E. Takasugi, Prog. Theor. Phys. 83 (Suppl) (1985) 1
6. M. Moe and P. Vogel, Ann. Rev. Nucl. Part. Science 44 (1994) 247
7. H. Ejiri, in Phys. Astronophys. Neutrinos, eds. M. Fukugita and A. Suzuki (Springer-Verlag, 1994) p500
8. Proc. Neutrino 98, Takayama, June 1998, eds. Y. Suzuki, & Y. Tostuka; and refs. therein. Nucl. Phys. to be published.
9. H. Ejiri et al., Nucl. Phys. A448 (1986) 271; J. Phys. G. Nucl. Phys. 13 (1987) 839
10. H. Ejiri et al., Nucl. Instr. Methods A302 (1991) 304
    H. Ejiri et al., Phys. Lett. B258 (1991) 17; J. Phys. G. Nucl. Phys. 17 (1991) 5155, Nucl. Phys. A611 (1996) 85
    H. Ejiri et al., J. Phys. Soc. Japan 64 (1995) 339; K. Kume et al., Nucle. Phys. A 577 (1994) 405c
11. R. Hazama, et. al, Proc. Int. Conf. WEIN (Weak and Electromagnetic Interactions in Nuclei, eds. H. Ejiri, T. Kishimoto, and T. Sato (World Scientific, 1995), p. 635
12. H. Ejiri and J. I. Fujita, Phys. Reports 38c (1978) 85
13. T. Tomoda, Rep. Prog. Part. Phys. 54 (1991) 53
14. H. Ejiri, Nucl. Phys. A599 (1996) 179c
15. H. Ejiri and H. Toki, J. Phys. Soc. Japan Lett. 65 (1996) 7
16. I. Daito, et al., Phys. Lett. B428 (1998) 27
17. H. Akimune, H. Ejiri et al., Phys. Lett. 394 (1997) 23
18. H. Ejiri, V. Beryaev, and H. Toki, Prog. Theor. Physics (1999)
19. H. Ejiri et al., Phys. Lett. B433 (1998) 257
20. H. Ejiri, Phys. Rev. C48 (1993) 1442
21. Y. Suzuki, et al. Phys. Lett. B311 (1993) 357
22. H. Ejiri and T. Yamada, Private Communication

23. R. Hazama, H. Ejiri, K. Fushimi, H. Ohsumi, Phys. Rev. C49 (1994) 2407
24. H. Ejiri and H. Toki, Phys. Lett. B306 (1993) 218
25. H. Ejiri, et al., Phys. Lett. 282 (1992) 281
26. T. Akama, et al., Phys. Rev. Lett. 68 (1992) 1826
27. M. Akei, H. Ejiri, et al., Nucl. Phys. A534 (1991) 478; Nucl. Instr. Methods A283 (1989) 46
28. H. Ejiri, T. Fukuda, T. Shibata, H. Bando, K-I. Kubo, Phys. Rev. C36 (1987) 1435
29. H. Ejiri, T. Kishimoto, H. Noumi, Phys. Lett. B225 (1989) 35
30. S. Ajimura, H. Ejiri et al., Phys. Lett. B282 (1992) 293
31. R. L. Jaffe, Phys. Rev. Lett. 38 (1977) 195
32. Y. Iwasaki, T. Yoshie and Y. Tsubou, Phys. Rev. Lett. 60 (1988)1371
33. H. Ejiri. E. Takasugi et al., Phys. Lett. 228 (1989) 24

NUCLEON SPIN ASYMMETRY AND NUCLEON AND MESON EFFECTIVE MASSES

T. NORO, H. AKIMUNE, I. DAITO, H. FUJIMURA, K. HATANAKA, F. IHARA, Y. MAEDA, N. MATSUOKA, E. OBAYASHI, K. TAMURA [a], H. YOSHIDA, M. YOSHIMURA

Research Center for Nuclear Physics, Osaka University, Osaka 567-0047, Japan
E-mail: noro@rcnp.osaka-u.ac.jp

T. ISHIKAWA, M. ITO, T. KAWABATA, M. NAKAMURA, H. SAKAGUCHI, H. TAKEDA, T. TAKI, A. TAMII, M. YOSOI

Department of Physics, Kyoto University, Kyoto 606-8502, Japan

H. AKIYOSHI, [b], S. MORINOBU, K. SAGARA

Department of Physics, Kyushu University, Fukuoka 812-8581, Japan

M. KAWABATA, H. YAMAZAKI

Laboratory of Nuclear Science, Tohoku University, Sendai 982-0826, Japan

A. OKIHANA

Kyoto University of Education, Kyoto 612-0863, Japan

Spin observables have been measured for $(p, 2p)$ reactions aiming at studying medium effects on NN interactions in nuclear field. Observed strong density-dependent reduction of the analyzing power is consistent with a model calculation where reduction of nucleon and meson masses are taken into account. The spin-transfer coefficients, which data are not reproduced by the model calculation, are found to be sensitive to reduction rate of each meson mass and have a possibility to test scaling lows in mass reductions.

1 Introduction

Modification of hadron properties in nuclear environment is one of the most interesting topics in current nuclear physics. Many authors have been discussing reductions of nucleon and meson masses in the framework of QCD [1,2,3] and Quantum Hadrodynamics [4]. At several hundreds MeV of nucleon energies, the nucleon-nucleon (NN) interaction is reasonably well described by a meson exchange model and thus such modification of nucleon and meson masses are expected to cause modification of the interaction in nuclear field [5,6].

The nucleon quasifree scattering is an NN scattering in nuclear field and provides a direct way to study medium effects on the NN interaction. Exclusive

[a]Present address: Fukui Medical University, Matsuoka, Fukui 910-1104, Japan

[b]Present address: RIKEN, Wako, Saitama 351-0198, Japan

measurement of this scattering, $(p, 2p)$ reaction, is particularly suitable for this study because two-body kinematics and the bound orbit of knocked out nucleons can be separately specified as experimental conditions. In addition, an averaged density of the nuclear region where the NN scattering takes place can be estimated for each measurement condition [7]. These advantages allow us to study the density dependence of the NN interaction less ambiguously.

In studying above nuclear medium effect, spin dependent amplitudes of the NN interaction, which cause nucleon spin asymmetries in NN scattering, are particularly important because they relate spin-parities of exchanged mesons directly. It is also worth mentioned that a spin-independent absorption is the dominant disturbance in the observation of those amplitude by using nuclear reactions and spin observables are less sensitive to it.

At RCNP, a program on measurements of spin observables in the $(p, 2p)$ reactions is ongoing aiming at investigating the medium effect in the NN interactions caused by mass modification of hadrons. In this report, we show our recent data on the analyzing power (A_y) and the spin transfer coefficients $(D_{I'J})$ to forward outgoing protons and comparisons of them with a model prediction where modification of nucleon and meson masses is taken into account. A few words will also be given on an extension of our study.

2 Experimental Results and PWIA and DWIA calculations

The experiment has been performed at RCNP by using a 392 MeV polarized proton beam. The two-arm spectrometer system, consisting of the Grand Raiden and the LAS, and a focal plane polarimeter mounted on the Grand Raiden have been used. Descriptions on this detector system is given elsewhere [8,9]. The targets used are ^{6}Li, ^{12}C and ^{40}Ca. Proton knockout reactions from the $1s_{1/2}$ orbit of former two targets and the $2s_{1/2}$ of the last have been observed in zero recoil kinematic condition, which corresponds to knockout of protons at rest in target nuclei. Since the s-knockout is the dominant process for this kinematics, ambiguities caused by nuclear structure and reaction mechanism are expected to be small.

The measurements has been performed for two kinds of angular settings. In asymmetric setting, the angle of Grand Raiden is kept at 25.5°. The angle of LAS and the field of the both spectrometers are set so as to satisfy the zero recoil condition at the relevant peak in each separation energy spectrum. In the case of symmetric setting, all of these parameters are uniquely determined from the zero recoil condition. A part of measured spin observables are plotted in Fig. 1 as functions of the averaged densities, mentioned above, which are estimated by using a factorized DWIA and a local density approximation.

Within the error bars, all of those data show monotonic decreases with the density and the decrease is especially distinct in the case of A_y.

In the same figure, conventional PWIA and DWIA calculations using a t-matrix in free space are also plotted by solid and dashed lines respectively. The calculations agree with the $D_{I'J}$ data reasonably. This shows the deviation of these data from the p–p values is dominantly a kinematical effect caused by finite Q-values of the reactions. On the other hand, the calculations completely fail to reproduce the distinct density dependence of A_y.

From the difference between those two calculations, it is found that the effect of the nuclear distortion is not large for the present condition. Therefore, ambiguities in optical-potential parameters are not likely the source of this reduction. Contributions of the multi-step processes have been estimated from a separation-energy spectrum, comparison of the recoil-momentum dependence of the cross section with a DWIA prediction [8] and a model analysis of pre-equilibrium processes in the case of ^{12}Ca$(p, p'p'')$ reaction [10]. The results consistently show that the mixture of the multi-step processes is only a few percent around the peaks of the one-step cross sections. Thus, they also do not

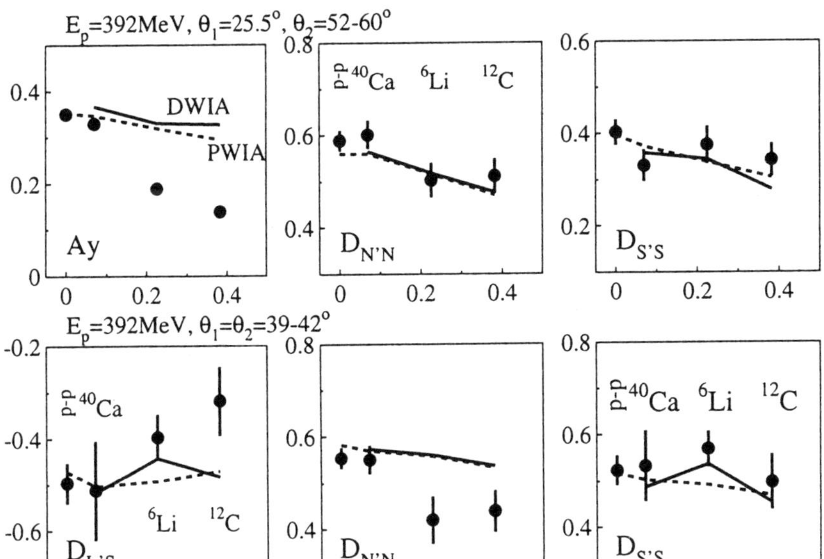

Figure 1: Comparison of A_y and $D_{I'J}$ data with theoretical predictions. The solid line and dashed line are results of DWIA and PWIA calculations with free t-matrix respectively.

cause observed distinct reduction of A_y. These facts suggest the A_y reduction is caused by some medium effect in the *NN* interactions.

3 Comparison with Calculations Including Nuclear Medium Effects

We tried to compare the data with a theoretical prediction where modifications of nucleon and meson massed are taken into account. In the relativistic framework, Horowitz have succeeded to reproduce the *NN* data by using the relativistic Love Franey model, a phenomenological meson-exchange model which dominant parameters are close to those of the Bonn potential [?]. We simply modified the meson-mass parameters and coupling constants which appear in their parameterization. The nucleon mass, in the lower component of the nucleon Dirac spinor, is also reduced [11]. The reduction rate of the effective masses of nucleon and σ-, ω- and ρ-mesons are taken to be 0.7 and that for σNN and ωNN coupling constants are set to be 0.75, at the saturation density. These rates are taken from Ref. [6] and optimization has not been made in the present calculations. For individual targets, a linear dependence of the effective mass on the nuclear density is assumed and the mass values corresponding to the averaged density are used in the calculation.

The results with PW approximation are shown in Fig. 2 by solid lines. The prediction explains the density dependent reduction of A_y very well. On the other hand, this modification of the *NN* interaction causes large deviations for some $D_{I'L}$'s, on which conventional calculations give reasonable predictions.

From these results, we conclude that the $(p, 2p)$ reactions are well sensitive to the change of the *NN* interaction caused by nucleon and meson mass modifications. In addition, it is also stated that such modification predict the density dependent reduction of A_y.

From some trial calculations, it was found that $D_{I'J}$'s are quite sensitive to modification of each meson mass while A_y is not so. Actually, only 10% reduction of σ-meson mass, at the saturation density, causes 15–20% changes of $D_{I'J}$'s for the ^{12}C case and considerable part of them is canceled out by modification of other mesons. This means present data give a good test to scaling lows as that proposed by Brown and Rho [1] and there is a possibility to obtain a parameter set which explain all the experimental data.

4 Extension of Present Study

One way to extend these study will be to obtain best fit mass parameters in the present framework. But the present model for the *NN* interaction is phenomenological parameterization and the data are not necessarily determinable enough. Instead of doing so, we are extending this study experimentally, aiming at complete determination of the NN amplitudes in nuclear field.

In the non-relativistic framework, the NN amplitudes are represented with five terms in spin space. These amplitudes are completely determined, excluding one common phase, by measurement of nine observables. In the relativistic model, the amplitude is again separated into five terms, each of which is presented by Dirac matrices corresponding to the spin-parity of exchanged (effective) mesons. Therefore, complete determination of the t-matrix enable us to study each meson exchange separately.

It is one of additional advantages of exclusive measurements that there are many kinds of experimental observables. We already observed six independent observables, the differential cross section, $A_y(=P)$, and $D_{I'J}$'s four of which are independent in the on-shell limit. Above game is realized by additional measurements of spin transfer coefficients for backward outgoing protons, $K_{I'J}$'s.

For the measurement of $K_{I'J}$'s, a new focal plane polarimeter is required

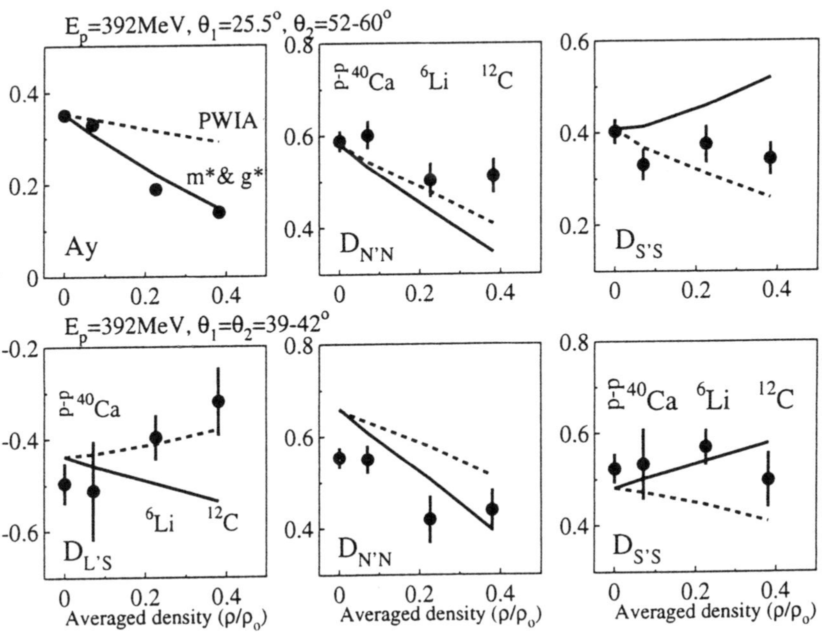

Figure 2: Comparison of A_y and $D_{I'J}$ data with a PW calculation where modifications of hadron masses and coupling constants are taken into account (solid lines). The dashed lines are a PWIA result with the original relativistic Love Franey interaction.

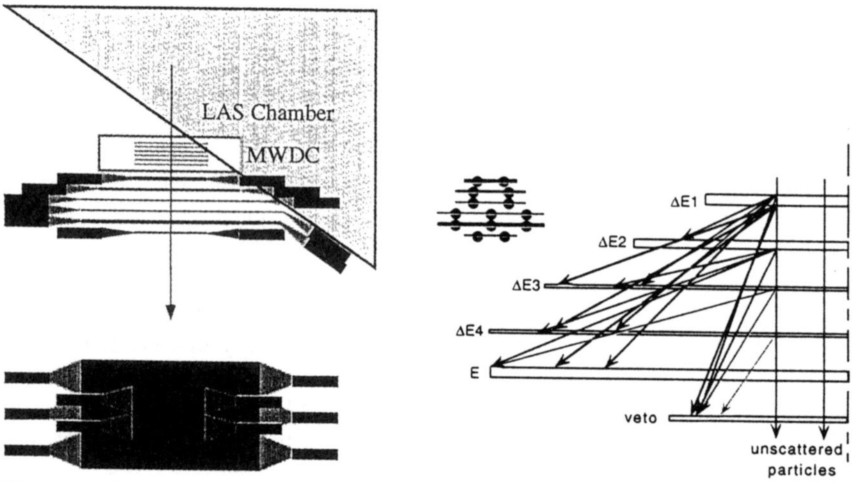

Figure 3: The focal plane polarimeter system for low energy protons under construction. Each plastic scintillator is used as both of scatter and catcher as shown in the right figure.

since the present one cannot be used for the backward outgoing protons, which energy is 85–90 MeV for our setting. Figure 3 shows the new polarimeter system which consists of a MWDC as a momentum analyzer and plastic scintillators used both as the second scatterer and the catcher with ray-tracing. The calibration works by using accelerated beams are ongoing for the new polarimeter.

References

1. G. E. Brown and M. Rho, *Phys. Rev. Lett.* **66**, 2720 (1991).
2. R. J. Furnstahl, D. K. Griegel, and T. D. Cohen, *Phys. Rev.* C **46**, 1507 (1992).
3. T. Hatsuda, *Nucl. Phys.* A **544**, 27c (1992).
4. B.D. Serot and J.D. Walecka, *Int. J. Mod. Phys.* E **6**, 515 (1997), and reterences therein.
5. G. E. Brown, A. Sethi, and N. M. Hintz, *Phys. Rev.* C **44**, 2653 (1991).
6. G. Krein *et al.*, *Phys. Rev.* C **51**, 2646 (1995).
7. K. Hatanaka *et al.*, *Phys. Rev. Lett.* **78**, 1014 (1997).
8. T. Noro *et al.*, *Nucl. Phys.* A **629**, 324c (1998).
9. M. Yosoi *et al.*, Proc. of the 11th Int. Sympo. on High Energy Spin Physics, Bloomington 1994, AIP Conf. Proc. **343**, p.157.
10. A. A. Cowley *et al.*, *Phys. Rev.* C **57**, 3185 (1998).
11. C. J. Horowitz and M. J. Iqbal, *Phys. Rev.* C **33**, 2059 (1986).

Gamow-Teller Strength in the Continuum Studied via the (p,n) Reaction

T. Wakasa, K. Hatanaka
Research Center for Nuclear Physics, Osaka University, Osaka 567-0047, Japan
E-mail: wakasa@rcnp.osaka-u.ac.jp

H. Sakai, S. Fujita, T. Nonaka, T. Ohnishi, K. Yako, K. Sekiguchi
Department of Physics, University of Tokyo, Bunkyo, Tokyo 113-0033, Japan

H. Okamura
Department of Physics, Saitama University, Saitama 338-8570, Japan

H. Otsu
Department of Physics, Tohoku University, Miyagi 980-8578, Japan

S. Ishida, N. Sakamoto, T. Uesaka, Y. Satou
The Institute of Physical and Chemical Research, Saitama 351-0198, Japan

M. B. Greenfield
International Christian University, Tokyo 181-8585, Japan

The double differential cross sections for $\theta_{\rm lab}$ between 0.0° and 14.7° and the polarization transfer coefficient $D_{NN}(0°)$ for the ^{27}Al$(\vec{p},\vec{n})$ reaction have been measured at a bombarding energy of 295 MeV. A multipole decomposition technique is applied for the cross section data to extract L = 0, 1, 2, and 3 contributions. The Gamow-Teller (GT) strength B(GT) deduced from the $L = 0$ contribution is compared with the B(GT) values calculated in a full *sd* shell-model space. The sum of B(GT) values up to 20 MeV excitation is $S_{\beta^-} = 4.0 \pm 0.1 \pm 0.1$. A fairly large L = 0 contribution is observed in the continuum region up to 50 MeV, which could be in part ascribed to the quenched GT strength. A limit on the effect that the Δ(1232)-isobar nucleon-hole admixture has upon the GT strength is estimated.

1 Introduction

The (p, n) reaction at intermediate energies (T_p > 100 MeV) has been shown to provide a highly selective probe of spin-isospin excitations in nuclei due to the energy dependence of the isovector part of nucleon-nucleon (N-N) t-matrices [1]. A striking feature of the 0° spectrum of this reaction is the dominance of the Gamow-Teller (GT) transition. A relatively simple relationship between the measured 0° ($L = 0$) cross section and the GT transition strength B(GT) has been derived [2,3]. This relationship has been successfully used to obtain the B(GT) values for the transitions which are energetically inaccessible to beta decay. Experimentally, only about 60% of the Ikeda's sum rule value [4,5] has been identified [6]. Part of this quenching might be ascribed to the $\Delta(1232)$-isobar nucleon-hole (ΔN^{-1}) admixture into the proton-particle neutron-hole ($1p1h$) GT state [7,8,9], however, part is due to the ordinary nuclear configuration mixing [10,11,12].

In our previous study [13], a multipole decomposition (MD) analysis [14] has been performed for the data of the ^{90}Zr($\vec{p},\vec{n}$) reaction, and we have found the following: (1) There is almost no contribution from $L \geq 1$ transitions in the GT resonance (GTR) region ($E_x < 16$ MeV). (2) A significant amount of the $L = 0$ component is located in the continuum up to 50 MeV excitation. (3) The quenching of the GT strength is mainly attributed to the mixing between $1p1h$ and two-particle two-hole ($2p2h$) states. In this report we present data for the ^{27}Al($\vec{p},\vec{n}$) reaction at $T_p = 295$ MeV. The data were analyzed with the same procedure used in the previous study to determine the contribution with each L transition in the continuum region including GTR. It is very interesting to extend our work to see whether or not there are features in common with the previous results.

2 Experiment

The data were taken with the neutron time-of-flight (TOF) facility [15] at the Research Center for Nuclear Physics (RCNP), Osaka University. The proton beam energy was 295 MeV and the neutron flight path length was 100 m. It should be noted that the ratio of the isovector spin-flip to non-spin-flip interactions is maximum around $T_p = 300$ MeV, while the effect of distortions becomes minimum. Thus the incident beam energy of 295 MeV is optimum for obtaining reliable nuclear structure information such as B(GT).

3 Multipole Decomposition Analysis

We have performed the MD analysis to analyze the angular distributions of the cross section and to obtain each L contribution to the cross section. The details concerning the MD analysis and the procedure for obtaining the B(GT) values are described elsewhere [13]. In the following, therefore, a brief description of the MD analysis relevant to the present analysis is presented.

We have calculated angular distributions for the following J^π transitions: 1^+ ($L = 0$); 0^-, 1^-, and 2^- ($L = 1$); 3^+ ($L = 2$); and 4^- ($L = 3$). The difference in shapes of the angular distributions for several $1p1h$ configurations for a given J^π transfer has been investigated. An active neutron hole is restricted to the $0s_{1/2}$, $0p_{3/2}$, $0p_{1/2}$, or $0d_{5/2}$ shell and an active proton particle is restricted to a shell up to the $0h_{11/2}$ shell. The $0s$ and $0p$ shells are excluded as a shell for the proton particle since they are assumed to be completely filled. It is found that the shapes of the angular distributions slightly depend on the $1p1h$ configurations. For each 2.0 MeV excitation energy interval, therefore, the optimal $1p1h$ configuration for a given J^π transfer has been selected to minimize the chi-square value of the fitting in the MD analysis as described below. The choice of a 2.0 MeV interval is comparable to the attained experimental energy resolution.

The experimentally obtained angular distribution $\sigma^{\exp}(\theta_{\text{c.m.}}, E_x)$ was fitted by means of the least square method with the sum of calculated angular distributions $\sigma^{\text{calc}}_{J^\pi}(\theta_{\text{c.m.}}, E_x)$ weighted with fitting coefficients a_{J^π} such that:

$$\sigma^{\exp}(\theta_{\text{c.m.}}, E_x) = \sum_{J^\pi} a_{J^\pi} \sigma^{\text{calc}}_{J^\pi}(\theta_{\text{c.m.}}, E_x) \ . \tag{1}$$

The fitting procedure has been performed for all possible 445500 combinations of the calculated angular distributions at each excitation energy. The combination of calculated angular distributions giving the minimum chi-square value was chosen.

Figure 1 shows the results of the MD analysis which are in excellent agreement with the measured cross sections for the whole excitation energy region for all angles. The present MD analysis clearly shows that the contribution from the $L = 0$ component is dominant in the GTR region. We have also carried out the MD analysis by using the angular distributions obtained in DWIA calculations using different optical model parameters. Furthermore, the effect of the Q value modification for 2^- transitions proposed by Mercer *et al.* [17] has been also examined. It is found that the results of the MD analysis in the GTR region are almost independent of the parameters used in DWIA calculations to generate the angular distributions. Beyond the GTR region, however, the results of the MD analysis are slightly sensitive to the modification of parameters used in DWIA calculations. This sensitivity is taken into account within the systematic uncertainty for each L contribution deduced from the MD analysis in the following discussion.

4 Gamow-Teller Strength

The $L = 0$ cross section $\sigma_{L=0}(q, \omega)$ is related to the corresponding $B(\text{GT})$ value as [3]

$$\sigma_{L=0}(q, \omega) = \hat{\sigma}_{\text{GT}} F(q, \omega) B(\text{GT}) \ , \tag{2}$$

where $\hat{\sigma}_{\text{GT}}$ is the GT unit cross section and $F(q, \omega)$ describes the dependence on momentum transfer q and energy loss ω. The momentum-transfer and energy-loss dependence has been determined by comparison with DWIA calculations for the GT transition. To obtain the $\hat{\sigma}_{\text{GT}}$ value, it is useful to use the $B(\text{GT})$

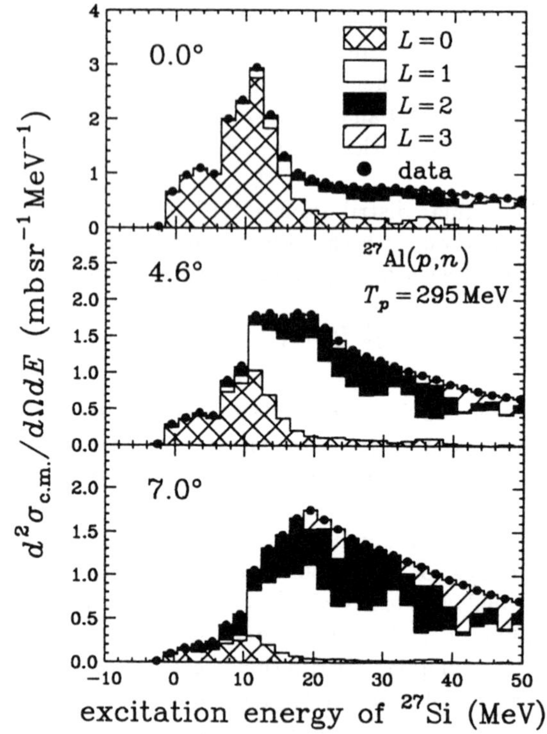

Figure 1: Results of the multipole decomposition analysis at 0.0° (top), 4.6° (middle), and 7.0° (bottom).

value for the g.s. transition which is known from the ^{27}Si(β^-) decay as 0.299 $\pm$ 0.006 [18]. We have obtained the cross section of the g.s. transition to be $\sigma_{\text{c.m.}}$ = 2.88 $\pm$ 0.03 mb/sr which includes the contribution from both GT and Fermi transitions. Thus the GT unit cross section can be deduced as

$$\begin{aligned}\hat{\sigma}_{\text{GT}} &= \frac{\sigma_{\text{c.m.}}(\text{g.s.})/F(q,\omega)}{B(\text{GT})+B(\text{F})/R^2} = \frac{(2.88\pm 0.03\ \text{mb/sr})/0.99}{(0.299\pm 0.006)+1.0/(14.7\pm 1.1)} \\ &= 7.9\pm 0.1\pm 0.2\ \text{mb/sr}\ , \qquad (3)\end{aligned}$$

where R^2 is the ratio of the GT unit cross section to the Fermi unit cross section which takes a value of 14.7 $\pm$ 1.1 at T_p = 300 MeV [19]. The first uncertainty comes from the statistical uncertainty of the cross section and the second uncertainty is due to the combined uncertainties of the B(GT) and R^2 values.

Figure 2 shows the B(GT) values obtained from the MD analysis as a function of excitation energy. The B(GT) values obtained from the shell-model

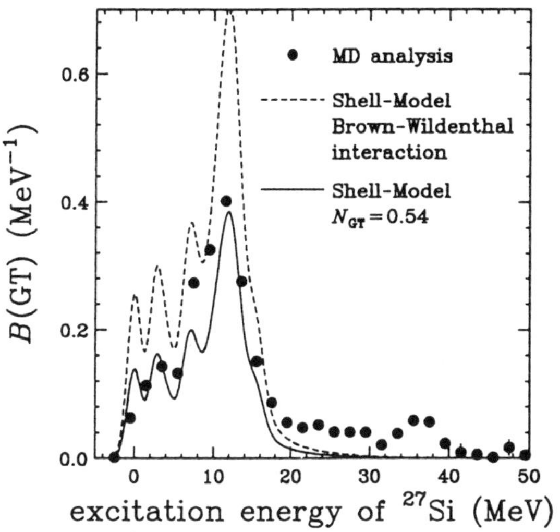

Figure 2: Gamow-Teller strength distribution (filled circles) obtained from the MD analysis. The dashed curve represents the shell-model calculation using the Brown-Wildenthal interaction. The solid curve represents the theoretical calculation normalized by a factor of 0.54 to reproduce the $B(\mathrm{GT})$ value for the g.s. transition obtained from the $^{27}\mathrm{Si}(\beta^-)$ decay.

(SM) calculation are presented by the dashed curve in Fig. 2. The theoretical $B(\mathrm{GT})$ values have been folded using a Gaussian distribution to simulate the experimental resolution. The centroid energies and widths of GT transitions are reproduced fairly well by the theoretical calculation. The SM calculation predicts the $B(\mathrm{GT})$ value for the g.s. transition to be 0.550, while the experimentally obtained value from the $^{27}\mathrm{Si}(\beta^-)$ decay is 0.299 ± 0.006 [18]. The solid curve in Fig. 2 shows the theoretical $B(\mathrm{GT})$ values normalized to reproduce the experimental $B(\mathrm{GT})$ value for the g.s. transition. The GT strength for the peaks at $E_x \simeq 3$ and 11 MeV is reproduced fairly well by the normalized theoretical calculation, while the normalized calculation underestimates the GT strength for the peak at $E_x = 8$ MeV.

The total experimentally observed GT strength summed over the region up to the GTR region ($E_x < 20$ MeV) becomes $S_{\beta^-} = 4.0 \pm 0.1 \pm 0.1$ where the first uncertainty comes from the statistical uncertainty of data and the second is due to the combined uncertainties for the $\hat{\sigma}_{\mathrm{GT}}$ value and the parameters used in DWIA calculations. The present MD analysis reveals the GT strength beyond 20 MeV, and the total GT strength in the region of 20 MeV $\leq E_x < 50$ MeV becomes $S_{\beta^-} = 0.9 \pm 0.1 \pm 0.1$ where the first and second uncertainties are the statistical and systematic uncertainties, respectively. A fairly large $B(\mathrm{GT})$ value beyond the GTR region can be ascribed to nuclear configuration mixing. However, it should be noted that there might be some contribution from the isovector spin-monopole (IVSM) strength which is indistinguishable from the GT strength in the present MD analysis. Thus the S_{β^-} value in this region is treated as an upper limit in the following discussion.

5 Ikeda's Sum Rule

Within the limitations of the MD analysis and the uncertainty in whether the $L = 0$ strength observed above 20 MeV is GT strength, it remains a useful exercise to place limits upon the extent to which the observed GT strength exhausts the Ikeda's sum rule. Recently the GT β^+ strength has been deduced from the 0°, ^{27}Al(d, ^{2}He) reaction at 270 MeV [20]. The S_{β^+} value in the region up to 10 MeV excitation is $S_{\beta^+} = 1.5 \pm 0.2 \pm 0.4$ where the first and second uncertainties are the statistical and systematic uncertainties, respectively. It is very interesting to examine whether or not the present S_{β^-} values and the S_{β^+} value deduced from the (d, ^{2}He) reaction satisfy the Ikeda's sum rule. The preceeding SM calculation predicts almost all of the GT strength ($\geq 90\%$) of the β^- and β^+ transitions in the regions of $E_x < 20$ MeV and $E_x < 10$ MeV, respectively. These energy regions correspond to the regions in which discrete GT transitions and GTR have been experimentally identified. Thus we define the two excitation energy regions: one is the low excitation energy region including GTR, and the other is the continuum region beyond GTR. If the ratios of the GT β^- strength in these two regions to the total GT β^- strength (f and $1 - f$) are the same as those for the β^+ transition, we can obtain the following equations

$$\begin{aligned} fS_{\beta^-} &= 4.0 \pm 0.1 \pm 0.1 \ , \\ fS_{\beta^+} &= 1.5 \pm 0.2 \pm 0.4 \ , \\ (1-f)S_{\beta^-} &\leq 0.9 \pm 0.1 \pm 0.1 \ , \end{aligned} \tag{4}$$

in which the inequality sign comes from the possible contribution from the IVSM strength in the continuum region beyond GTR. From Eq. (4), S_{β^-}, S_{β^+}, and $S_{\beta^-} - S_{\beta^+}$ values become

$$\begin{aligned} 4.0 \pm 0.1 \pm 0.2 \leq S_{\beta^-} &\leq 5.0 \pm 0.1 \pm 0.2 \ , \\ 1.5 \pm 0.2 \pm 0.4 \leq S_{\beta^+} &\leq 1.8 \pm 0.2 \pm 0.5 \ , \\ 2.5 \pm 0.2 \pm 0.4 \leq S_{\beta^-} - S_{\beta^+} &\leq 3.1 \pm 0.2 \pm 0.5 \ , \end{aligned} \tag{5}$$

where the first and second uncertainties are the statistical and systematic uncertainties, respectively. The lower limit of the $S_{\beta^-} - S_{\beta^+}$ value corresponds to $84 \pm 5 \pm 15\%$ of the Ikeda's sum rule value of $3(N - Z) = 3$. It should be noted that the present upper limit of this value exhausts the Ikeda's sum rule.

Thus the effect of the ΔN^{-1} admixture into the GT state on the Ikeda's sum rule has been estimated to be less than 16%.

We would like to thank T. Niizeki for valuable discussions and for providing the experimental data before publication, K. Muto for the calculation of one-body transition amplitudes, and Toru Suzuki for helpful discussions and suggestions. This work is supported in part by the Grants-in-Aid for Scientific Research Nos. 6342007 and 0442004 of the Ministry of Education, Science, Sports and Culture of Japan.

References

1. F. Osterfeld, Rev. Mod. Phys. 64 (1992) 491.
2. C.D. Goodman et al., Phys. Rev. Lett. 44 (1980) 1755.
3. T.N. Taddeucci et al., Nucl. Phys. A 469 (1987) 125.
4. K. Ikeda, S. Fujii, and J.I. Fujita, Phys. Rev. Lett. 3 (1963) 271.
5. C. Gaarde et al., Nucl. Phys. A 334 (1980) 248.
6. C. Gaarde, Nucl. Phys. A 369 (1983) 127c.
7. M. Ericson, A. Figureau, and C. Thévenet, Phys. Lett. B45 (1973) 19.
8. E. Oset and M. Rho, Phys. Rev. Lett. 42 (1979) 47.
9. I.S. Towner and F.C. Khanna, Phys. Rev. Lett. 42 (1979) 51.
10. K. Shimizu, M. Ichimura, and A. Arima, Nucl. Phys. A226 (1974) 282.
11. A. Arima and H. Hyuga, in *Mesons in Nuclei*, edited by M. Rho and D. H. Wilkinson, (North-Holland, Amsterdam, 1979), p. 683.
12. G.F. Bertsch and I. Hamamoto, Phys. Rev. C 26 (1982) 1323.
13. T. Wakasa et al., Phys. Rev. C 55 (1997) 2909.
14. M.A. Moinester, Can. J. Phys. 65 (1987) 660.
15. H. Sakai et al., Nucl. Instrum. Methods Phys. Res. A 369 (1996) 120.
16. J. Lisantti et al., Phys. Rev. C 37 (1988) 2408.
17. D.J. Mercer et al., Phys. Rev. C 49 (1994) 3104.
18. S. Raman et al., At. Data Nucl. Data Tables 21 (1978) 567.
19. W.P. Alford et al., Phys. Lett. B 179 (1986) 20.
20. T. Niizeki et al., in Proceedings of the 3rd. Japan-China Joint Nuclear Physics Symposium, Sendai and Niigata, Japan, 1997, to be published; T. Niizeki (private communication).

ROLE OF ISOBAR COMPONENTS IN THE LOW-LYING LEVELS IN LIGHT NUCLEI

C.RANGACHARYULU

Department of Physics, University of Saskatchewan,
Saskatoon, SK, Canada, S7N 5E2
E-mail: chary@sask.usask.ca

In this talk, I present the current status of the search for the evidence of isobar components in the low-lying levels of light nuclei. The on-going work on the pion electroproduction off the ^3He nucleus will be described. Also arguments will be presented to show that the mirror symmetry breaking in the conjugate nuclei may have origins in isobar components of the wave functions.

1 Introduction

From the very early times, nuclear systems have been sources of inspiration for physicists to look for symmetry arguments and organize the systematics of nuclear level properties and transition probabilities. Heisenberg [1], in view of the fact that both proton and neutron have nearly the same masses, invoked a symmetry argument with neutron and proton as two charge states of a hypothetical particle "nucleon". This concept, along with the approximation that nuclear forces are of two-body type, leads to the symmetry principles such as charge symmetry and charge independence. The observation that level properties of many nuclei follow closely the expectations from these symmetry arguments are attestations that these principles are good guides of nuclear structures.

Nevertheless, one finds departures from the mirror symmetry arguments which appear as binding energy differences in mirror nuclei to be more than what one could account for by Coulomb energy differences. One also finds differences in the electromagnetic and weak decay properties of mirror transitions in conjugate nuclei. Further improvement in the description of these properties might suggest the following avenues:
a) introducing the 3-, 4- and many body forces in addition to conventional 2-body force approximation,
b) invoking non-nucleonic degrees of freedom with excited nucleon configurations as part of nuclear wavefunctions,
c) incorporating multipion and/or heavier meson exchange forces
d) eliciting the quark degrees of freedom and
e) some combinations of the above four.

It should also be remarked that the above extensions are not totally independent and the results from one approach might mimic that of others.

Our work has been motivated to take the option b), i.e. invoking non- nucleonic degrees of freedom as part of the nuclear wavefunctions. The rationale is that as the nuclear force is due to the exchange of pions between nucleons, one might expect that this exchange would have a finite probability that the interacting nucleons will acquire the quantum numbers of excited nucleons, since the latter are simply nucleon-pion configurations. In this context, the lowest lying Delta configuration with a strong M1 coupling to the nucleons is of particular interest. However, the delta-like component in a nucleus should not be considered to be on-mass shell, rather it is a virtual nucleon-pion composite of the quantum numbers of a Delta. In this paper, I will present two lines of work a) search for isobar components in the ground state wavefunction of the ^3He nucleus and b) to account for the asymmetries in the log ft values of mirror decays by non-zero isobar components.

2 The studies on ^3He

For nearly over a decade, we have been involved in attempts to find an evidence for isobar components in the ground state wavefunction of ^3He by examining the production of pions by real and virtual photons over the delta resonance region. The impetus for this work came from the publication of Lipkin and Lee [2]. The real photon work was completed and published a while ago [3]. In this experiment, we measured the $\gamma +^3 He \rightarrow p\pi^+nn$ and $\gamma +^3 He \rightarrow p\pi^-pp$ in the same setting simultaneously for identical kinematical conditions. From this experiment, we deduced the Δ component to be $\leq 2\%$ [3,4].

As part of the A1 collaboration at Mainz, we[5] initiated $^3He(e,e'\pi^\pm)$ measurements and subsequently $^3He(e,e'p\pi^+)$ triple coincidence experiments to obtain a quantitative estimate of the Δ component [5,6]. Here we make use of the three spectrometer arrangement with the high duty, low energy spread ($\Delta E/E \approx 10^{-4}$) electron beams at the Mainz Microtron (MaMi) facility. As is shown in Figure 1, the high resolution arrangement permitted us to clearly separate the 2-body channel ($t\pi^+$) from the disintegration channels ($nd\pi^+$ and $nnp\pi^+$) in the $^3He(e,e'\pi^+)$ measurement. In the same figure, we plot the model calculations for the disintegration channels (high missing mass). Dashed and solid curves are calculations without and with a Delta component in the model descriptions. As to be expected, the role of Delta component becomes increasingly important at higher missing masses. Also, one sees here that the four-body final state ($nnp\pi^+$) is the most sensitive for Delta components.

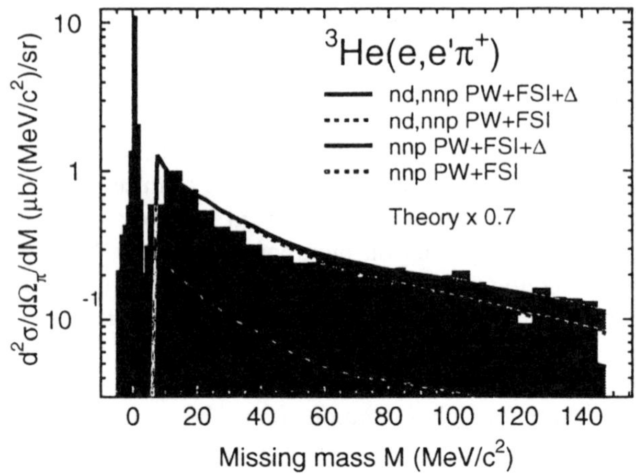

Figure 1. The coincidence spectrum of the $^3He(e,e'\pi^+)$ reaction for incident electron energy of 855 MeV. The zero on the missing mass scale corresponds to the undetected triton. Note the logarithmic scale for the cross section. The histogram is the experimental data and the curves are model calculations for the breakup channels.

As we could resolve the $t\pi^+$ channel from the other channels, it was possible to make a detailed study of the coherent pion production. We were able to determine the Transverse(T) and Longitudinal(L) components of the cross section by Rosenbluth plots of the momentum transfer matching measurements[7]. Also, we measured the angular distributions of π^+ for both coherent channel and the disintegration channels. From these angular distributions, the TL interference term can be extracted. The first results of these measurements along with the model calculations are shown in the Figure 2. As is seen there, the models predict that the interference term is large at the forward angles. More interesting thing to note is that medium modification contributions, which reproduce the transverse and longitudinal components of the cross section [7], also predict a negative interference effect. This is in contrast to the calculations for the production off the free nucleon, which predict positive contributions. In the figure, the circles are the data so far analyzed. It should be remarked that the data preceded the model calculations and recently, we have taken more data at forward angles. The data analyses are in progress.

The next step in the experiment was to perform a more exclusive $^3He(e,e'p\pi^+)nn$ measurement to mimic the Δ-knockout process. The experiment was designed to look for signatures where the incoming virtual photon interacts with the Delta in the nucleus, with the two neutrons acting as spectators. Furthermore, one expects that the Delta component, in ^3He nucleus must belong to a high momentum component, since it must be in the D- state for reasons of angular momentum conservation. However, one could not ascertain that the D-state component in ^3He is solely due to Delta component. Nevertheless, one expects to see a signature of the knockout candidates by measuring the yield of $p\pi^+$ events as a function of invariant mass of the pair, the missing mass and missing momentum and pion emission angle, etc. In view of the finite acceptance of the spectrometers, the choice of kinematical settings is not a trivial task. We have finished a part of the data-taking for this process and I am quite happy to report that we have observed triple coincidence events. Data analyses and further plans of the measurements are in progress.

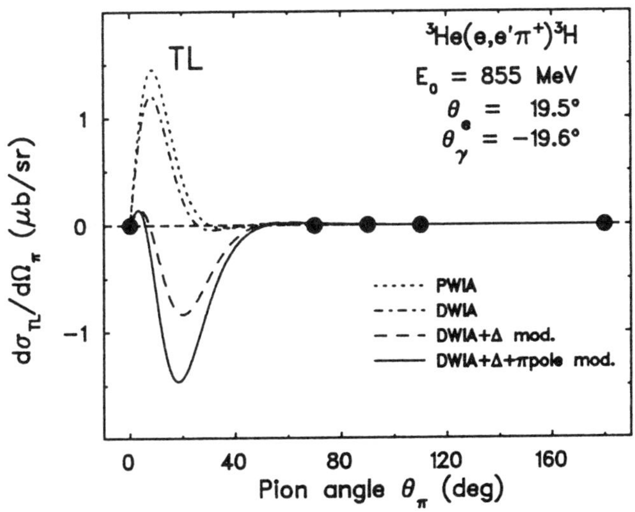

Figure 2. The angular distribution of pions for the electron kinematics indicated in the figure. The curves are the theoretical predictions of the TL interference terms for the Coherent channel. See text for details.

3 Mirror $\beta^\pm$ Decays in Conjugate Nuclei.

A consequence of charge symmetry is that the corresponding $\beta^\pm$ strengths in mirror decays must be of identical strengths. While this rule holds good for many cases, there are a few exceptions. Our attempt was to see if a meaningful isobar component (a few %) can account for the observed asymmetries. I

Table 1. Mirror Gamow-Teller beta transitions between levels of $T_i = 3/2$ and $T_f = 1/2$ and log ft values of interest to the present work. The J^π are indicated. Also included are the asymmetry parameters for the corresponding transitions and deduced isobar components (see text for details).

Transition	log ft	Asymmetry	% Delta
^{13}B(g.s.)→^{13}C(g.s.)	4.034 ±0.006		
^{13}O(g.s.)→^{13}N(g.s.)	4.081 ±0.011	0.114±0.030	3.9±1.0
$3/2^- \to 1/2^-$			
^{17}N(g.s.)→^{17}O(4.55)	4.41±0.02		
^{17}Ne(g.s.)→^{17}F(4.64)	4.59±0.034	0.197±0.04	6.5±1.1
$1/2^- \to 3/2^-$			

briefly sketch the rationale below and details can be found in Baker etal [8]. The cases of our interest are listed in Table 1.

These nuclei are of particular interest since they can be described as a few particle-hole configurations and are amenable to detailed shell model calculations. We model the transitions to be of two parts: a) nucleonic transitions and b) nucleonic transitions involving a single isobar spectator component. These are first order calculations, where Delta component is treated as a spectator with its role limited to that of providing the optimal angular momentum and isospin couplings with the nucleons making the transitions. For each mass chain, an inert core of $J^\pi = 0^+, T = 1$ with valence contributions of either 3-nucleons or 2-nucleons and a Delta forming the makeup of the nucleus. The $\beta^\pm$ transitions arise from $n \leftrightarrow p$ as in the conventional aproaches. It is conceivable and certainly desirable to include $\Delta^{++} \leftrightarrow \Delta^+$, $\Delta^+ \leftrightarrow \Delta^0$ amplitudes for a complete description. Our aim was, however, to examine if small but finite isobar components provide a reasonable physics description and thus we limited to a toy-model approach. It is heartening to note the numbers we obtain (last column in Table 1) are what one considers as reasonable admixtures of the isobar configurations in nuclear states.

In conclusion, one can say that isobar components in nuclear wavefunctions provide a useful, may not be unique, prescription for the nuclear phenomenology.

Acknowledgments

Over the years, I have benefitted very much from my collaborations in Germany and elsewhere. While I thank all my collaborators, I am especially grateful to Achim Richter for a long-standing support, extending for about 20 years. This paper is dedicated to Hiroyasu Ejiri on the occasion of his retirement from the Director and Professor positions of Research Center for Nuclear Physics, Osaka University. I have had the privilege of knowing him for nearly 27 years and I have always been impressed with his passion and dedication for Physics, especially for Electro-Weak Nuclear Physics.

References

1. W. Heisenberg, Zeit. Für Physik **77**, 1 (1932).
2. H.J. Lipkin and T.S-H. Lee, Physics Letters **B183**, 22 (1987).
3. T. Emura et al, TAGX collaboration, Physics Letters **B306**, 6 (1993).
4. B. Lasiuk, M.Sc. Thesis, University of Saskatchewan, 1993 (unpublished).
5. A. Richter et al, Proposals to the MaMi Program Advisory Committee, A1/4-90, A1/2-94, and A1/2-96.
6. I. Blomqvist et al, Phys. Rev. Lett. **77**, 2396 (1996); M. Kuss, Ph. D. Thesis (TU-Darmstadt) 1996 (unpublished).
7. I. Blomqvist et al, Nuclear Physics **A626**, 871 (1997).
8. T. Baker, B. Lasiuk and C.Rangacharyulu, Int. J. of Modern Physics **8**, 11 (1999).

WHAT DO WE LEARN ABOUT HADRONIC INTERACTIONS AT ULTRAHIGH ENERGIES FROM EXTENSIVE AIR SHOWER OBSERVATIONS?

H.REBEL

Forschungszentrum Karlsruhe -Institut für Kernphysik
KASCADE collaboration
D 76021 KARLSRUHE-Germany
E-mail: rebel@ik3.fzk.de

The interpretation of extensive air shower (EAS) observations needs a sufficiently accurate knowledge of the interactions driving the cascade development in the atmosphere. While the electromagnetic and weak interaction parts do not provide principal problems, the hadronic interaction is a subject of uncertainties and debates, especially in the ultrahigh energy region extending the energy limits of man made accelerators and experimental knowledge from collider experiments. Since the EAS development is dominantly governed by soft processes, which are presently not accessible to a perturbative QCD treatment, one has to rely on QCD inspired phenomenological interaction models, in particular on string-models based on the Gribov-Regge theory like QGSJET, VENUS and SYBILL. Recent results of the EAS experiments KASCADE are scrutinized in terms of such models used as generators in the Monte Carlo EAS simulation code CORSIKA.

1 Introduction

In cosmic ray investigations, in addition to the astrophysical items of origin, acceleration and propagation of primary cosmic rays, there is the historically well developed aspect of the interaction of high-energy particles with matter. Cosmic rays interacting with the atmosphere as target (on sea level it is equivalent to a lead bloc of 1m thickness) produce the full zoo of elementary particles and induce by cascading interactions intensive air showers (EAS) which we observe with large extended detector arrays distributed in the landscapes, recording the features of different particle EAS components. The EAS development carries information about the hadronic interaction (but it has to be disentangled from the unknown nature and quality of the beam). When realizing the present limits of man made accelerators, it is immediately obvious why there appears a renaissance of interest in cosmic ray studies from the point of view of particle physics. EAS observations of energies 10^{15} eV represent an almost unique chance to test theoretical achievements of very high energy nuclear physics.

Actually the astrophysicist is faced with the situation that reliable interpretations of the features of the secondary particle production, and of their relation to the characteristics of the primary particle are necessarily related to our understanding of the hadronic interactions. This aspect is particularly stimulating for high-energy physicists, since there is not yet an exact way to calculate the properties of the bulk of hadronic interactions.

This lecture is directed to review some relevant aspects of hadronic interactions affecting the EAS development, illustrated with recent results of EAS investigations of the KASCADE experiment [1], especially of studies of the hadronic EAS component using the iron sampling calorimeter of the KASCADE central detector.

2 EAS development and hadronic interactions

The basic ingredients for the understanding of EAS are the total cross sections of hadron air collisions and the differential cross sections for multiparticle production. Actually our interest in the total cross section is better specified by the inelastic part, since the elastic part does not drive the EAS development. Usually with ignoring coherence effects, the nucleon-nucleon cross section is considered to be more fundamental than the nucleus-nucleus cross section, which is believed to be obtained in terms of the first. Due to the short range of hadron interactions the proton will interact with only some, the so-called wounded nucleons of the target. The number could be estimated on basis of geometrical considerations, in which size and shape of the colliding nuclei enter. All this is mathematically formulated in the Glauber multiple scattering formalism, ending up with nucleon-nucleus cross sections.
Looking for the cross features of the particle production, the experiments show that the bulk of it consists of hadrons emitted with limited transverse momenta ($< P_t >\sim 0.3\,\mathrm{GeV/c}$) with respect to the direction of the incident nucleon. In these "soft" processes the momentum transfer is small. More rarely, but existing, are hard scattering processes with large P_t-production.

It is useful to remind that cosmic ray observations of particle phenomena are strongly weighted to sample the production in forward direction. The kinematic range of the rapidity distribution for the Fermilab proton collider for 1.8 TeV in the c.m. system is equivalent to a laboratory case of 1.7 PeV. Here the energy flow is peaking near the kinematical limit. That means, most of the energy is carried away longitudinally. This dominance of longitudinal energy transport has initiated the concept, suggested by Feynman: The inclusive cross sections are expressed by factorizing the longitudinal part with an universal transverse momentum distribution G(P_t) and a function scaling with the dimensionless Feyman variable x_F, defined as the ratio of the longitudinal momentum to the maximum momentum. Though this concept, expressing the invariant cross sections by

$$E \cdot d^3\sigma/dp^3 \sim x_F \cdot d^3\sigma/dx_F dp_T \qquad (1)$$

provides an orientation in extrapolating cross sections, it is not correct in reality, and the question of scaling violation is a particular aspect in context of modeling ultrahigh-energy interactions.

3 Hadronic interaction models as generators of Monte-Carlo simulations

Microscopic hadronic interaction models, i.e. models based on parton-parton interactions are approaches, inspired by the QCD and considering the lowest order Feyman graphs involving the elementary constituents of hadrons (quarks and gluons). However, there are not yet exact ways to calculate the bulk of soft processes since for small momentum transfer the coupling constant α_s of the strong interaction is so large that perturbative QCD fails. Thus we have to rely on phenomenological models which incorporate concepts from scattering theory.

A class of successful models are based on the Gribov-Regge theory. In the language of this theory the interaction is mediated by exchange particles, so-called Reggeons. At high energies, when non-resonant exchange is dominating, a special Reggeon without colour, charge and angular momentum, the Pomeron, gets importance. In a parton model the Pomeron can be identified as a complex gluon network or generalised ladders i.e. a colourless, flavourless multiple (two and more) gluon exchange. For inelastic interactions such a Pomeron cylinder of gluon and quark loops is cut, thus enabling colour exchange ("cut cylinder") and a re-arrangement of the quarks by a string formation. Fig.1 recalls the principles by displaying some parton interaction diagram's.

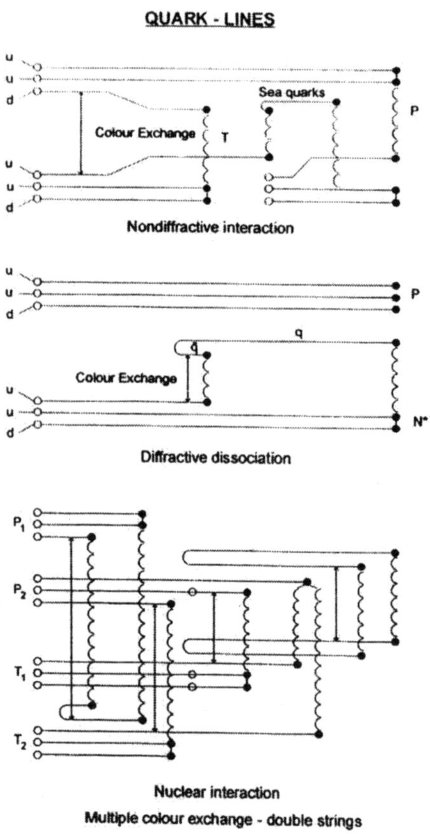

Figure 1. Parton interaction lines

- The interacting valence quarks of projectile and target rearrange by gluon exchange the color structure of the system (the arrow indicates the colour exchange by opening the cylinder). As a consequence, constituents of the projectile and target (a fast quark and slow di-quark e.g.) for a colour singlet string with partons of large relative momenta. Due to the confinement the stretched chains start to fragment (i.e. a spontaneous $q\bar{q}$-production) in order to consume the energy within the string. We recognize a target string (T) and a projectile string (P), which are the only chains in pp collisions. In multiple collision processes in a nucleus, sea quarks are additionally excited and may mediate nucleon-A interactions. While in the intermediate step the projectile diquark remains inert, chains with the sea quark of the projectile are formed.

- Most important are diffractive processes, signaled in the longitudinal momentum (x_F) distribution by the diffractive peak in forward directions. Here the interacting nucleon looks like a spectator, in some kind of polarisation being slowed down a little bit due to a soft excitation of another nucleon by a colour exchange with sea quarks (quark-antiquark pairs spontaneously created in the sea).

- There is a number of such quark lines, representing nondiffractive, diffractive and double diffractive processes, with single and multiple colour exchange.

The various string models differ by the types quark lines included. For a given diagram the strings are determined by Monte Carlo procedures. The momenta of the participating partons are generated along the structure functions. The models are also different in the technical procedures, how they incorporate hard processes, which can be calculated by perturbative QCD. With increasing energy hard and semihard parton collisions get important, in particular minijets induced by gluon-gluon scattering.

In summary, the string models VENUS [2], QGSJET [3] and DPMJET [4] which are specifically used as generators in Monte-Carlo EAS simulations are based on the Gribov-Regge theory.They describe soft particle interactions by exchange of one or multiple Pomerons. Inelastic reactions are simulated by cutting Pomerons, finally producing two color strings per Pomerons which subsequently fragment into color-neutral hadrons.All three models calculate detailed nucleus-nucleus collisions by tracking the participants nucleons both in target and projectile.The differences between the models are due to some technical details in the treatment and fragmentation of strings. An important difference is that QGSJET and DPMJET are both able to treat hard processes, whereas VENUS, in the present form, does not. VENUS on the other hand allows for secondary interactions of strings which are close to each other in space and time. That is not the case in QGSJET and DPMJET. SYBYLL [5] and HDPM [6] extrapolate experimental data to high energies guided by simple theoretical ideas. SIBYLL takes the production of minijets into account. These models are implemented in the Karlsruhe Monte Carlo simulation programm CORSIKA [6,7] to which we refer in the analyses of data.

4 The KASCADE apparatus

From the very beginning, when planning the KASCADE experiment [1] the setup of an calorimeter for efficient studies of the hadronic component in the shower center has been foreseen with the intention of checking the predictions of hadronic interaction models.

The KASCADE detector array consists of an field array of 252 detector stations, arranged in a regular way in an area of $200 \cdot 200\,\mathrm{m}^2$, and of a complex central detector with a sampling calorimeter for hadron detection. The field detectors identify the EAS event, they provide the principal trigger (a coincidence in at least eight stations), the basic characterisation (angle of incidence, shower axis and core location) and do sample the lateral distribution of the electron-photon and muon component from which the shower size and quantities characterising the intensity

and muon content of the showers are determined. In the array stations the muon detectors are positioned directly below the scintillators of the electron-photon detectors, shielded by lead and iron corresponding to 20 radiation lengths, imposing a energy detection threshold of about 300 MeV.
The central detector combines various types of detector installation with with an iron sampling calorimeter of eight layers of active detectors.
The iron absorbers are 12-36 cm thick, increasingly in the deeper parts of calorimeter. Therefore the energy resolution does not scale as $1/\sqrt{E}$, but is rather constant, slowly varying from $\sigma\sqrt{E} = 20\%$ at 100 GeV to 10% at 10 TeV. In total (including the concrete ceiling) the calorimeter thickness corresponds to 11 interaction lengths ($\lambda_I = 16.7\,\text{cm Fe}$) for vertical muons. On top, a 5 cm lead layer absorbs the electromagnetic component to a sufficiently low level.

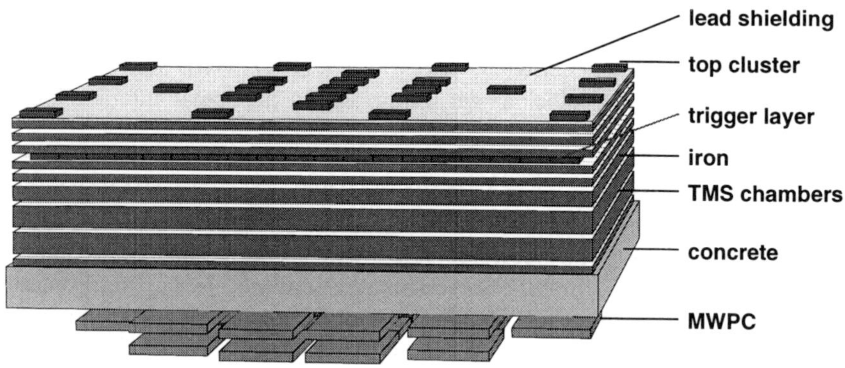

Figure 2. Scheme of the KASCADE central detector

The active detectors are 10.000 ionisation chambers using room temperature liquids tetramethylsilan (TMS) and tetramethylpentane (TMP) operated with a large dynamical range (5.10^4). This ensures that the calorimeter measures linearly the energy of single hadrons up to 15 TeV. The third layer of the calorimeter setup is an "eye" of 456 plastic scintillator, which deliver a fast trigger signal. Independently from hadron calorimetry, it is used as additional muon detector and as timing facility for muon arrival time measurements. In the basement of the iron calorimeter there are position sensitive multiwire proportional chamber (MWPC) installed for specific studies of the structure of the shower core and of the EAS muon component with an energy threshold of 2 GeV.
The energy calibration of the energy deposit of single ionisations chambers is made by means of the through-going muons, and the transition curves, i.e. the longitudinal profiles of the energy deposition are compared with simulations (using the detector simulation code GEANT [8] with the FLUKA description).

5 Test of EAS observables [9]

The general scheme of the analysis of EAS observations involves Monte Carlo simulations constructing pseudo experimental data which can be compared with the real data [10]. The king-way of the comparison is the application of advanced statistical techniques of multivariate analyses of nonparametric distributions [11].
The mass composition of cosmic rays in the energy region above 0.5 PeV is poorly known. Hence the comparison of simulation results based on different interaction models has to consider two extreme cases of the primary mass: protons and iron nuclei, and the criteria of our judgment of a model is directed to the question, if the data are compatible in the limits of the predicted extremes of protons and iron nuclei. The hadronic observables, which we consider in dependence from shower parameters and characterize the registered EAS, in particular indicating the primary energy, are:

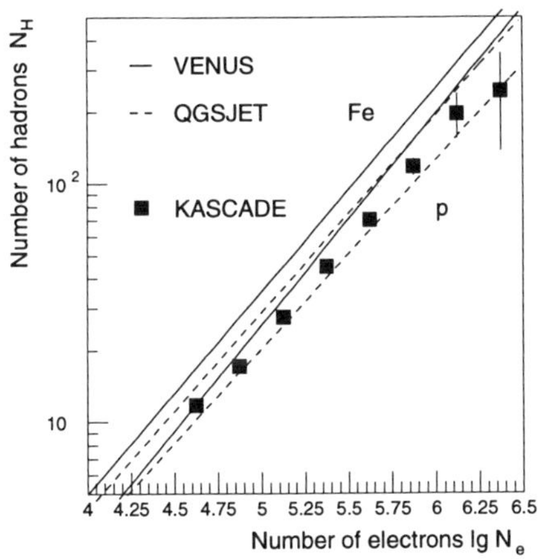

Figure 3. Hadron number N_H - shower size N_e correlation

- The shower size N_e, i.e. the total electron number
- The muon content N_μ^{tr} which the number of muons obtained from an integration of the lateral distribution in the radial range from 40 to 200 m. It has been shown that this quantity is approximately an mass independent energy estimator for the KASCADE layout, conveniently used for a first energy classification of the showers [12].

First, the dependence of the average number of hadrons N_H with an energy $E_H > 100\,\mathrm{GeV}$ from the shower size is shown and compared with the predictions of the VENUS and QGSJET model. The energy range covers the range from 0.2 PeV to 20 PeV. The result shows some preference for the QGSJET model, and such an indication is corroborated by other tests.
There is another feature obvious. When shower observables are classified along the electromagnetic shower sizes N_e, a proton rich composition is displayed. This effect is understood by the fact that at the same energy protons produce larger electromagnetic sizes than iron induced showers, i.e. with the same shower size iron primaries have higher energies, where the steeply falling primary induces the dominance of protons in the sample.
Another example considers the frequency distributions of the energy of each single hadron E_H with respect to the energy of the most energetic hadron E_H^{max}. The data are compared with predictions of SYBILL and QGSJET for iron and proton induced showers.

- For a primary proton one expects that the leading particle is accompanied by a swarm of hadron of lower energies. For a primary iron nuclei the energy distribution appears narrowed.
- The two upper curves display the case for a primary energy below the knee (about 3 PeV). The deficiencies of SYBYLL are obvious and have been also evidenced by other tests, especially with the muon content [9]. SYBILL seems to produce a wrong EAS muon intensity, and it fair to mention that just this observation has prompted the authors to start a revision of the SYBILL model.
- At energies well above the knee (about 12 PeV) also the QGSJET exhibits discrepancies, at least in the energy distribution of the hadrons of the shower core. Other observables like lateral distribution and the total number of hadrons, however are appear more compatible with the model.

How to interpret this results? Tentatively we may understand that in the simulations E_H^{max}, the energy of the leading hadrons is too large. Lowering E_H^{max} would lead to a redistribution of the E/E_H^{max} distribution shifting the simulation curves in direction of the data.

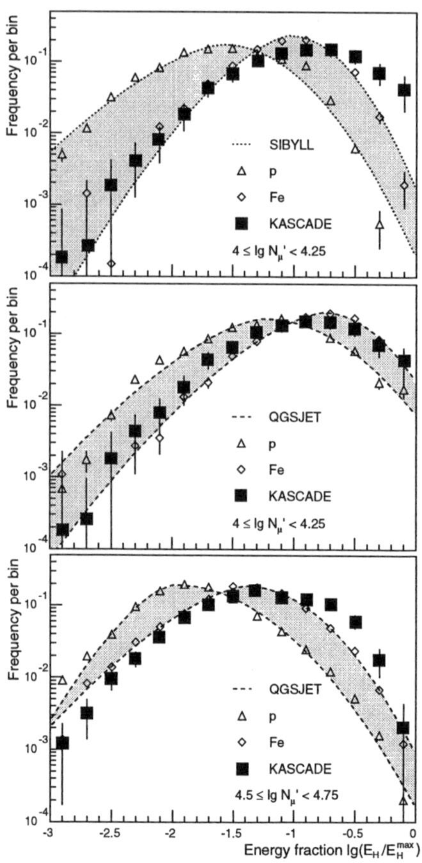

Figure 4. Distribution of the energy fraction of the EAS hadrons

A further test quantity is related to the spatial granularity of hadronic core of the EAS. The graph (Fig.5 left) shows the spatial distribution of hadrons for a shower induced by a 15 PeV proton. The size of the points represents the energy (on a logarithmic scale). For a characterisation of the pattern a minimum spanning tree is constructed. All hadron points are connected by lines and the distances are weighted by the inverse sum of energies. The minimum spanning tree minimizes the total sum of all weighted distances. The test quantity is the frequency distribution of the weighted distances d_{MST}. Results are shown for two different bins of the truncated muon size or of the primary energy (2 and 12 PeV), respectively (Fig.5 right). Again we are lead to the impression that either the distribution pattern is not reproduced or the high-energy hadrons are missing in the model.

Tentatively we may deduce from these indications, that the transfer of energy to the secondaries - what we phenomenologically characterize with the

not very well defined concept of the inelasticity of the collision - appears to be underestimated.
In order to underline this feature we may inspect the variation of some other observables with the quantity $log_{10}(N_\mu^{tr}) \propto log_{10}(E_{prim})$: The so-called shower age s, which characterizes the stage of the EAS development, the number of observed hadrons N_h with $E_H > 100\,\text{GeV}$, the energy sum $\sum E_h$ of this hadrons and the energy of the highest energy hadrons E_H^{max}. Fig.6 compares with predictions with the QGS model (with the limit log $N_\mu^{tr} < 4.6$). The predictions of the VENUS display the same features. Globally we realize the tendency that the experimental data approach the predictions for iron induced showers, i.e. for faster developed EAS. But this may be hardly interpreted as consequence of a heavier mass composition, rather as arising from a larger inelasticity of the hadronic collisions e.g..

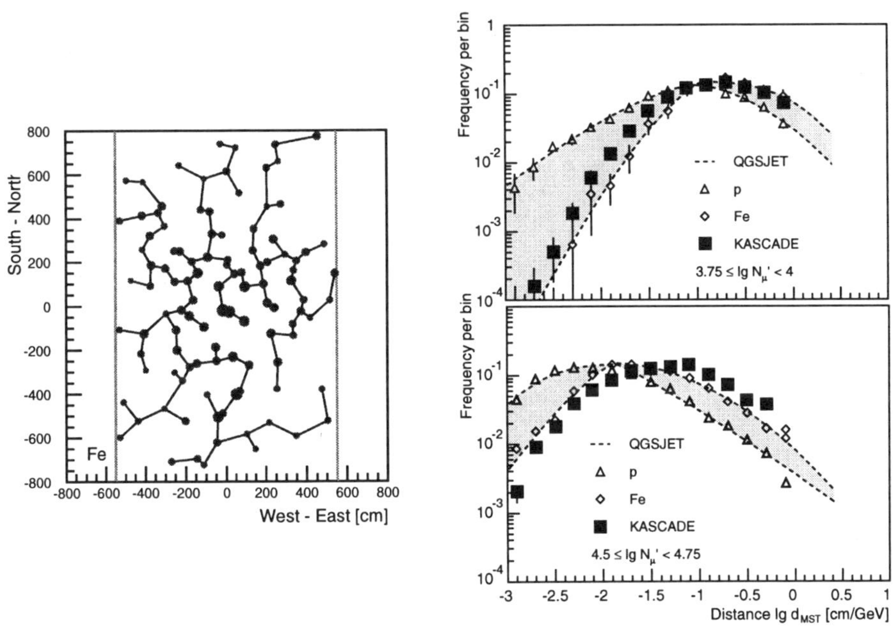

Figure 5. Left: Example of a hadronic core observed in the calorimeter (top view). Right: Frequency distributions of the distances of the minimum-spanning-tree

6 Concluding remarks

From the investigation of a series EAS observables and comparisons with different hadronic interaction models, en vogue for ultrahigh energy collisions, we conclude with following messages:

- The model SYBILL, in the present release, has problems, in particular when correlations with the muon content of the showers are involved.

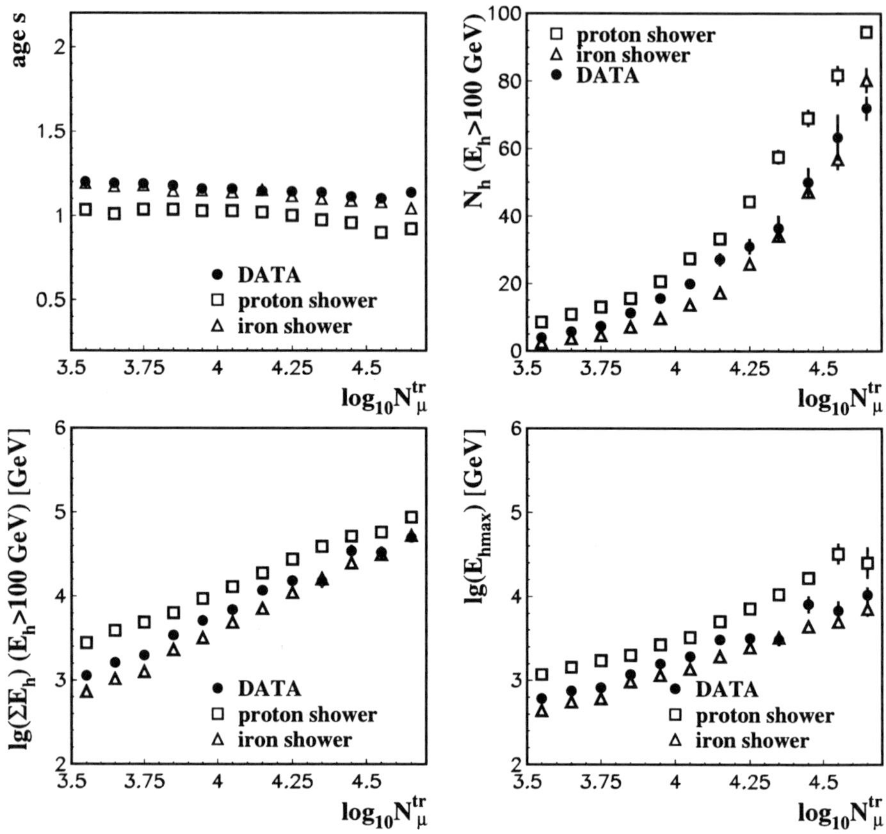

Figure 6. Comparison of various experimentally accessible EAS variables with predictions of the the QGSJET model [13]

- The model VENUS is in fair agreement with the data, but it indicates also some problems at high energies, when correlations with the shower sizes are considered.
- In the moment the model QGSJET, which includes the minijet production - in contrast to VENUS - reproduces sufficiently well the data, though it underestimates the number of high energy hadrons for high energies.
- In general there are tentative indications that the inelasticity in the fragmentation region is underestimated especially with increasing energy.

All models are in a process of refinements. Actually somehow triggered by the experimental indications, there is a common enterprise of VENUS and QGSJET towards a combined model descriptions: NEXUS [14]. That is a unified approach combining coherently the Gribov-Regge theory and perturbative QCD. Faced with the experimental endeavour to set up giant arrays for astrophysical observations at extremely high energies, the Monte Carlo simulations need certainly a safer ground of model generators. Hence our efforts in KASCADE are directed to extend the array and to refine the present studies with results towards primary energies of 10^{17} eV.

Acknowledgments

The experimental results are based on a KASCADE publication [9]. In particular, I would like to thank Dr. Andreas Haungs for contributions and clarifying discussions.

References

1. H.O.Klages et al. (KASCADE collaboration), Nucl.Phys.B(Proc.Suppl.) 52B (1997) 92
2. K.Werner, Phys.Rep.232 (1993) 87
3. N.N. Kalmykov and S.S. Ostapchenko, Phys.At.Nucl.56(1993) 346
4. J.Ranft, Phys.Rev. D 51 (1995) 64
5. R.S.Fletcher el al. Phys.Rev.D 50 (1994) 5710
6. J.N.Capdevielle et al., KfK Report 4998 (1992), Kernforschungszentrum Karlsruhe
7. D.Heck et al., FZKA Report 6019 (1998), Forschungszentrum Karlsruhe
8. GEANT 3.15, Detector Description and Simulation Tool, CERN Program Library Long Writeup W5015, CERN (1993)
9. T.Antoni et al. (KASCADE collaboration), Journal of Physics G: Particles and Nuclei (submitted)
10. H.Rebel, in FZKA Report 6215 (1998), Forschungszentrum Karlsruhe
11. A.Chilingarian, "ANI - Nonparametric Statistical Analysis of High Energy Physics and Astrophysics Experiments", Users Guide, 1998; M.Roth, FZKA Report 6262 (1999), Forschungszentrum Karlsruhe
12. R.Glasstetter for the KASCADE collaboration, 25th ICRC(Durban, South Africa) Vol 6,157
13. A.Haungs, private communication
14. H.J. Drescher et al., preprint hep-ph/9903296 (March 1999)

Neutrinos in Explosive Nucleosynthesis: Big-Bang and Supernovae

Toshitaka Kajino[1,2]

[1] National Astronomical Observatory, Mitaka, Tokyo, 181-8588

[2] Also, Department of Astronomy, University of Tokyo, Tokyo, 113-0033; and Department of Astronomical Science, Graduate University for Advanced Studies, Tokyo, 181-8588

Abstract. We study the neutrino-nucleus interactions in explosive nucleosynthesis of light-to-heavy mass elements in various astrophysical conditions. In neutrino-driven winds of gravitational core collapse of SNeII, several light nuclei as well as heavy unstable nuclei paly significant roles in producing r-process elements. We discuss also that many reactions in the r-process are relevant to a Big-Bang nucleosynthesis.

1 Introduction

Big-Bang nucleosynthesis in the early Universe is thought to produce the light elements D and ^{3,4}He with small amounts of ^{7}Li, ^{9}Be and ^{11}B. Since these abundances are used as unique observables to determine the cosmological parameter Ω_b, careful studies of the few body reactions of light nuclei are highly desired. Stars with various masses provide a variety of production sites for ^{4}He, ^{12}C and heavier elements. Very massive stars $\geq 10M_\odot$ culminate their evolution by supernova (SN) explosions which are also presumed to be most viable candidate for the still unknown astrophysical site of r-process nucleosynthesis. Even in the nucleosynthesis of heavy elements, initial entropy and density at the surface of proto-neutron stars are so high that nuclear statistical equilibrium (NSE) favors production of abundant light nuclei. In such explosive circumstances few-body reactions play a significant role.

The study of the origin of r-process elements is also critical in cosmology. It is a potentially serious problem that the cosmic age of the expanding Universe derived from cosmological parameters may be shorter than the age of the oldest globular clusters. Since both age estimates are subject to the uncertain cosmological distance scale, an independent method has long been needed. Thorium, which is a typical r-process element and has half-life of 14 Gyr, has recently been detected along with other elements in very metal-deficient stars. If we model the r-process nucleosynthesis in these first-generation stars, thorium can be used as a cosmochronometer completely independent of the uncertain cosmological distance scale.

In this article, we first demonstrate theoretically that supernova explosions of very massive stars could be a viable site for r-process nucleosynthesis. We study the role of few-body reactions and calculate the sensitivity of the r-process yields to the cross section for ^{4}He$(\alpha n,\gamma)^9$Be, this rate is poorly known experimentally but is expected to affect the result strongly. Next we note several key reactions in the r-process which are relevant to the Big-Bang nucleosynthesis.

2 Explosive Nucleosynthesis in Supernovae

Recent mesurements using high-dispersion spectrographs with large Telescopes or the Hubble Space Telescope has made it possible to detect minute amounts of heavy elements in faint metal-deficient ([Fe/H] $\leq$ -2) stars. The discovery of r-process elements in these stars has shown that the relative abundance pattern for the mass region 120 $\leq$ A is surprisingly similar to the solar system r-process abundances independent of the metallicity of the star as shown in Fig. 1. Here metallicity is defined by [Fe/H] = log[N(Fe)/N(H)] - log[N(Fe)/N(H)]$_\odot$. It obeys the approximate relation t/10^{10}yr $\sim 10^{[Fe/H]}$. The observed similarity strongly suggests that the r-process occurs in a single environment which is independent fo progenitor metallicity. Massive stars with $10M_\odot \leq M$ have a short life $\sim 10^7$ yr and eventually end up as violent supernova explosions, ejecting material into the intersteller medium early on quickly from the history of the Galaxy. However, the iron shell in SNe is excluded from being the r-process site because of the observed metallicity independence.

Hot neutron stars just born in the gravitational core collapse of SNeII release most of their energy as neutrinos during the Kelvin-Helmholtz cooling phase. An intense flux of neutrinos heat the material near the neutron star surface and drive matter outflow (neutrino-driven winds). The entropy in these winds is so high that the NSE favors a plasma which consists of mainly free nucleons and alpha particles rather than composite nuclei like iron. The equilibrium lepton fraction Y_e is determined by a delicate balance between $\nu_e + n \rightarrow p + e^-$ and $\bar{\nu}_e + p \rightarrow n + e^+$, which overcomes the difference of chemical potential between n and p, to reach $Y_e \sim 0.45$. R-process nucleosynthesis occurs because there are plenty of free neutrons at high temperature. This is possible only if seed elements are produced in the correct neutron to seed ratio before and during the r-process.

Although Woosley et al. [1] demonstrated a profound possibility that the r-process could occur in these winds, several difficulties were subsequently identified. First, independent non relativistic numerical supernova models [2] have difficulty producing the required entropy in the bubble S/k $\sim$ 400. Relativistic effects may not be enough to increase the entropy dramatically [3, 4, 5]. Second, even should the entropy be high enough, the effects of neutrino absorption

$\nu_e + n \rightarrow p + e^-$ and $\nu_e + A(Z, N) \rightarrow A(Z+1, N-1) + e^-$ may decrease the neutron fraction during the nucleosynthesis process. As a result, a deficiency of free neutrons could prohibit the r-process [6]. Third, if neutrinos are massive and have approximate mixing parameters, energetic ν_μ and ν_τ are converted into ν_e due to flavor mixing. This activates the $\nu_e + n \rightarrow p + e^-$ process and results in a deficiency of free neutrons.

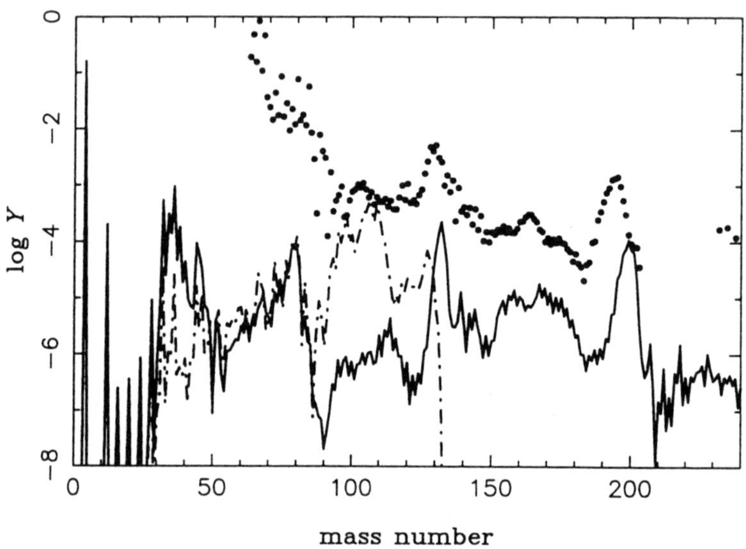

Figure 1. Calculated r-process abundance yields [5] in the neutrino-driven winds as a function of atomic mass number. Filled circles are the solar system r-process abundances [7], shown in arbitrary units. Solid line displays the calculated result obtained by using the cross section estimated by Woosley and Hoffman [9], assuming the ^{8}Be + n structure of ^{9}Be, and long-dashed line is the result for that cross section multiplied by 10. See text for details.

In order to resolve these difficulties, we have studied neutrino-driven winds in a Schwarzschild geometry under the reasonable assumption of spherical steady-state flow [5]. Then, we carried out r-process nucleosynthesis calculations in our wind model [8]. The nuclear reaction network used for light-mass region is shown in Fig. 2. The full network consists of $\sim$ 3000 elements including radioactive unstable nuclei up to the plutonium isotopes.

We found [5] that the general relativistic effects make expanding dynamic time scale τ_{dyn} much shorter, although the entropy per baryon increases by about 40 % from the Newtonian value of S/k $\sim$ 90. By simulating many supernova explosions, we have found some interesting conditions which lead to successful r-process nucleosynthesis [5, 8], as shown in Fig. 1. The best case is for a relatively large neutron-star mass $M > 1.7 M_{\odot}$ and a neutrino luminosity $L_{\nu} > 5 \times 10^{51} erg\ s^{-1}$.

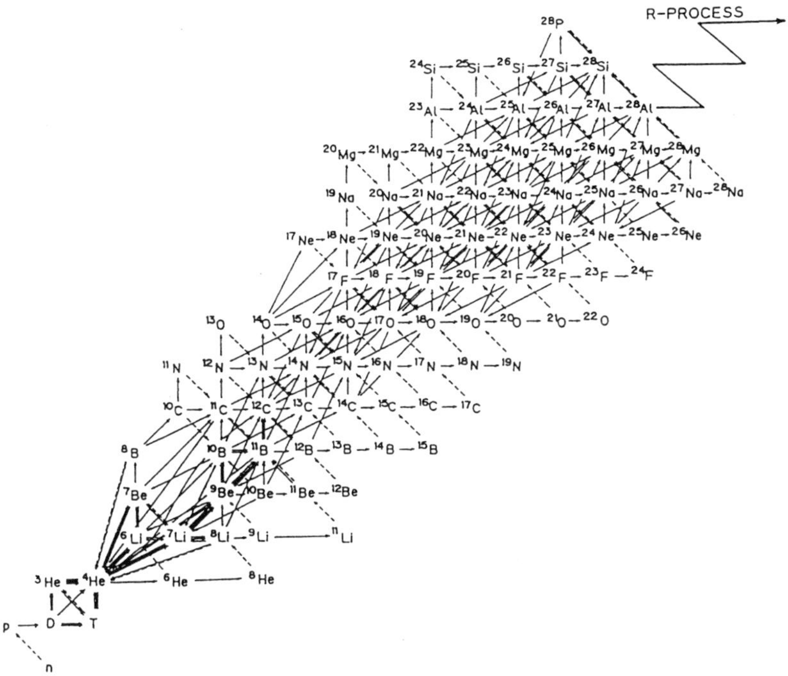

Figure 2. Light mass region of the nuclear reaction network used in the calculations. Solid arrows indicate nuclear reactions in the direction of positive Q–values. Dashed arrows show beta decays. Solid arrows indicate nuclear reactions in the direction of positive Q–value, and dashed arrows show beta decays. The full network consists of $\sim$ 3000 elements including radioactive unstable nuclei up to plutonium isotopes [8].

The key to resolve the first difficulty noted above is the short dynamic time scale $\tau_{dyn} \sim 10$ ms. As the initial nuclear composition of the relativistic plasma consists of neutrons and protons, the α-burning begins when the plasma temperature cools below T $\sim$ 0.5 MeV. The ^{4}He$(\alpha\alpha,\gamma)^{12}$C reaction is too slow at this temperature, and alternative nuclear reaction path ^{4}He$(\alpha n,\gamma)^9$Be$(\alpha,n)^{12}$C triggers explosive α-burning to produce seed elements with A $\sim$ 100. Therefore, the time scale for nuclear reactions is regulated by the ^{4}He$(\alpha n,\gamma)^9$Be. It is given by $\tau_N \equiv \left(\rho_b^2 Y_\alpha^2 Y_n \lambda(\alpha\alpha n \rightarrow ^9\text{Be})\right)^{-1}$. If the neutrino-driven winds fulfill the condition $\tau_{dyn} < \tau_N$, then fewer seed nuclei are produced during the α-process with plenty of free neutrons left over when the r-process begins at

T $\sim$ 0.2 MeV. The high neutron-to-seed ratio, $n/s \sim 100$, leads to appreciable production of r-process elements, even for low entropy S $\sim$ 130, producing both the 2nd ($A \sim 130$) and 3rd ($A \sim 195$) abundance peaks and the hill of rare-earth elements ($A \sim 165$) (cf. Fig.1).

The three body nuclear reaction cross section for ^{4}He$(\alpha n,\gamma)^9$Be is poorly determined experimentally. Because two different channels, ^{8}Be + n and ^{5}He + α, contribute to this process it is also a theoretical challenge to understand the reaction mechanism and the resonance structure. We show two calculated results in Fig. 1: The solid line displays the result obtained by using the cross section estimated by Woosley and Hoffman [9], assuming a ^{8}Be + n structure for ^{9}Be. The cross section estimate may only be good to one order of magnitude. Therefore, we also calculated the r-process by multiplying this cross section by a factor of 10 (long-dashed line). This makes a drastic change in the r-process abundance yields. Both theoretical and experimental studies of the ^{4}He$(\alpha n,\gamma)^9$Be reaction are highly desired.

The specific collision time for neutrino-nucleus interactions is given by

$$\tau_\nu \approx 201 \times L_{\nu,51}^{-1} \times \left(\frac{\epsilon_\nu}{\text{MeV}}\right)\left(\frac{r}{100\text{km}}\right)^2\left(\frac{\langle\sigma_\nu\rangle}{10^{-41}\text{cm}^2}\right)^{-1} ms. \quad (1)$$

$\tau_{dyn} \approx 10$ ms $\ll \tau_\nu \approx 200$ ms holds at the α-burning site, r $\approx$ 100 km. This resolves the second difficulty: Because there is not enough time for ν's to interact with n's in such rapidly expanding winds, the neutron fraction is very insensitive to the neutrino absorption.

We have recently shown that the neutrino flavor oscillation could destroy the r-process condition if the mixing parameters satisfy 0.3 eV$^2 \leq \Delta m^2$ in our wind model. Recent experiments of the atmospheric neutrinos and the missing solar neutrinos have indicated much smaller Δm^2, while the LSND experiment suggests larger Δm^2. We should wait for more experiments.

3 Big-Bang Nucleosynthesis and Galactic Chemical Evolution

We now found [10] that new possible nuclear reaction paths, which were neglected in the previous calculations. In addition to ^{4}He$(\alpha n,\gamma)^9$Be at relatively late epoch in the r-process, there are also ^{4}He$(t,\gamma)^7$Li$(\alpha,\gamma)^{11}$B and ^{7}Li$(n,\gamma)^8$Li$(\alpha,n)^{11}$B and ^{7}Li$(t,n)^9$Be$(t,n)^{11}$B, followed by (n,γ) and beta-decays. It is also likely that three-body reactions like ^{9}Li$(2n,\gamma)$, ^{15}B$(2n,\gamma)$, ^{24}O$(2n,\gamma)$, ^{27}F$(2n,\gamma)$, etc., may be important. This has already been pointed out in the inhomogeneous Big-Bang nucleosynthesis model (IBBN) [11, 12] although the physical conditions are different.

These new paths produce ^{9}Be, ^{11}B and even intermediate-mass nuclei 12 $\leq$ A in the IBBN, the abundances of which are many orders of magnitude higher than those of SBBN, as displayed in Fig. 3. This figure shows also the Galactic chemical evolution. The rise of the ^{9}Be abundance with increasing [Fe/H] arises from spallation of Galactic cosmic rays after Galaxy formation. Each solid curve reaches a different abundance plateau for Big-Bang nucleosynthesis (SBBN or IBBN) in the limit of [Fe/H] = -∞, i.e. t = 0.

The nuclear physics of the few-body reactions of light neutron-rich nuclei is thus very important not only in studies of r-process nucleosynthesis but in Big-Bang nucleosynthesis and Galactic chemical evolution as well.

This work has been supported in part by the Grant-in-Aid for Scientific Research (10640236, 10044103, 11127220) of the Ministry of Education, Science, Sports and Culture of Japan and also by JSPS-NSF Grant of the Japan-U.S. Joint Research Project. The author deeply acknowledges his collaborators's invaluable contrinution to this study: Dr. K. Otsuki, Dr. S. Wanajo, Dr. M. Orito Dr. A. Tokuhisa, Ms. M. Terasawa, Mr. T.-K. Suzuki, Prof. Y. Yoshii, Dr. H. Tagoshi, Dr. K. Sumiyoshi, Dr. I. Tanihata, Prof. G. J. Mathews, and Prof. R. N. Boyd.

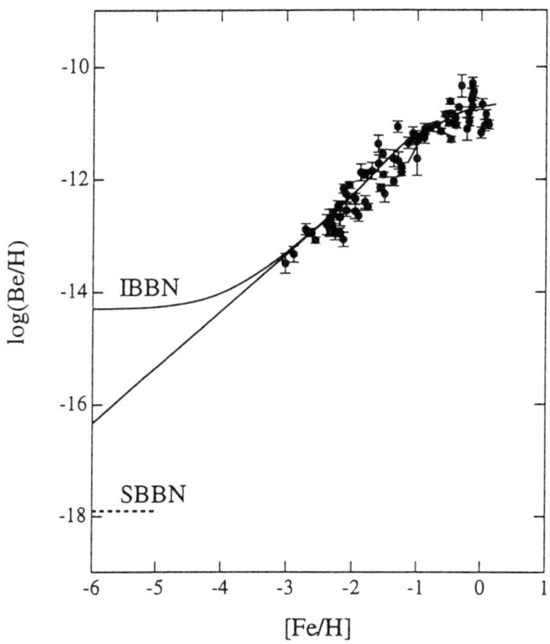

Figure 3. Predicted evolution of N(^{9}Be)/N(H) vs. [Fe/H] with the standard Big-Bang nucleosynthesis model (SBBN; lower dashed plateau) and the inhomogeneous Big-Bang nucleosynthesis model (IBBN; upper solid plateau). The rise of ^{9}Be abundance with increasing [Fe/H] arises from spallation production of Galactic cosmic rays after the Galaxy formation [13]. Each solid curve reaches different abundance plateau of Big-Bang nucleosynthesis (SBBN or IBBN) in the limit of [Fe/H] = $-\infty$. ght mass region of the nuclear reaction network used in the calculations.

References

1. Woosley, S.E., et al. 1994, ApJ 433, 229
2. Witti, J., Janka, H.-Th. & Takahashi, K. 1994, A&Ap 286, 842.
3. Qian, Y.Z. & Woosley, S.E. 1996, ApJ 471, 331.
4. Cardall, C.Y. & Fuller, G.M. 1997, ApJ 486, L111.
5. Otsuki, K., Tagoshi, H., Kajino, T. & Wanajo, S. 2000, ApJ, in press.
6. Meyer, B.S. 1995, ApJ 449, L55.
7. Käppeler, F., Beer, H. & Wisshak, K. 1989, Rep. Prog. Phys. 52, 945.
8. Wanajo, S., Otsuki, K., Kajino, T. & Mathews, G.M. 1999, ApJ, submit.
9. Woosley, S.E. & Hoffman, R.D. 1992, ApJ 395, 202.
10. Terasawa, M., Sumiyoshi, K., Kajino, T., Tanihata, T., and Mathews, G. J. 1999, ApJ, submitted.
11. Kajino, T. & Boyd, R.N. 1990, ApJ 359, 267; Kajino, T., et al. 1990, ApJ 364, 7.
12. Orito, M., Kajino, T. & Mathews, G.M. 1997, ApJ 488, 515.
13. Tokuhisa, A. & Kajino, T. 1999, in preparation.

Electromagnetic production of hyperons

K. Maeda
Physics Department, Graduate School of Science, Tohoku University, Sendai 9808578, Japan
E-mail: maeda@mail.cc.tohoku.ac.jp

The experimental studies of Λ photoproductions on ^{3}He and ^{12}C near the threshold region are presented. They were measured at Tokyo Electron Synchrotron Laboratory using the TAGX spectrometer. They are compared with numerical estimates with different models. We found the angular distribution of the K^0 photoproduction is much different from that of K^+. We discuss the use of polarized γ beams for the elementary (γ, K) reactions. It enables us to study the detailed elementary kaon production amplitudes.

Λ photoproduction on nuclei

The recent intensive study related to the hyperon production on nuclei is the Λ-hypernuclear spectroscopy using (π^+, K^+) reactions.[1] They have shown that a Λ hyperon production in a nucleus provides a unique opportunity to study fundamentals on the nucleon-hyperon and the nucleus-hyperon interactions. The photoproduction of hyperons becomes important concern because of the comparison with these accumulated data.[2] The Λ-producing (γ, K^+) and (γ, K^0) reactions are believed to be complementary to the hadoronic reaction.[3] The dominance of the spin-flip amplitudes in (γ, K) near the threshold region enables us to study the spin dependent behavior of Λ in nuclear potential. Since the mean free path of γ in nuclear medium is much longer than any hadrons, we can probe Λ in nuclei with less distortion. In spite of the importance of the electromagnetic production of hyperons on nuclear targets, suitable γ beams have not been available. High duty and high intensity γ beams in GeV-region are required to study Λ photoproductions because of the small cross sections and the requirements of photon tagging.

In this article, we introduce the ^{12}C(γ, K^+) and ^{3}He(γ, K^0) data near the threshold region. These measurements are the first observation of the K^+ and K^0 photoproduction on nuclei using the tagged photon beam. In addition, we notice the importance of the detailed study of the elementary kaon production amplitudes, which can be investigated by using the polarized γ beams. They can be done at the Laser-Electron-Photon facility at SPring8.[4] They will give us a complete data set for the elementary hyperon photoproduction amplitudes.

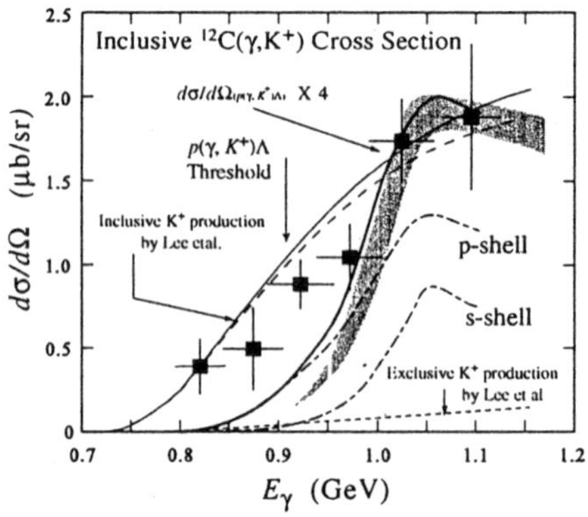

Figure 1: Differential cross section for ^{12}C(γ, K^+) reaction. TAGX data were compared with the elementary cross section and the numerical estimates in terms of the quasifree kaon photoproduction.

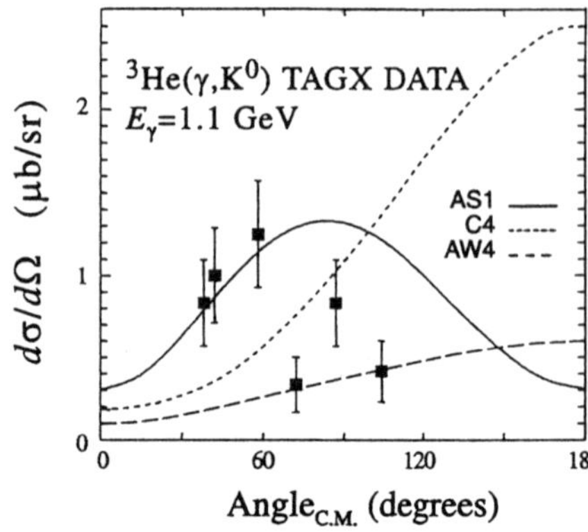

Figure 2: Differential cross section for ^{3}He(γ, K^0). Lines show the numerical estimates with different models.

K^+ photoproduction at TAGX

The ^{12}C(γ, K^+) reaction was carried out using the TAGX spectrometer[5] at the 1.3-GeV Tokyo Electron Synchrotron Laboratory.[6] The incident γ energy range was $E_\gamma = 0.78 - 1.1$ GeV. The charged particle events were momentum analyzed from the hits of Si-Strip Counters and Cylindrical Drift Chambers in the magnetic field. The timing and triggering counters were inner- and outer-scintillation-counter hodoscopes, and a time-of-flight wall. They covered the angular ranges of 10 to 40 degrees on the left-hand side and 0 to 30 degrees on the right-hand side. A typical momentum resolution was 6% for protons of 0.4 GeV/c.

Figure 1 shows the obtained differential cross section. Solid squares show the TAGX data compared with an average of the elementary cross sections multiplied by a factor of four. Thick curves show the Monte Carlo calculations, in which the quasifree process is assumed. The proton momentum distribution were taken from the ^{12}C$(e, e'p)$ data[7]. The dot-dashed curves represent the contributions of two protons in the p shell and four protons in the s shell, respectively. The thick solid line shows the sum of these two contributions. We can understand that the nuclear Fermi motion causes the K^+ yield to rise more slowly near the threshold than the elementary process.

Theoretical investigations of ^{12}C(γ, K^+) have been done in terms of the quasifree process leading to an unbound Λ with a three body model,[8,9] and an intranuclear cascade model[10]. Lee *et al.* [8] predicted both the inclusive and exclusive ^{12}C(γ, K^+) cross sections by using the recently developed amplitudes[11] of $\gamma p \to K^+\Lambda$ and wave functions from the relativistic mean-field theory[12]. They obtained 0.19μb/sr for the sum of the cross sections for all bound Λ states, which is close to the experimental value $0.21 \pm 0.05\mu$b/sr. The inclusive cross section in the energy region up to $E_\gamma = 1.1$ GeV was calculated based on the three body model. Their numerical estimates are shown in Figure 1. The thin-dotted line is the calculated cross section of the exclusive ^{12}C(γ, K^+)$^{12}_{\Lambda}$B. The thin-dashed curve is the cross section of the quasifree ^{12}C(γ, K^+)X process. The thin-solid curve is the sum of these two contributions. The quasifree Λ production dominates the (γ, K^+) process in the threshold region. A reduction factor $\sim$ 2.2 was used to fit the data. It is mainly due to the medium effects of γ, K^+ and Λ. When we use the experimental value of the total γN and K^+N cross section, the reduction factor become $R \sim 1.6$. It suggests that the nuclear medium effects on the Λ propagation in nuclei must be significant.

K^0 photoproduction at TAGX

The K^0 photoproduction on^{3}He were measured using the TAGX spectrometer. The invariant mass spectrum was constructed from two pion events. The γ energy range is above the reaction threshold for the elementary K^0 photoproduction and the vertex cut was employed outside of the target cell. In order to estimate the experimental cross section, the TAGX Monte Carlo was used to determine the detection efficiency.[5] We obtained the cross section $\sigma = 0.69\mu$b for K^0-short photoproduction on ^{3}He. Therefore the σ_{γ,K^0} becomes to about 1.4μb, which is similar to σ_{γ,K^+} at $E_\gamma \sim 1$ GeV.

The differential cross section is shown in Figure 2. Although the K^+ cross sections are strongly peaked in forward direction, the TAGX data does not show the forward peaking. It is understood by a lack of the t-channel diagram in the elementary K^0 photoproduction. The curves in Figure 2 are the numerical calculations using a computer code developed by Sotona.[13] Input parameters for the production amplitudes are obtained from K^+ photoproduction models.[14,15,16] It is reported that the predicted (γ, K^+) cross section using these different models show rather similar behavior in the 1 GeV energy region. In Figure 2, different model predictions denoted by AS1[14], C4[15] and AW4[16] are compared with experimental data. The numerical values of the (γ, K^0) cross sections are strongly model dependent in contrast to the K^+ photoproduction.

Elementary kaon photoproduction

The knowledge about the wide-range elementary $(\gamma, K^+(K^0))$ amplitude is essential for the substantial data analysis. When we detect Λ producing K^+ and K^0 photoproduction, their basic processes are, $\gamma + p \rightarrow K^+ + \Lambda$ and $\gamma + n \rightarrow K^0 + \Lambda$. In these cases, a spin 0 kaon is photo-produced on a spin 1/2 nucleon, leading to a spin 1/2 Λ particle. There are eight possible spin states in the system. Since a real γ does not have a longitudinal component, eight possible spin states correspond to four independent amplitudes. The (γ, K) process can be expressed with a photon polarization vector $\boldsymbol{\varepsilon}$,the nucleon spin σ, the photon $\boldsymbol{p}_\gamma$, the kaon momentum $\boldsymbol{p}_K$ and these amplitudes. The defined observables are differential cross sections, three single polarization asymmetries of P(recoil), Σ(beam) and T(target), and twelve double polarization asymmetries. Their definitions are shown in Figure 3. It is sufficient to measure eight of the polarization observables to complete the data sets for the elementary $N(\gamma, K)\Lambda$ processes. Until now, there are no beam polarization data and quite few data for target polarization, and several recoil polarization data of Λ are available with large experimental error bars [14,17,18]. They are shown in Figure 4. These exploration, which can be obtained by using the polarized γ beams, is very important to understand the hyperon photoproduction process.

CrossSection

$$\frac{d\sigma}{d\Omega}$$

1

Single Polarization Observables

$$P = \frac{(d\sigma/d\Omega)^+ - (d\sigma/d\Omega)^-}{(d\sigma/d\Omega)^+ + (d\sigma/d\Omega)^-} \quad \cdots\text{(Recoil polarization asymmetry)}$$

$$\Sigma = \frac{(d\sigma/d\Omega)^\perp - (d\sigma/d\Omega)^\parallel}{(d\sigma/d\Omega)^\perp + (d\sigma/d\Omega)^\parallel} \quad \cdots\text{(Polarized photon asymmetry)}$$

$$T = \frac{(d\sigma/d\Omega)^+ - (d\sigma/d\Omega)^-}{(d\sigma/d\Omega)^+ + (d\sigma/d\Omega)^-} \quad \cdots\text{(Polarized target asymmetry)}$$

3

Double Polarization Observables

$$X = \frac{(d\sigma/d\Omega)^{++} + (d\sigma/d\Omega)^{--} - (d\sigma/d\Omega)^{+-} - (d\sigma/d\Omega)^{-+}}{(d\sigma/d\Omega)^{++} + (d\sigma/d\Omega)^{--} + (d\sigma/d\Omega)^{+-} + (d\sigma/d\Omega)^{-+}}$$

$$= \frac{(d\sigma/d\Omega)^{++} - (d\sigma/d\Omega)^{+-}}{(d\sigma/d\Omega)^{++} + (d\sigma/d\Omega)^{+-}}$$

$$= \frac{(d\sigma/d\Omega)^{--} - (d\sigma/d\Omega)^{-+}}{(d\sigma/d\Omega)^{--} + (d\sigma/d\Omega)^{-+}}$$

Beam-Target X4

Beam-Recoil X4

Target-Recoil X4

12

Figure 3: Observables in $N(\gamma, K)\Lambda$ reaction.

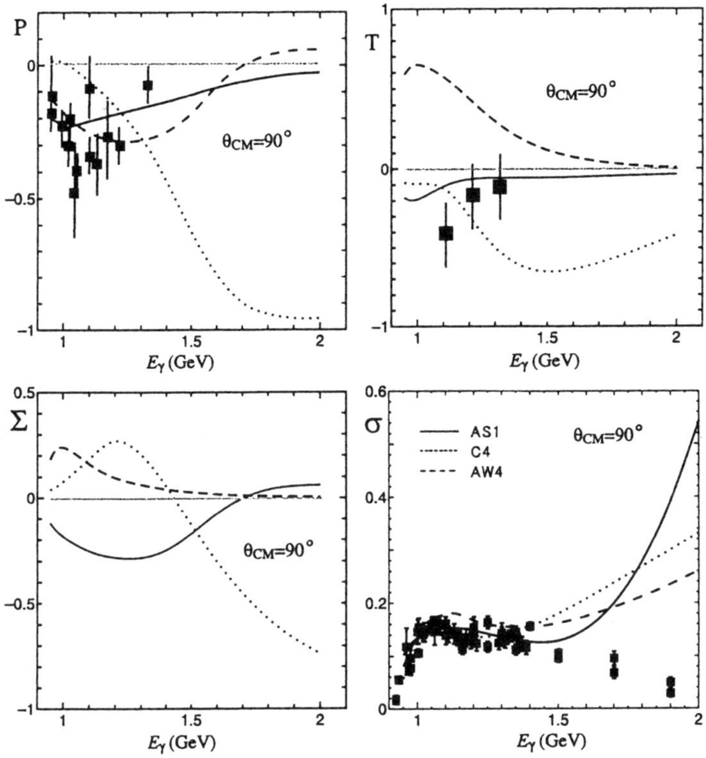

Figure 4: The present data status of experimental observables in $N(\gamma, K)\Lambda$ reaction. Lines show the numerical estimates using different models (see text).

Polarized photon beam at SPring8

In order to produce high-energy polarized γ, the use of Backward-Compton-Scattering (BCS) method in a laser-electron colliding process was proposed in 1962 [19]. This technique has been tried and materialized at several laboratories [20]. The advantages of the BCS method are in high polarization and low backgrounds. The magnitude of the γ polarization is determined by the polarization of laser light and the scattering angle. We can obtain almost 100% polarized γ near the maximum energy region.

The SPring-8 (Super Photon Ring 8 GeV) was built as a synchrotron radiation source.[21] It consists of a 1-GeV electron linear accelerator, an 8-GeV booster synchrotron and an 8-GeV storage ring. A dedicated beam line is ensured for the Laser-Electron-Photon at SPring8 (LEPS) project. It has a straight section of 7.8 m between two dipole magnets. Laser light is injected along the straight section, and it collides with the 8-GeV electron. We use a dipole magnet just after the colliding region as a magnetic spectrometer to analyze Compton-recoil electron energy. The electron counter consists of 0.1-mm pitch silicon strip detectors and plastic hodoscopes. They detect the electron energy range of 4.6 GeV to 6.5 GeV. The corresponding γ-ray energy region is 1.5 GeV to 3.4 GeV with $\Delta E_\gamma / E_\gamma \leq 1\%$ resolution.

When a 351 nm line of Ar laser is used, the maximum BCS photon energy is 2.4 GeV. The obtainable γ intensity, which is limited by the requirement of the light source operation, is $I \leq 10^7$ γ/sec. A magnetic spectrometer, which is completed by the LEPS project, will be used to measure K^+ and two charged pions from K^0 decay. The tagged photon luminosity (L_γ) will be $L_\gamma = N_\gamma N_T \cdot 10^{-30} \approx 0.6$ [μb^{-1}sec^{-1}] with $I = 1 \cdot 10^6$ γ/sec and one mol target. Therefore, the LEPS facility has enough power to carry out the complete measurements for the elementary hyperon photoproduction.

Summary

We measured the differential cross sections of the ^{12}C(γ, K^+) and ^{3}He(γ, K^0) reaction in the threshold region at 1.3-GeV Tokyo electron synchrotron laboratory. They are compared with numerical estimates and theoretical interpretations. We found that the accurate study of the elementary (γ, K) process and nuclear medium effects of γ, kaons and Λ propagation must be important for the future investigation of the hyperon photoproduction on nuclei. We noted that the measurements of the spin observables in (γ, K) is very effective to understand the reaction process. They can be done at the new Laser-Backward-Compton facility in the near future. They will give us important information to study the Λ-nucleus interaction and the structure of the Λ-hypernuclei.

The author would like to acknowledge the continuing guidance and kind encouragement of Professor H. Ejiri.

1. O.Hashimoto, *et al.*, *Nucl.Phys.* **A629**, (1998) 405c and references therein.
2. S.S. Hsiao and S. R. Cotanch, Phys. Rev. C**28**, (1983) 1668, M. Sotona, *et al. Nucl. Phys.* **A547**, 63c (1992), and C. Bennhold, *Nucl. Phys.* **547** (1992) 79c.

3. C. B. Dover, *Nucle. Phys.* **A547**, 27c (1992).
4. T.Nakano, H.Ejiri, M.Fujiwara, T.Hotta, K.Takanashi, H.Toki, S.Hasegawa, T.Iwata, K.Okamoto, T.Murakami, J.Tamii, K.Imai, K.Maeda, K.Maruyama, S.Date, M.M.Obuti, Y.Ohashi, H.Ohkuma, N.Kumagai, *Nucl.Phys.* **A629**, 559c (1998).
5. TAGX Collaboration: *Nucl. Instrm. Methods* **A376**, 335 (1996).
6. K.Maeda, *et al.*, *Nucl. Phys.* **A577**, 227c (1994), and H.Yamazaki, *et al.*, *Phys. Rev.* C **52**, R1157 (1995).
7. J. Mougey, M. Bernheim, A. Bussière, A. Gillebert, Phan Xuan Hô, M Priou, D. Royer, I. Sick and G. J. Wagner, *Nucl. Phys.* **A262**, 461 (1976).
8. T.S. Lee, Z.Y. Ma, B. Saghai and H. Toki, *Phys. Rev.* C**58**, 1558 (1998).
9. E. Ya. Paryev, private communication.
10. S. de Pina, E. C. de Oliveira, E. L. Medeiros, S. B. Duarte and M. Gonalves, *Phys. Lett.* **B434**, 1 (1998).
11. J. C. David, C Fayad, G. H. Lamot and B. Saghai, *Phys. Rev.* C**53**, 2613 (1996).
12. Z. Ma, J. Speth, S. Krewald, B. Chen, and A. Ruber, *Nulc. Phys.* **A608**, 305 (1996).
13. M. Sotona, K. Itonaga, T. Motoba, O. Richter and J. Žofka, *Nucl. Phys.* **A547**, 63c (1992), and M. Sotona and S. Frullani, *Prog. Theor. Phys.* **117**, 151 (1994).
14. R. A. Adelseck and B. Saghi, *Phys. ReV.* C**42**, 108 (1990).
15. R. Williams, Ch. R. Ji and S. R. Cotanch, *Phys. ReV.* C**46**, 1617 (1992).
16. R. A. Adelseck and L.E. Wright, *Phys. ReV.* C**38**, 1965 (1988).
17. S.S. Hsiao and S.R. Cotanch, *Phys. Rev.* C**28**, 1668 (1983).
18. C. Bennhold, *Nucl. Phys.* **A547**, 79c (1992).
19. R. H. Milburn, *Phys. Rev. Lett.* **10**, 75 (1963).
20. C. Bensporad *et al. Phys. Rev.* B**138** 1548 (1965), O. F. Klikov *et al.*, *Phys. Lett.* **13**, 344 (1964), C. E. Thorn *et al.*, *Nucl. Instrm. Methods* **A285**, 447 (1989)
21. URL:www.spring8.or.jp

BAND CROSSING IN THE O(6) SYMMETRY REGION AROUND A=130

A. Gelberg[1,4], N. Yoshida[2], T. Otuska[1], A. Arima[3], A. Dewald[4], and P. von Brentano[4]

[1] *Department of Physics, University of Tokyo, Bunkyo-ku, Tokyo 113-0033, Japan*
[2] *Faculty of Informatics, Kansai University, Ryozenji, Takatsuki 569-1095, Japan*
[3] *Ministry of Education, Science, Sports and Culture, Chiyoda-ku, Tokyo 100-0013, Japan*
[4] *Institut für Kernphysik der Universität zu Köln, 50937 Köln, Germany*

The band crossing in ^{128,132}Ba has been studied by using the model in which two neutrons and two protons have been coupled to a proton-neutron boson model core. The properties of the isomeric 10^+ state in ^{132}Ba have been well described. The most striking features of ^{128}Ba are the three-band mixing and the fact that both 10^+ and 12^+ states are mixed. Lifetimes in the yrast cascade and E2 branching ratios have been calculated. The calculation is in satisfactory agreement with the experimental data.

1 Introduction

In most even-even nuclei of the Ba region, the presence of two parallel *s-bands* has been observed, [1] which have been interpreted as the crossing of the ground band with two quasiparticle bands built on the $\pi h^2_{11/2}$ and $\nu h^{-2}_{11/2}$ configurations, respectively. This phenomenon is a direct consequence of the gamma- softness of these nuclei, i.e. the existence of the O(6) symmetry in the framework of the Interacting Boson Model (IBM). [2]

In ^{132}Ba only one *s-band* based on an isomeric 10^+ state with $\tau = 12ns$ is known. A striking feature of the band crossing pattern in ^{128}Ba is the presence of three nearly degenerate 12^+ states. A three-band mixing calculation has been carried out in, [3] under the assumption that the yrast 10^+ state is purely collective. Recent lifetime measurements[4] have shown that also the 10^+ states are mixed.

In the present report, we use the model of[5] which is able to handle the simultaneous crossing of a collective band with both a two proton and a two neutron bands; the pairs of protons and neutrons are coupled to an IBM-2 core. As a matter of fact, [5,6] contain also has become much richer. The aim of the present work is to apply the boson core plus two quasiparticle model of[5] to the band crossing in ^{128,132}Ba. Special attention will be paid to reduced E2 transition probabilities B(E2) and branching ratios.

2 The model

The calculations are based on the proton-neutron interacting boson model(IBM-2).[5] The model space is divided into three subspaces:

I. Pure boson core states with total angular mometum $I: |I> = |L = I>$, where L is the boson angular momentum;
II. Two-proton states: $|L, j_\pi^2(J_\pi); I>$, where the protons with j_π are coupled to J_π;
III. Two-neutron staes: $|L, j_\nu^2(J_\nu); I>$, where the neutrons with j_ν are coupled to J_ν.

The sum of the boson numbers and the fermion-pair numbers is always kept constant.

The Hamiltonian $H = H^B + H^F + V^{BF}$ is assumed in which

$$H^B = \epsilon(n_{d_\pi} + n_{d_\nu}) + \kappa Q_\pi^B \cdot Q_\nu^B + ... \tag{1}$$

is the Hamiltonian of the boson core, [2] where n_{d_π} and n_{d_ν} are the proton and neutron boson number operators respectively, and $Q_\rho^B = d_\rho^\dagger s_\rho + s_\rho^\dagger \tilde{d}_\rho + \chi_\rho [d_\rho^\dagger \tilde{d}_\rho]^{(2)}$ is the boson quadrupole operator ($\rho = \pi$ or ν). The fermion Hamiltonian is

$$H^F = E_\pi n_\pi + E_\nu n_\nu + V_\pi^F + V_\nu^F \tag{2}$$

where $E_\pi(E_\nu)$ is the single particle energy of the proton (neutron), $n_{\pi(\nu)}$ is the number of protons (neutrons), and $V_\pi^F(V_\nu^F)$ is the two-body interaction between the protons (neutrons). In this calculation the delta interaction has been used. A quadrupole type interaction of the fermions with the core is assumed, as in the well known particle to vibration coupling [7]:

$$V^{BF} = \kappa Q_\pi \cdot Q_\nu - \kappa Q_\pi^B \cdot Q_\nu^B - \sum_{\rho=\pi,\nu} \frac{\Lambda_\rho}{\sqrt{2j_\rho + 1}} : \left[[a_{j_\rho}^\dagger \tilde{d}_\rho]^{(j_\rho)} [\tilde{a}_{j_\rho} d_\rho^\dagger]^{(j_\rho)}\right]_0^{(0)} : \tag{3}$$

where the total quadrupole operator is defined as

$$Q_\rho = Q_\rho^B + \alpha_\rho [a_\rho^\dagger \tilde{a}_\rho]^{(2)} + \beta_\rho [[a_\rho^\dagger a_\rho^\dagger]^{(4)} \tilde{d}_\rho]^{(2)} - \beta_\rho [d_\rho^\dagger [\tilde{a}_\rho \tilde{a}_\rho]^{(4)}]^{(2)} \tag{4}$$

with $\rho = \pi, \nu$ and $\tilde{a}_{jm} = (-1)^{j-m} a_{j-m}$. The third and fourth terms in eq. (3) are responsible for the mixing of purely boson states with two-particle ones. These terms mix states belonging to the subspace I and others. The direct mixing between the II and III is quite weak.

In the calculation of E2 matrix elements, only core contributions are taken into account. [8] This, of course, includes collective transitions inside a two-quasiparticle band. The E2 transition operator is

$$T(\mathrm{E2}) = e_\pi Q_\pi + e_\nu Q_\nu \tag{5}$$

where $e_{\pi(\nu)}$ is the proton (neutron) boson effective charge.

3 Experimental data

Although several bands have been observed in ^{132}Ba, [9,10] there is only one s-band, based on an isomeric 10^+ state. The measurement of the magnetic moment of the isomeric 10^+ state[11] showed that it has a $\nu h_{11/2}^{-2}$ configuration. A peculiarity of this nucleus is the small energy difference between the 10_1 and 8_1 states. The structure of ^{128}Ba is pretty complex. [12,13] In the present work we will concentrate our attention on the crossing of the ground band with the two quasiparticle bands build upon the $\pi h_{11/2}^2$ and $\nu h_{11/2}^{-2}$ configurations, respectively.

The ground band can be followed up to $I^\pi = 12^+$. The 12^+ levels to which both *s-bands* decay have been observed, and their energies are pretty close to that of the third 12^+ state. Only in one of the *s-bands* a 10^+ level at $E_x =$ 3.522 *MeV* has been observed.

The results of lifetime measurements up to the yrast 10^+ state can be found in. [4] The lifetimes of the 16^+, 18^+ and 20^+ have been given in. [14] Recently a more accurate lifetime experiment [15] has been carried out, and the reduced transition probabilities are in good agreement with those of [4]; however, the standard deviations are smaller. One can notice that the B(E2;$10_1^+ \to 8_1^+$) is strongly reduced in comparison to the collective values, i.e. this state is also mixed.

4 IBM-2 Clculation

4.1 ^{132}*Ba*

The core parameters for the subspaces I and III were fitted to the ground bands of ^{132}Ba and ^{134}Ba, respectively. The single fermion energy was treated as a free parameter. The peculiar backbending plot shown in Fig. 1 has been partially reproduced by the calculation. The isomeric 10_1^+ state has only a weak ≈ 1 % core component, and this reduces strongly the B(E2;$10_1 \to 8_1$), in agreement with the experiment. At the same time, contrary to the situation in lighter Ba, there is a weak $10_1 \to 8_2$ branch, which indicates a slight breaking of the O(6) symmetry (see Table 1). The fit represent a compromise, through

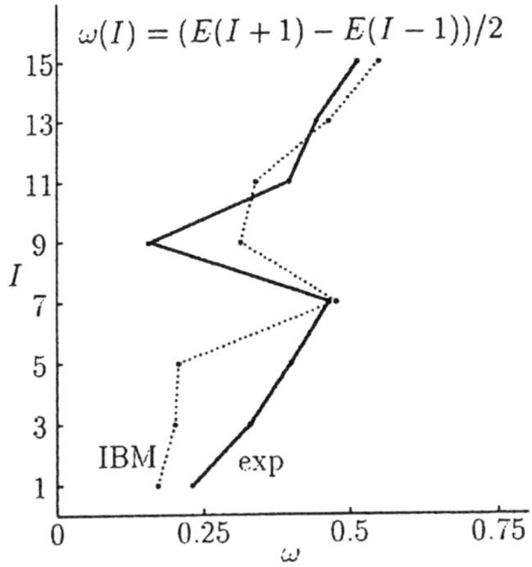

Figure 1: The backbending plot in ^{132}Ba.

Table 1: B(E2) values in ^{132}Ba ($e^2 fm^4$)

Observable	Exp.	IBM-2
B(E2;$2_1 \rightarrow 0_1$)	1720(172)	1720
B(E2;$10_1 \rightarrow 8_1$)	19(5)	33
$\frac{B(E2;10_1 \rightarrow 8_2)}{B(E2;10_1 \rightarrow 8_1)}$	0.13	0.29

Table 2: Gyromagnetic factors

$g(I)$	Exp.	IBM-2
$g(2_1)$	0.34(3)	0.31
$g(10_1)$	-0.145	-0.146

which a small B(E2;$10_1 \rightarrow 8_1$) was achieved without increasing too much the B(E2;$10_1 \rightarrow 8_2$). As can be seen in Table 2, the agreement of the experimental and theoretical g-factors is quite good.

4.2 ^{128}Ba

The parameters for the three cores were first fitted to the excited states of ^{128}Ba, ^{126}Xe and ^{130}Ba, respectively. These parameters are close to those of.[16] However, at the moment when the mixing interaction is switched on, the boson states are puched down in energy. In order to bring them in agreement with the experimental excitation energies, all parameters which have the dimension of an energy have been multiplied by a common scaling factor. The fact that not only the 10^+, but also the 12^+ states are mixed, imposed a supplementary condition on the parameters of the (N_π, N_ν) = (2,5) and (3,4) cores. Since the E(12^+) - E(10^+) energy differences are larger than the excitaion energies of

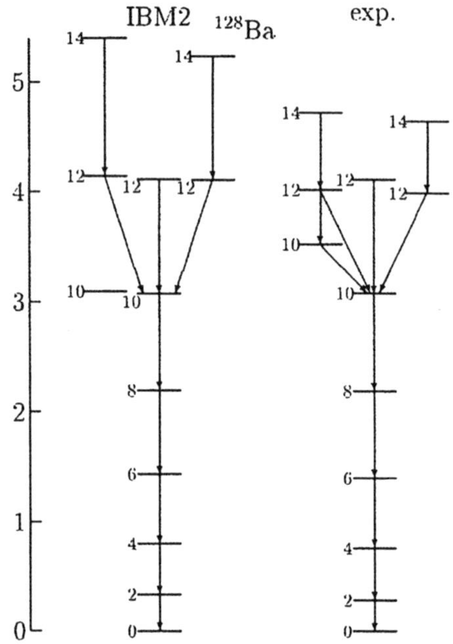

Figure 2: Experimental and calculated excitation energies in ^{128}Ba.

the corresponding 2_1^+ states, it is obvious that we can not keep all parameters at the values fitted to the neighbouring nuclei.

A comparison of experimantal and calculated excitaion energies of the ground band and the *s-bands* is shown in Fig.2. The closeness of the three 12^+ states is well described by the model. As a matter of fact, the degeneracy of the calculated levels is even more pronounced than in the experiment. The calulated 10^+ levels are also nearly degenerate, while in reality they are separated by ≈ 400 *keV*

Calculated and experimental values of B(E2) are shown in Fig.3. The agreement is quite good. The B(E2)'s show the usual increase with spin, up to $I = 8_1^+$, which is typical for E2 collective transitions. On the contrary, the $10_1^+ \rightarrow 8_1^+$, transition has B(E2) $= 0.196\ e^2b^2$, which is definitely lower than the collective value. This show that the 10_1^+ state is mixed. The low value of B(E2;$12_1^+ \rightarrow 10_1^+$) indicates that the yrast 12^+ state should be mixed too.

One should be very careful in ascribing a calculated state to a certain band. As mentioned above, the 10^+ and 12^+ states are very close in energy.

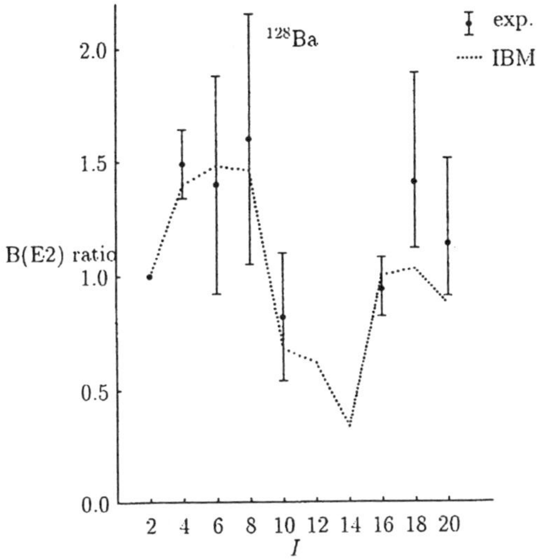

Figure 3: Experimental and Calculated B(E2) ratios B(E2; $I \to I-2$)/B(E2; $2 \to 0$) of the yrast states.

Therefore, one has to follow the strong B(E2)'s in orde to identify the states which belong to a certain band. In order not to confuse the reader, calculated states will be numbered accrording to the experimental energy ordering.

Experimental and calculated branching ratios have been compared. Relative B(E2) values have been calculated for the decay of the 14^+ and 12^+ levels. With the exception of the decay of the 14_1 state, the calculated branching ratios are in satisfactory agreement with the experiment.

As expected, the states belonging to the ground band are dominated by a boson component. The nature of the states changes dramatically in the corssing region. Above the crossing, i.e. for $I \geq 14$, the states of an *s-band* are dominated either by a two proton, or by a two neutron configuration, in agreement with [1]. As stated above, we still do not have a convincing experimental criterion for assigning configurations to the *s-bands*. In the present calculation the yrast band has a $\nu h_{11/2}^{-2}$ configuration, and the yrare band has a $\pi h_{11/2}^{2}$ configuration. The differences between the calculated B(E2)'s of corresponding transitions in the two *s-bands* is too small for making a configuration assignment.

According to a recent analysis [14] carried out on the basis of the B(E2)'s of transitions from states with $I = 20$, 18, 16 respectively, the yrast band is likely to have a proton character. Since the experimental errors are quite large, and the calculation based on the rotor plus particle model represents only an approximation, the comparison with the present calculation should be made with caution.

In order to exhibit the complex character of the mixing, we show in Fig. 4 the squared amplitudes of the boson, two proton and two neutron components respectively for the cascade originating from the two neutron band. The strong mixing of boson, two proton and two neutron components in the crossing region can be easily seen. As a matter of fact, the structure of the wave functions is even more complex, owing to the innumerable coupling possibilities.

5 Conclusions

Excitaion energies and transition strengths in the O(6) type nuclei ^{128}Ba and ^{132}Ba have been calculated by the IBM plus two quasiparticles model. In ^{132}Ba the energy backbending and the properties of the isomeric 10_1^+ state have been satisfactorily reproduced.

In ^{128}Ba the model is able to reproduce the main features of the band corssing, namely: the strong reduction of B(E2) starting from 10_1 and the three band mixing. The branching ratios have been moderately well reproduced.

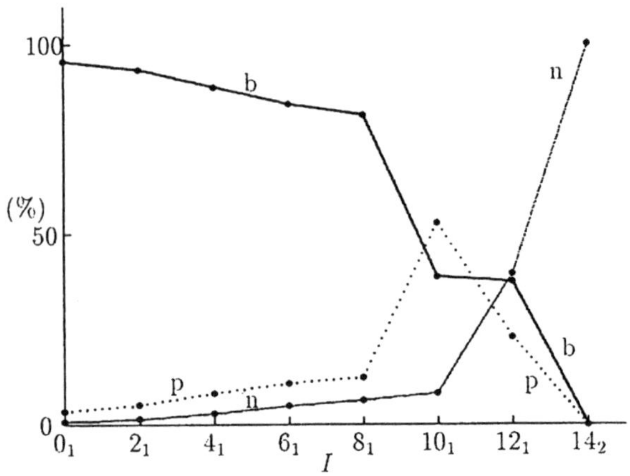

Figure 4: The squared amplitudes of the pure boson (b), two-proton (p) and two-neutron (n) components for the cascade from the two-neutron $I = 14$ state.

References

1. R. Wyss et al., Nucl. Phys. **A505**, 337 (1989) and references within.
2. F. Iachello and A. Arima, The Interacting Boson Model (Cambridge University Press, Cambridge 1987).
3. S. Frauendorf, W. Lieberz, D. Lieberz, P. von Brentano and A. Gelberg. Phys. Lett. **274B**, 149 (1992).
4. P. Petkov et al., Nucl. Phys. **A543**, 589 (1992).
5. N. Yoshida, A. Arima and T. Otuska, Phys. Lett. **114B**, 86 (1982).
6. A. Faessler et. al., Nucl. Phys. **A438** 78 (1985).
7. A. Bohr and B. Mottelson, Nuclear Structure, vol.II, W. A. Benjamin, Reading, Mass., 1975.
8. Brief Manual of the TW (Boson plus Two particles) program, Kansai University, Takatsuki, Japan 1997 (unpublished).
9. A. Dewald et. al., Phys. Rev. **C37**, 289 (1988).
10. E. S. Paul et. al., Phys. Rev. **C40**, 1255 (1989).
11. S. Harissopulos et. al., Phys. Rev. **C52**, 1796 (1995).
12. K. Schiffer, A. Dewald, A. Gelberg, R. Reinhardt, K. O. Zell, Sun Xiang-Fu, and P. von Brentano, Nucl. Phys. **A458**, 337 (1986).
13. U. Neuneyer et al., Z. Phys. **A336**, 245 (1990).
14. P. Petkov et al., Nucl. Phys. **A640**, 293 (1998).
15. P. Petkov et al. (to be published; private communication).
16. G. Puddu, O. Scholten and T. Otuska, Nucl. Phys. **A348**, 109 (1980).

ELECTRON ACCUMULATOR RING FOR THE PEARL PROJECT

K. Hatanaka
Research Center for Nuclear Physics, Osaka University,
10-1 Mihogaoka, Ibaraki, Osaka 567-0047, Japan
E-mail: hatanaka@rcnp.osaka-u.ac.jp

Results of design studies are presented for an electron accumulator ring proposed for the PEARL project. Maximum energy of electrons is 10 GeV. Electrons are polarized and the luminosity for internal target experiment reaches 10^{33} to 10^{34} nucleons $cm^{-2}s^{-1}$. High energy real photons are produced with Compton backscattering technique. The maximum photon energy is expected be 5 GeV at the intensity of 5×10^5 photons $GeV^{-1}A^{-1}W^{-1}s^{-1}$.

1 Introduction

The Research Center for Nuclear Physics(RCNP), Osaka University, is a laboratory complex. The main facility is the cyclotron laboratory located in this Suita-campus. It consists of coupled cyclotrons, K140 AVF and K400 ring cyclotrons, and unique experimental apparatuses including dual magnetic spectrometers and a neutron detector/polarimeter. Others are Laser Electron Photon Laboratory at the SPring-8 site, and the underground laboratory at Oto Cosmo Observatory. A lot of excellent results and perspectives are presented in this volume of proceedings.

Recently, a committee was organized to discuss medium range plans of RCNP. The committee reviewed the present RCNP activities and important physics subjects to be studied by RCNP. On the basis of the review, following suggestions were made on RCNP medium range plans. [1]

Medium range plans of RCNP should be based on the present RCNP activities and the present potential power, the current subjects of nuclear physics and on the national and international situations of nuclear particle physics. RCNP is strongly encouraged to promote the following subjects.

1. Quark lepton nuclear physics with multi-GeV electron and proton/nucleus collider.
2. Sensitive frontiers with high quality/sensitive spectrometers and detectors.

RCNP will contribute to the progress of nuclear physics in first four among those current subjects in nuclear physics.

1. Standard theory and beyond
2. Neutrinos and weak interactions

3. Hadrons and non-perturbative QCD Flavor nuclear Physics
4. Quark confinement
5. QGP and exotic nuclear structures
6. Origin of universe and astrophysics problems

A conceptual design of an electron-nucleus collider was performed according to the suggestion. [2] In the design, beam energy is 5 GeV for both electrons and protons. For this combination of energies, the luminosity was found to be limited to $10^{31}\ cm^{-2}s^{-1}$. The limitation comes from the space charge effect in the proton accelerator. An asymmetric collider geometry can provide the luminosity of $10^{33}\ cm^{-2}s^{-1}$ for the collision of a few GeV electrons and and a few tens GeV(/u) protons(ions). Such kind of collider is proposed at IUCF as discussed by Prof. J. Cameron in this Symposium. [3]

As the first stage to extend the present RCNP activities to quark nuclear physics, an action program was started to design an electron accumulator ring at 8-10 GeV proposed as the Photon Electron Accumulator Ring Laboratory (PEARL). The non-perturbative regime of QCD is current challenging task on the border line between nuclear and particle physics. Outstanding questions are the quark-gluon structure of baryons and mesons, the existence or non-existence of glueballs which are predicted by QCD and the origin of the confinement which prevents colored objects propagating freely in space. The PEARL project will give the opportunity to investigate these subjects.

The results of the design and feasibility studies of the 10 GeV electron accumulator ring are presented in the following sections.

2 Electron Accumulator Ring

2.1 General Description of the Ring

General features of the electron accumulator ring are summarized below.

1. Maximum energy is 10 GeV.
2. Expected beam current is 0.5 A and will be increased to 1 A in future.
3. Polarized photons are produced by the Compton backscattering of Laser lights.
4. Electrons are polarized and internal gas targets are used.
5. Circumference is around 600 m in order to accommodate the ring in a proper area.

Figure 1 shows kinematic regions of the electron energy loss, ν, and the squared electron four momentum transfer, Q^2, accessible with 10 GeV beam. It enables

1. the access to high momentum transfers, implying fine spatial resolution
2. the broad access to the deep inelastic scattering (DIS) regime
3. the access to baryons up to 4 GeV mass, and mesons up to 3 GeV mass. For both baryons and mesona, a large fraction of the spectrum is missing as discussed by H. Toki in this Symposium.[4]

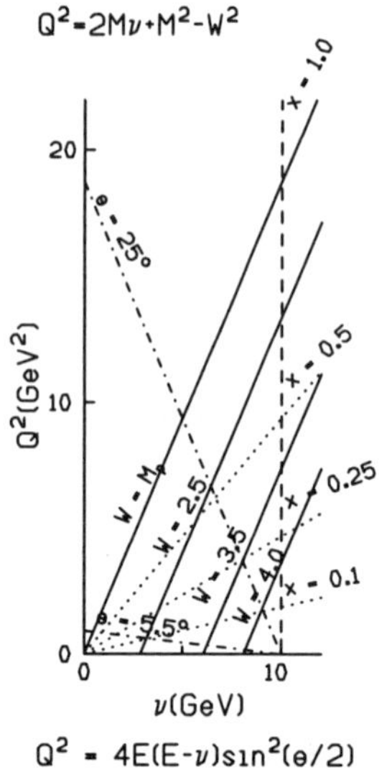

Figure 1: Kinematic regions of the 10 GeV accumulator ring.

2.2 Lattice Parameters

The general parameters of the accumulator ring are listed in Table 1 and the optical functions for the arc section are shown in Fig.2. The injection energy is 8 GeV, and the maximum energy 10 GeV. Arcs are composed of simple FODO arrays and bending radius is 27.5 m. Dispersion suppressors ensure the transition between arcs and the straight sections where the dispersion is zero. Length of a long and a short straight section is 100 m and 40 m, respectivelly.

2.3 RF System

The RF paramerters of the ring are listed in Tabe 2. Radio frequency is 508.58 MHz, which is same as that of SPring-8. Shunt impedance of a cavity is assumed 6 MΩ. There are 4 stations in the ring, and 6 units per station. 4 cavities are installed at each unit. In total, 96 cavities are used to supply 42 MV and 16 MW continuously to a beam of 0.5 A. Power dissipated in the cavities is 3.4 MW assuming 10 % shunt impedance drop. Total power is 21 MW assuming 10 % loss in transmission lines. Input power per cavity is 200 kW, and power dissipated per cavity is 35 kW. There is a limit of 300 kW for the power that can be carried by each input coupler.

Table 1: Parameters of a 10 GeV Accumulator Ring.

Parameter	Value	Unit
Circumference	586	m
Injection energy	8	GeV
Maximum energy	10	GeV
Lattice type	FODO	
Bending radius	27.52	m
Number of dipole magnets	48	
Number of normal cells	20	
Number of dispersion suppression cells	8	
Super periodicity	2	
Momentum spread	1.5×10^{-3}	
Nominal tune (ν_x/ν_y)	17.9/12.7	
Natural chromaticity	-26.3/-18.5	
Dispersion in straight cells (D $_x$)	0.0	
Long straight section (100 m)	2	
Short straight section (40 m)	2	
Radiation loss	32	MeV/turn
Harmonic number	994	
Damping time τ_x	2.0	ms
τ_y	2.0	ms
τ_s	1.0	ms

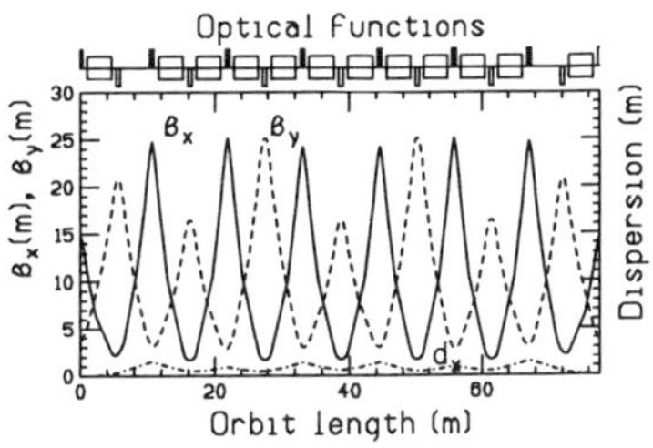

Figure 2: Optical functions for the arc section.

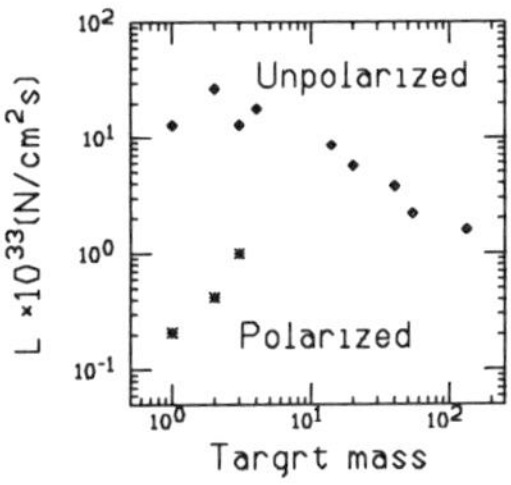

Figure 3: Maximum obtainable luminosity in nucleons/cm^2/s for gaseous elements assuming a beam life time of 1 hour and a beam current of 0.5 A.

3 Internal Target Luminosities

The luminosity that can be obtained with internal targets depends strongly on the atomic number Z. The beam lifetime is reduced by atomic bremsstrahlung of electrons in the target, In the absence of other beam-loss mechanisms, the lifetime is given by [5]

$$T_{target} \simeq T_0[Z(Z+1)(\ln \frac{183}{Z^{1/3}})(\frac{n}{8 \times 10^{25} atoms/cm^2})]^{-1},$$

Table 2: RF parameters of a 10 GeV Accumulator Ring.

Radio frequency	508.58	MHz
Maximum RF voltage	42	MV
Number of RF stations	4	
Number of units/station	6	
Number of cavities/unit	4	
Total number of cavities	96	
Power transferred to the beam at 0.5 A	16	MW
Power Dissipated in the cavities*	3.4	MW
Total Power**	21	MW
Shunt impedance of a cavity	6	MΩ
Input power per cavity	200	kW
Power dissipated per cavity	35	kW

* Assuming 10 % shunt impeddance drop. ** Assuming 10 % loss in transmission lines.

where n is the target thickness and $T_0 = 2\mu s$ is the revolution time in the accumulator ring. The maximum thickness for internal targets are obtained for gaseous elements corresponding to $T_{target} = 1hour$. Figure 3 shows the corresponding luminosity L_A for a beam current of 0.5 A as a function of the target mass. It is found the luminosity for internal target experiment can reache 10^{33} to 10^{34} nucleons $cm^{-2}s^{-1}$.

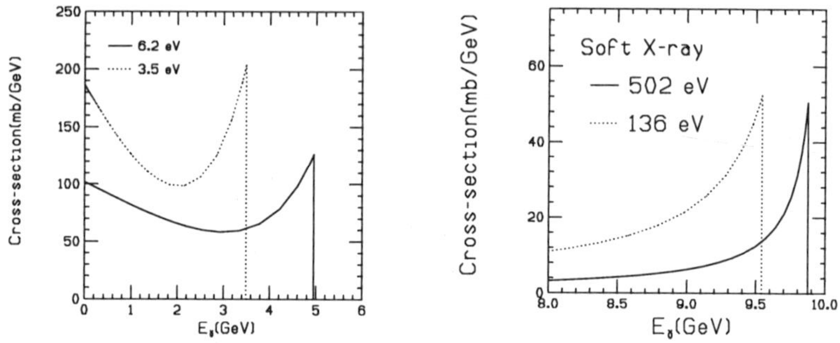

Figure 4: Energy spectrum of γ-rays for incoming electron beam energy 10 GeV.

4 Photon Beams

In 1963 it was pointed out for the first time that the backward scattering of Laser light against high energy electrons could produce polarized γ-rays ($\vec{\gamma}$). [6] There are several working facilities based on the Compton backscattering technique and the RCNP LEPS facility will be in opration in 1999. [7] With the electron energy of 10 GeV, the maximum photon energy is expected be 5 GeV at the intensity of 5×10^5 photons $GeV^{-1}A^{-1}W^{-1}s^{-1}$ using a Laser light of 200 nm wavelength. If X-rays or soft X-rays from SPring-8 are available in stead of a Laser light, Compton backscatterd γ-rays have almost monochromatic energy. Figure 4 shows energy spectrum of γ-ray for incoming electron beam energy 10 GeV and Laser energy 3.5 eV, 6.2 eV, 136 eV and 502 eV, respectively.

5 Conclusion

The results of the design and feasibility studies of the 10 GeV electron accumulator ring for the PEARL project are discussed. The ring will give the opportunity to challenge the non-perturbative regime of QCD with polarized electrons and real photons.

Acknowledgments

The author would like to thank Profs. H. Ejiri, H. Toki, T. Nakano, M. Nomachi, H. Yonehara and S. Date for useful information and discussion.

References

1. T. Kajino *et al*, Report in Japanese(1997).
2. K. Hatanaka *et al*, RCNP Annual Report (1997) 312.
3. J. Cameron, to be published in these Proceedings.
4. H. Toki, to be published in these Proceedings.
5. M. Düren, DESY-HERMES 90-01 (1990).
6. F.R. Arutyunyan and V.A. Tumanian, *Phys. Lett.* **4**, 176 (1963).
   R.H. Milburn, *Phys. Rev. Lett.* **10**, 75 (1963).
7. T. Nakano, to be published in these Proceedings.

CONCLUSION REMARK

S. KULLANDER
Department of Radiation Sciences, Uppsala University, S-751 21 Uppsala, Sweden

1 Introduction

It is my privilege to make a few remarks to a very exciting conference covering a wide range of topics in modern subatomic physics.

On the threshold to a new millennium it is of interest to remind of some of the most important developments of modern physics that was founded from the theory of relativity and quantum mechanics in the beginning of the century. The unification of the weak and electromagnetic interactions due to theoretical work by Glashow, Salam and Weinberg but also due to theoretical work by Yang, Mills and t'Hooft and its experimental verification due to discoveries of weak neutral currents (neutrino scattering) and of the intermediate Z and W bosons are among the major achievements. Discoveries of the atomic nucleus in the beginning of the century, of the proton and the neutron, of new elements, of antimatter, of quarks and gluons are other important mile-stones of modern physics.

The strong interaction is best described at low energy by meson exchanges and at high energy by quarks and gluons. The development of Quantum Chromo Dynamics (QCD) by Feynman, Bjorken and Nambu and its experimental verification especially in deep inelastic scattering have given us a clear understanding on how to interpret high energy collisions. Of particular importance is the role of the gluon (a co-discoverer was the late Bjorn Wiik) that takes part explicitly at high energy and is hidden in the quark environment at low energy. A main challenge to physicists in the next millennium is to understand the nucleus in terms of quarks. Important aids for our understanding will be conservation laws and symmetries, in the focus of NEWS99, and novel instruments and experimental methods.

It is out of range of my possibilities to give proper justice to the many high-class talks that have been presented during this conference. Instead I shall select some examples, on the border between nuclear and particle physics, of interesting problem areas which may mature in the next millennium. In the focus of this presentation is of course professor Hiro Ejiri who has had a very pronounced ambition to cross-fertilize these two sub-fields of modern physics. He will leave the institute, RCNP, that will be in a very strong position to compete successfully in future physics research. A key problem for the application of the quark picture to nuclear physics is to understand the transition from perturbative to non-perturbative QCD and it is appropriate to present ongoing work in this direction. Attention will also be paid to violation of symmetries and the oscillation of neutrinos in a meeting with the theme Symmetries and Electro-Weak Nuclear processes. The evolution of physics beyond the problems discussed today will be guided by new instruments and methods as has been shown in some exciting talks here. Finally I would like to make a few statements on how to make the role of physics more appreciated in society.

2 Hiro Ejiri and RCNP

It is appropriate on this occasion when professor Ejiri is about to leave his office as the Director of RCNP to remind you of the rapid transition, under his leadership, of the science profile of the Institute from low-energy nuclear physics to intermediate energy physics; from nucleon-meson physics to quark-lepton physics. Six years ago, when professor Ikegami had finished his term of office, the RCNP cyclotron and the associated experimental facilities were just about ready. We now have witnessed the scientific exploitation of these facilities. For example, the beautiful excitation energy spectra of the residual silicon nucleus resulting from the $^{27}Al(^{3}He,t)^{27}Si$ at an incident energy of 450 MeV were possible thanks to the excellent energy resolution of the Grand Raiden magnetic spectrometer. Y. Fujita in his talk showed how it was possible to compare resolved states in the giant resonance region, excitation energies between 8 and 12 MeV, with states in the mirror nucleus ^{27}Al excited in inelastic electron and proton scattering and to demonstrate that the $M1/GT$ strengths are equal revealing isospin symmetry. Another example, discussed by T. Kishimoto, of the potential of the RCNP cyclotron for studies of very rare phenomena is the forthcoming measurements of weak production of Λs in *pn* interactions whereby parity violation in the baryon-baryon system can be directly exploited.

Thanks to the research at the Oto Cosmo Observatory highly relevant studies of double beta decay using the ELEGANT detectors are carried out as reported in the talk by Hiro Ejiri. ELEGANT VI will also be used for searches of dark matter candidates using 25 modules of CaF_2 scintillators measuring the ^{19}F nuclear recoil from its interaction with a dark matter particle. In his talk, Hiro Ejiri showed how the decay time limit for the case of the ^{100}Mo had been pushed to $0.22 \cdot 10^{23}$ years implying a limit on the neutrino mass of 1.9 eV. These limits will be further improved thanks to comparatively low background conditions.

The new photon beam facility (LEPS) at SPring-8 will provide polarized monoenergetic photons due to the backscattering of laser photons from the electrons circulating in the synchrotron. A number of interesting physics issues may be probed at this new facility. T. Hatsuda discussed how medium effects may lead to a σ/π degeneracy and to spectral shifts of the masses of ρ, ω and ϕ. H. Shimizu pointed out that Primakoff production of σ could be useful for tests of the Chiral perturbation theory. A. Titov considered production of ϕ for investigations of exchange mechanisms and of the strange sea content of the nucleon. According to T. Nakano test experiments on photoproduction using 1.5-2.4 GeV tagged photons ($5 \cdot 10^6$) will start in September 99 with an energy resolution of 15 MeV. These examples show that physicists are well prepared to start a programme in which effects of quarks can become fairly visible. Hiro Ejiri's aim of having LEPS as a facility for quark-nuclear physics is certainly very important for the extension of the RCNP physics programme.

3 From perturbative to non-perturbative QCD

Perturbative QCD has been shown to be valid for reactions in which momenta greater than one GeV/c are transferred. The quarks and the gluons may be sensed in deep inelastic lepton scattering off nucleons or produced as quark-antiquark pairs or/and gluons in high energy collisions. The interactions with the quarks and gluons give rise to inelastic form factors, structure functions, that remain essentially constant as a function of the four-momentum transfer Q. The production of quarks and/or gluons give rise to jets containing several hadrons within a narrow solid angle.

The picture that has emerged from high energy experiments is that the nucleon consists of about 50 per cent quarks and 50 per cent gluons. A few per cent of the quark content is made up of quark-antiquark pairs usually called sea quarks to be distinguished from the three valence quarks that are necessary to give the nucleon its baryon identity. The masses of the "free" point quarks are not well defined and are about 10 MeV for the d quarks and 5 MeV for the u quarks. These point quarks measured in high energy experiments are often referred to as current quarks. The theoretical interpretation of different experiments using perturbative QCD is very satisfactory and for example the strong coupling constant α_s has been measured with very high precision for Q greater than 1 GeV/c.

Another picture of the quark constituents comes from hadron spectroscopy. The results of hadron spectroscopy are that the u quarks and the d quarks share the energy content of the nucleon, each carrying an energy of about 300 MeV. These quarks that are extracted from spectroscopy are often referred to as constituent quarks and they are considered each one to have a spatial extension so that they together fill up the hadron.

It is important to remember that the results from perturbative QCD are based on interactions with current quarks. For example, the so called spin crisis detected in deep inelastic scattering and explained in many models by contributions to the nucleon spin from the gluons, would not be seen at lower energy where a constituent quark model is applicable; the gluons and their spin content are not separable from the constituent quarks.

Concerning the content of strange sea quarks in the nucleon, there are *ep* elastic scattering experiments on parity violation in progress or planned at MIT-Bates, Mainz and Jefferson Lab as reported by J. Arvieux. The vector current contributions of photon and Z exchanges due to strange quarks in the nucleon are expected to give parity violations showing up on the order of 10^{-5} in the asymmetry. These experiments are very important since they will connect information about the sea quark content in the nucleon between the region of high Q^2 where perturbative QCD is valid to the region of Q^2 below 1 $(\mathrm{GeV/c})^2$ which is more typical for the momentum transfers in nuclear physics.

Interesting theoretical approaches on how to adapt the quark picture to a nuclear physics environment were presented. L. Glozman suggested a chiral constituent quark model based on an analogy with the phonons and electrons in a crystal to explain differences in Λ and N baryon spectra. In the model, the pions couple to the constituent quarks rather than to the nucleon. In the limit of zero momentum transfer, the interaction is smeared out so that the calculated value for the πqq coupling constant becomes equal to 0.6 instead of 14 for the πNN coupling.

The dual Ginzburg-Landau theory as an effective theory of low energy QCD was presented by H. Toki and also by H. Suganuma. The essence of the approach is an Abelian gauge and a QCD monopole. The result of the Abelian dominance is a diagonal gluon instead of a colour charged gluon, and the former has a more extended pion-like range function than the latter. The theory leads to Chiral symmetry, Regge behaviour and a QCD phase transition and results in a prediction of a QCD Higgs particle of mass 1.5 GeV similar to the mass of glueball candidates.

A. Thomas presented how Chiral corrections to lattice QCD lead to a rapidly converging expansion. He stressed that a pion cloud around a bare nucleon is essential for the understanding of masses, form factors and deep inelastic scattering where an excess of sea anti-d quarks is seen. A physical neutron is expanded in a bare neutron and a neutral pion, a bare proton and a negative pion and, in order to have a rapid convergence, a bare Δ and a pion. The inclusion of the $\Delta\pi$ term in the expansion leads to a marginal change of

$f_{\pi NN}$ by 10 per cent as compared to a factor 2 to 3 without the term. The importance of the $\Delta\pi$ term has been noted earlier but it was rediscovered with the progress of Chiral Perturbation Theory.

As seen from these examples, there are many interesting ideas in the air for a better understanding of how to connect the world of quarks and gluons with the world of mesons and nucleons. The beginning of the next millennium will very likely be an exciting time for this important field of physics which will pave the way for a better understanding of nuclear structure and nuclear reactions in terms of basic constituents.

4 Neutrino oscillations come and go

The field of neutrino oscillations continues to be of great interest at scientific conferences. One reason for this is the positive evidence seen in the experiments at Kamioka, and this evidence is becoming firmer as reported by Y. Suzuki. The neutrinos created in the interactions of cosmic rays have typically energies in the GeV range. It seems more and more clear that the deficit of muon neutrinos is due to neutrino oscillations. The results from the Superkamiokande experiment as of December 1998 are that, on the 90% confidence level, the mass difference Δm between the mass of a muon neutrino and that of another appearing neutrino is $\Delta m^2 = (1.5 - 6) \cdot 10^{-3}$ $(\mathrm{eV})^2$. The result for the mixing angle θ is that $sin^2 2\theta > 0.9$.

A controversial issue has for a rather long time been the positive evidence, announced by the Los Alamos group (LSND), of electron neutrinos appearing in a muon - neutrino beam produced in the decays of positive pions at rest. J. Kleinfeller reported on new analysed data from the ISIS - KARMEN experiments which are nearly identical to the LSND experiment. In the data from February 1997 to April 1998, the seven electron neutrinos found are consistent with the background. The data are defined by $sin^2 2\theta < 1.3 \cdot 10^{-3}$ and cover practically the full range of LSND.

R. Raghavan reviewed the status of solar neutrino measurements and discussed in particular the problem with the deficit of neutrinos from beryllium-7. Since boron-8 cannot be made without beryllium-7, it is hard to understand the deficit unless there are neutrino oscillations. A very sensitive experiment that has now commenced, is Borexino in Gran Sasso. The detector medium is an organic scintillator. Since the solubility for radioactive elements that cause background is very low the experimental conditions are favourable. The threshold is as low as 50 keV. A flavour independent determination of the flux of 862 keV neutrinos, originating from beryllium-7, will be made by measuring the recoil electrons from neutrino - electron scattering.

Raghavan has suggested a low energy neutrino spectrometer (LENS). It will tag a neutrino - electron reaction by the gamma ray decaying from an excited state of the product nucleus. This latter differs by one unit of charge from the initial nucleus. Suitable target nuclei are ^{176}Yb, ^{160}Gd and ^{82}Br. Interactions with them result in product nuclei having levels that are populated with a high probability for the case of 862 keV neutrinos from beryllium. In this way LENS will be both a source specific and flavour specific detector. Experimentally, a few per cent of gadolinium mixed with a liquid scintillator will be used.

M. Fujiwara has used the (^{3}He,t) charge exchange reaction at 450 MeV to investigate how different states of residual nuclei are populated. It is assumed that this reaction leads to the same relative population as the weak (ν, e) interaction. From these investigations it is found that the rates of absorbed beryllium neutrinos is enhanced and is comparable to the rates of absorbed neutrinos of the pp type and far above absorbed neutrinos of other types. The absorption rates of beryllium neutrinos per year in 1.3 tons of ^{176}Yb and in 4.4 tons of ^{160}Gd are estimated to be 85 and 135 respectively.

These examples show how knowledge of nuclear physics and of methods and instruments used in subatomic physics are very important for the advancement of our understanding of neutrino properties and of the nuclear fuel cycle in the sun. This understanding is also of importance for astrophysics and cosmology where neutrino observatories gain in importance.

5 Violations of the Standard Model?

The question of massive neutrinos has been substantiated by the Kamiokande experiments. Whether there are Majorana or Dirac neutrinos was discussed by J. Vergados and according to him there now seems to be consensus among theorists about the Majorana choice.

Physicists would like to see a more complete theory than the present Standard Model (SM) which has too many free parameters. Attempts to falsify the SM are therefore followed with great interest. Several speakers stressed the importance of the double beta decay experiments for studies of effects beyond the Standard Model; Higgs scalars, SUSY, R parity violation. Discovery of neutrino - less double beta decay could be a signal of WIMPs or SUSY particles.

Could the Fermi constant G_F vary with time? The rate of double beta decay is now 50% higher than 2 billion years ago according to A Barabash when present results from NEMO are compared with geochemical experiments on ^{82}Se, ^{96}Zr and ^{130}Te. The planned NEMO3 detector having ^{100}Mo as a source will provide better data and may improve the present 3σ effect. A geochemical experiment, also on molybdenum-100, would allow a direct comparison and shed more light on this interesting effect.

Y. Kuno reviewed two different types of lepton flavour violation experiments involving muons; $\mu \rightarrow e\gamma$ decay and $\mu \rightarrow e$ conversion. The present limits are around 10^{-12}. New attempts to improve the decay branching ratios using polarised muons will be done at PSI - SINDRUM and to improve the conversion factors at BNL - MECO and KEK - PRISM. The hope is to reach figures of 10^{-16}. These very rare processes are important since they give bounds on SUSY and grand unified theories.

W. T. H. van Oers suggested that a high precision measurement of the asymmetry A in np elastic scattering be done at 320 MeV to have results with the same significance as the measured asymmetry for pp elastic scattering. A difference between the two asymmetries would mean that P_{even}/T_{odd} components, not allowed in the SM, are present.

These different proposed experiments on how to detect any cracks in the SM are difficult but challenging. If any of them would give a positive result, the consequences would certainly lead to rethinking among theorists. Even negative results are of importance since our present understanding will further mature.

6 Novel instrumentation

Progress in science is often linked to the development of new instruments and methods. During this meeting, some in my mind extraordinary developments and ideas have been presented. One of these by E. Fiorini is a bolometer to be used for double beta decay experiments. A bolometer detects heat and has thus the advantage, compared with detectors sensing light or ionisation, that it registers all the energy absorbed in the detector medium. The secret to obtain a good performance is a low heat capacitance. In Milano, at a temperature of 7 mK, a resolution of 5 eV for 6 keV X rays has been obtained, the theoretical limit being 0.1 eV. Using 500 mg of $AgReO_4$ as a medium, 4 Kurie plots have been taken of the beta decay of ^{187}Re with an upper end point energy of 2.7 keV. In the future, the technology is of interest for double beta decay experiments and in the CUORE experiment, 230 kg of TeO_2 as a medium is being considered.

S. Matsuki presented an idea of how to search for axions of extremely small mass. The axion is a particle that has been suggested as a means of understanding the CP problem of the strong interaction. The idea is to place Rb in an external magnetic field and to use the Primakoff process $a \rightarrow \gamma$ for the creation of atomic Rydberg states by a hypothetical axion, a. By selective field ionisation, the atom would be ionised and the electron detected by means of channel plates.

Studies of how to make a powerful source of polarised neutrons at RCNP have been done by Y. Masuda on a suggestion by Hiro Ejiri. Using protons of 400 MeV energy and 3 mA intensity from the RCNP cyclotron, neutrons are produced by the spallation process. After moderation and cooling in superfluid helium, a source of $2.1 \cdot 10^5$ ultra cold neutrons per cm^3 is obtainable. This is 2000 times more intense than a present source in Grenoble. Having such a source of polarised neutrons would make a number of very interesting symmetry tests possible; T symmetry in n-nucleus scattering, Electric Dipole Moment of the neutron and asymmetry in neutron beta decay.

7 The future role of physics?

We have assisted in a very exciting meeting with topics in the frontier of modern physics. It may seem inappropriate on this occasion to question the role of basic physics. Nevertheless the resources for physics research are being reduced in many countries, and investments are instead steered to applied research and to the life sciences. It is easier for politicians to see investments in these areas of more direct use to society. We have therefore a common duty to convey our appreciations of physics to the public and to the political establishment.

Physics is on the same time a philosophical science, a base science and a practical science. Concerning the philosophical aspects of physics, the secrets of the Microcosmos and the Macrocosmos have fascinated mankind from the days when spirits were thought to control the world to our days when physics represents a hard core of observational evidence of Nature and natural phenomena. Involved in this transition from Magic to Science were philosophers in the early Middle East, Greece, South America and China. The secrets of Cosmos have fascinated us at this meeting and we can share this fascination in our role as teacher and science educator at our home universities.

Physics is also an important base, together with mathematics, for many other disciplines in science, technology and medicine. In these disciplines, knowledge of the properties and interactions of a great variety of radiation types besides the visible light is required. This knowledge comes from physics and physics research. It is also necessary to apply the laws of quantum mechanics increasingly to everyday life as materials become miniaturised and manufactured down to the atomic scale. Last but not least is the challenge to provide the world with clean and environment safe energy resources where the nuclear option seems to be one of the few realistic possibilities for big cities.

Physics is also a practical science since the advancement of working methods and instrumentation often have direct practical applications. Two examples can illustrate this practicality of physics. The GPS satellite navigation system has evolved from space physics and depends critically on the exactness of atomic clocks. The WWW was developed in a high energy physics laboratory, CERN. The conference participants are well aware of these and many other applications resulting from physics research where accelerators, sensors, data acquisition systems, simulations, modelling and numerical computations are frontier developments that are of interest to a high tech society.

It is my hope that the assets represented by physics researchers will be well funded even in the future. I am sure that we have to make more propaganda than ever for a sacred case. Nobody else can do it or will do it better than ourselves.

8 Acknowledgments

It is my great pleasure to thank for the hospitality extended to us by our hosts. It is a particular pleasure to thank professor Hiro Ejiri for his hospitality, congratulate him for all his achievements and wish him a successful future as an emeritus professor. Also I would like to thank the organising committee and in particular H. Toki, T. Kishimoto and K. Hatanaka. Finally I look forward to see a continued successful development of RCNP under the leadership of Y. Nagai.